Genome Mapping and Sequencing

Edited by

Ian Dunham

The Wellcome Trust Sanger Institute, Cambridge, UK

horizon scientific press

Copyright © 2003
Horizon Scientific Press
32 Hewitts Lane
Wymondham
Norfolk NR18 0JA
U.K.

www.horizonpress.com

British Library Cataloguing-in-Publication Data

A catalogue record for this book is available from the British Library

ISBN: 1-898486-50-6

Printed and bound in Great Britain by Biddles Limited

Contents

Contributors

David M. Beare
The Wellcome Trust Sanger Institute
Wellcome Trust Genome Campus
Hinxton
Cambridgeshire CB10 1SA, UK

Asif T. Chinwalla
Washington University Genome
Sequencing Center
4444 Forest Park Boulevard
St. Louis, MO 63108, USA

Marguerite Ciancio
Genomics Department
Exelixis, Inc.
170 Harbor Way
P.O. Box 511
South San Francisco
CA 94083-0511, USA

John E. Collins
The Wellcome Trust Sanger Institute
Wellcome Trust Genome Campus
Hinxton
Cambridgeshire CB10 1SA, UK

Alan Coulson
The Wellcome Trust Sanger Institute
Wellcome Trust Genome Campus
Hinxton
Cambridgeshire CB10 1SA, UK

Hugues Roest Crollius
Genoscope & CNRS UMR8030
2 rue Gaston Crémieux
91057 Evry Cedex, France
hrc@genoscope.cns.fr

Panos Deloukas
The Wellcome Trust Sanger Institute
Wellcome Trust Genome Campus
Hinxton
Cambridgeshire CB10 1SA, UK

Evan E. Eichler
Case Western Reserve University
Department of Genetics BRB 720
Cleveland, OH 44106, USA

Fred Engler
Arizona Genomics Computational
Laboratory
University of Arizona Tucson
Tucson, AZ 85721, USA
fred@genome.arizona.edu

Jeffrey Garnes
Genomics Department
Exelixis, Inc.
170 Harbor Way
P.O. Box 511
San Francisco, CA 94083-0511, USA

Andreas Gnirke
Genomics Department
Exelixis, Inc.
170 Harbor Way
P.O. Box 511
San Francisco, CA 94083-0511, USA
gnirke@genome.wi.mit.edu

Susan Gribble
The Wellcome Trust Sanger Institute
Wellcome Trust Genome Campus
Hinxton
Cambridgeshire CB10 1SA, UK

LaDeana W. Hillier
Washington University Genome
Sequencing Center
4444 Forest Park Boulevard
St. Louis, MO 63108, USA

Julie E. Horvath
Case Western Reserve University
Department of Genetics BRB 720
Cleveland, OH 44106, USA

Sean J. Humphray
The Wellcome Trust Sanger Institute
Wellcome Trust Genome Campus
Hinxton
Cambridgeshire CB10 1SA, UK

Adrienne R. Hunt
The Wellcome Trust Sanger Institute
Wellcome Trust Genome Campus
Hinxton
Cambridgeshire CB10 1SA, UK

M.C. Jones
The Wellcome Trust Sanger Institute
Wellcome Trust Genome Campus
Hinxton
Cambridgeshire CB10 1SA, UK

Devin P. Locke
Case Western Reserve University
Department of Genetics BRB 720
Cleveland, OH 44106, USA

Andrew J. Mungall
The Wellcome Trust Sanger Institute
Wellcome Trust Genome Campus
Hinxton
Cambridgeshire CB10 1SA, UK

Michael A. Quail
The Wellcome Trust Sanger Institute
Wellcome Trust Genome Campus
Hinxton
Cambridgeshire CB10 1SA, UK

Harold Riethman
The Wistar Institute
3601 Spruce St
Philadelphia, PA 19104, USA

Gregory D. Schuler
National Center for Biotechnology
Information
National Institutes of Health
Bethesda, MD 20894, USA
schuler@ncbi.nlm.nih.gov

S.K. Sims
The Wellcome Trust Sanger Institute
Wellcome Trust Genome Campus
Hinxton
Cambridgeshire CB10 1SA, UK

Cari Soderlund
Arizona Genomics Computational
Laboratory
University of Arizona Tucson
Tucson, AZ 85721, USA

Michael C. Wendl
Washington University Genome
Sequencing Center
4444 Forest Park Boulevard
St. Louis, MO 63108, USA

David L. Willey
The Wellcome Trust Sanger Institute
Wellcome Trust Genome Campus
Hinxton
Cambridgeshire CB10 1SA, UK

Preface

The sequence of the human genome has recently been completed to a high standard, at least for 99% of the euchromatin. The last five years have also seen at least draft sequences of each of the common model organisms including mouse, rat, *Drosophila*, *C. elegans*, Zebrafish, *Fugu* and yeast, plus a bewildering array of micro-organisms and other eukaryotes. This flood of DNA sequence data has amply served to demonstrate the value of sequencing so that it now seems that rates of sequence accumulation will continue to increase, rather than decrease. In the genome sequence age, study of any organism requires a genome sequence, whether it is for enumerating the basic gene catalogue or studying similarity and variation within and between species. If costs continue to decrease, perhaps down to the $1000 human genome sequence, genome sequencing will become a ubiquitous research tool.

Despite the successes of whole genome shotguns for smaller genomes, and partial successes for mammalian genomes, it seems to me that map and clone based strategies will continue to be necessary at least for the first pass at a new genome, and especially if that genome is to be finished. This is the motivation behind this book. I have tried to bring together a collection of chapters which cover the processes required to map and sequence a genome to a finished standard, and to provide a gene annotation. The book starts with the initial considerations of strategy and the up front information required to determine the strategy. It then moves step by step through establishing long range maps and clone resources, assembling clone maps, dealing with problem genomic regions including repeats and telomeres, and selecting the clones to sequence. Sequencing itself is dealt with at both the shotgun and finishing stages. While these chapters concentrate on the clone by clone approach, the methods are also broadly applicable to whole genome shotgun as well. At the end we deal with annotating genes, and getting the information out to the scientific community through the international sequence databases and the new generation of genome browsers. At each stage the book provides representative protocols, and detailed coverage of the software required. I make no apologies for the selection of authors predominantly from the International Human Genome Sequencing Consortium, and particularly from past or current members of the Sanger Institute. I think this provides a level of integration between the chapters that can not normally be achieved with a multi-author format, since the processes that are described, and sometimes even the people, are used to talking to each other. I hope that together this information will be useful for genomes yet to be sequenced.

Finally I'd like to thank all the authors for their cooperation and goodwill during the production of this volume, the staff at Horizon Press, and Charlotte Cole for extensive help with proof reading and indexing.

Ian Dunham
Hinxton
May 2003

Other Titles of Interest

From: *Genome Mapping and Sequencing*
© 2003 Horizon Scientific Press, Wymondham, UK

1

Parameters of a New Genome

Hugues Roest Crollius

Abstract

Genome sequencing is a young but evolving field. At the beginning, the possibility of sequencing entire eukaryotic genomes was naturally applied to those for which biological data was already the most abundant. Today however, genome sequencing is turning towards species that are of interest because of their key position in the evolutionary tree or because of their economic value. The corollary to this situation is that little information may be available on the genome of a species before a sequencing project is initiated. This chapter nevertheless describes some parameters that are worth looking into before time and money is invested in a large project, such as determining precisely the size of the genome, estimating the level of polymorphism in a population or exploring the repeat content. Depending on each particular case, knowledge of some of these basic properties may help lay the foundation for a successful sequencing project.

1. Introduction

The first eukaryotic genomes that were sequenced other than human (International Human Genome Sequencing Consortium, 2001; Venter *et al.*, 2001), such as that of yeast (Goffeau *et al.*, 1996), worm (The C. elegans Sequencing Consortium, 1998), fly (Adams *et al.*, 2000), mustard weed (The Arabidopsis Genome Initiative, 2000), and mouse were natural first choices for several good reasons. In particular, a large amount of genetic and functional information had been accumulated on these organisms after decades of laboratory experiments. Hence, it was expected that the knowledge of the entire genome sequence would help bring together this information into one conceptual framework. The very first genomes to be sequenced (yeast, worm, fly and weed) had the added advantage of being relatively small compared to species of the same phylum. This practical consideration meant that sequencing the entire genome was a realistic goal with the technologies of the time. A second practical advantage of these model organisms was the availability of detailed genetic and physical maps that enabled the layout of the sequencing project and the reconstruction of the genome sequence along the chromosome length.

The sequence of a genome provides a physical basis for the integration of previous data, but it is also a platform for gene discovery, genome comparisons, and a reference for measuring genetic variation. With this in mind species that are candidates for genome sequencing are these days selected on different grounds. Those that hold a key position in an evolutionary tree may for instance be of interest to understand fundamental aspects of the evolution of that lineage, and thus enriching the "vertical" catalogue of intermediate species in the tree. Also of great interest from the point of view of evolution is the sequencing of closely related species ("horizontal catalogue") to better understand the forces that drive genome evolution across shorter time scales or between species of similar physiology. For instance, such a study was recently conducted in yeasts (Souciet et al., 2000). More generally, genome sequence comparisons enable the identification of coding and non-coding conserved regions to pinpoint sequences of functional importance. But outside of evolution and comparative genomics, genome sequencing also holds promises for much more practical applications. Organisms of economic value may be better studied, improved and cared for if their genetic background is totally known. This is clearly the reason behind the sequencing of the maize or the rice genomes (Goff et al., 2002; Yu et al., 2002) by international collaborations and/or private companies. Finally, some species are being chosen because their genome sequence may hold the key to the treatment of important human diseases. This is the driving motivation for the sequencing of the *Anopheles gambiae* genome (mosquito vector of malaria).

From these observations, it appears that instead of representing the culmination of decades of biological research (as for yeast, worm or fly), the sequence of a genome now increasingly represents the foundation over which the first biological and bioinformatics investigations will be performed.

This conclusion implies that, in contrast to well known model organisms, those that will be selected in the future increasingly start off with few known genomic features. Although some characteristics will naturally be derived from the sequence itself, it may be important to investigate some key characteristics before an expensive full scale sequencing project commences. This chapter describes the reasons and the methods for delineating a number of basic parameters of a "new" genome that is entering a sequencing project. These include the genome size, the karyotype, the phylogenetic position and molecular signature in comparison to closely related species or even genus, the degree of polymorphism within a population and some basic features of genome architecture, repeat content and distribution.

2. Determining Genome Size

The differences in genome size in different species is the subject of much debate among scientists interested in genome evolution and dynamics. The reasons for the discussion are the still elusive search for a general relation between the size of the haploid genome in base pairs (C-value), and a potential environmental, developmental, or metabolic constraint that would influence this size. Indeed, the problem is obvious when considering even closely related species that possess genomes of vastly different sizes, with apparently no correlation of evolutionary position, organismal complexity, or any other simple parameter (the so called C-value paradox (Gregory and Hebert, 1999; Thomas, 1971). The classic example of the 200 fold difference between the unicellular *Amoeba dubia* genome (an enormous 670 000 Mb) and the human genome (3000 Mb) illustrates this conundrum (Li, 1997).

In fact the C-value can itself be the main argument for sequencing a genome. For instance the freshwater pufferfish *Tetraodon nigroviridis* possesses the smallest known vertebrate genome with a C-value of 0.35 pg (Hinegardner, 1972; Lamatsch *et al.*, 2000). Considering that the mouse and human genomes - which are about 8 times larger - were the only two vertebrate genomes initially earmarked for sequencing, this observation promoted an interest in the genome of both the marine pufferfish *Takifugu rubripes* (Brenner *et al.*, 1993; Elgar *et al.*, 1999) and the freshwater pufferfish *Tetraodon nigroviridis* (Crnogorac-Jurcevic *et al.*, 1997; Roest Crollius *et al.*, 2000). It was argued that sequencing such compact genomes would be an efficient way to access most of the genes in a vertebrate, and this reasoning led to the beginning of a full scale sequencing project of the

Tetraodon genome in 1997 and of Takifugu in 2000, despite the fact that little else than their C-value was known as a biological feature.

In general however genomes are selected for sequencing based on different criteria. In those cases the determination of the C-value is of secondary importance to the biologist but crucial to the sequencing centre. It is of practical importance because large genomes are generally rich in repeats, which means that sequencing is longer, more difficult, more expensive, and analysis more complicated. The genome size is thus the main factor that will determine the time frame of the project, the sequencing capacity that must be devoted to it and the funding necessary to bring it to completion.

As the C-value paradox illustrated, genome size is difficult to predict given a species of interest. In closely related organisms, even different strains from an inbred species (e.g. mice (Capparelli et al., 1997)), can display noticeable differences in genome sizes. More dramatically, a survey of 79 species of polychaete worms showed a > 70 fold variation in genome sizes (from 0.1 to 7.2 pg) with large variation in closely related species (Gambi et al., 1997; Sella et al., 1993). Other examples include C-value measurements in two strains of the fungus Histoplasma capsulatum that show a 40% difference (Carr and Shearer, 1998) and two different stocks of the parasite Trypanosoma bruccei ssp (Kanmogne et al., 1997) that have been measured with a 29% difference. These examples underline the necessity of measuring genome size specifically and avoiding inferences from other species, however closely related. The most straightforward technique for measuring C-values is by flow cytometry, where cellular DNA is labelled with a fluorogenic dye and its intensity compared to known standards, i.e. genomes of known sizes (Meistrich et al., 1978; Vindelov and Christensen, 1990). It is also currently the method of choice because of its high precision (up to 0.005 pg with good quality DNA) and hence its narrow standard deviation values. Other methods include Feulgen microspectrophotometry, biochemical extraction and reassociation kinetics but these are more laborious and less accurate. Strictly diploid organisms rarely exist (Brodsky and Uryvaeva, 1985) and the body of an organism is often a mixture of different ploidy levels depending on the tissue being examined, and this applies to plants as well as animal species. It may therefore be important to test nuclei from several tissues depending on the species. Erythrocytes are nucleated in fishes and provide a good source of cells in large animals although fins are also an alternative (Lamatsch et al., 2000). Splenocytes are often used in mammals as well as haploid spermatocytes. In plants any nuclei containing tissue such as seedlings, roots, flower or seeds can be used, although leaves are most commonly chosen.

3. Determining the Karyotype

Because in most cases genes are self contained functional units, knowing which chromosome they belong to is of secondary importance when studying single genes. This is now increasingly the case when genes are "discovered" by sequence comparisons between two species, and the target gene is immediately focused upon. However in many other cases, genes are discovered because they are physically linked either to a polymorphism or to another gene. This notion of physical linkage is central to genetic studies and is the basis for understanding the dynamics of genome evolution through the conservation of synteny. So although it is possible to establish a useful catalogue of individual genes on the draft sequence of a genome, such an endeavour takes its full meaning when genes are ordered and assigned to a specific chromosome, *i.e.* when the genome is reconstructed as it is in the nucleus. For many eukaryote genomes that have been sequenced, this was not an issue because genetic and physical maps of the genome had long been available and provided the necessary long range order and physical framework. Such maps may alleviate the initial need for a precise karyotype, as is the case in the zebrafish *Danio rerio*. The haploid zebrafish genome has 25 chromosomes, most of which are difficult to distinguish (Daga *et al.*, 1996; Gornung *et al.*, 1997). Because of the small size and similarity of zebrafish chromosomes, linkage groups identified by meiotic mapping were not assigned to cytogenetically identifiable chromosomes at a time when the sequencing of the genome was already well advanced. To circumvent this problem, single copy genes from each linkage group are being mapped on chromosomes, using genomic clones as probes in Fluorescent In Situ Hybridizations (FISH) experiments (reviewed in (Sola and Gornung, 2001)). Genomes which are now being considered for sequencing often do not benefit from such prior knowledge concerning gene linkage (although see chapter 6 on the establishment of physical maps), and the only way to associate the sequence to its physical foundation is by first determining the karyotype and then directly mapping probes from sequenced contigs to chromosomes using FISH. Classical cytogenetics methods available to distinguish at least a subset of chromosomes include silver staining of the Nucleolus Organizer Regions (NOR) (Howell and Black, 1978; Sumner, 1990), C-banding of constitutive heterochromatin (Sumner, 1972; Sumner, 1990) and the use of base-specific fluorochromes that preferentially stain AT- or GC-rich regions (Fischer *et al.*, 2000; Gornung *et al.*, 1997). These approaches can be supplemented by molecular cytogenetics methods. In particular, the use of repeated sequences in FISH experiments is an efficient way to localise the major heterochromatic blocks in the chromosome complement and thus to potentially identify some pairs specifically (see section on repeats in the genome below). Finally, once the karyotype is well characterised, FISH can then be employed to map single-copy sequences to each chromosome. This arduous enterprise is best performed when long sequence contigs or scaffolds are available, and is greatly facilitated by the use of double colour labelling techniques. For instance, one approach may consist in selecting one genomic clone from each one of the largest scaffolds of the assembly and cross-hybridising them in pairs using a different label for each (often biotin and digoxygenin respectively). After exhausting all possible pair combinations, and in conjunction with cytogenetics data obtained previously, this approach makes it possible to assign specific molecular probes - as well as a good fraction of the assembly - to individual chromosomes.

4. Characterising the Species at the Molecular Level

Well characterised genetic models or species of agricultural importance often exist as inbred strains, cultivars or germoplasm collections that ensure that any experiment is always carried out on the same species, often with exactly the same genotype. Some recent species of interest for genome sequencing do not benefit from such prior knowledge: the first DNA samples are often prepared from individuals sampled in a wild population. This was the case for the pufferfish *Tetraodon nigroviridis*, and the urochordate *Ciona savignyi* sequencing projects for example. In these cases the reproductive cycle may not be known or controlled in the laboratory, or the generation time is too long to envisage the creation of lines, implying that any future sampling will be done again from the wild population. This may be a problem if different subspecies have few morphological differences or if the geographical distribution of the population is not well known and/or overlaps with closely related species. The consequence of this may be that new samplings of individuals, for the purpose of further biological experiments during or after the genome sequencing phase might accidentally originate from a different subspecies.

It is thus of practical importance to establish an unambiguous signature for the species selected for sequencing that is easy to assay by other laboratories. Inferences concerning the historical relationships between organisms (phylogenetics) are also a prerequisite for accurate comparative analyses. Molecular markers provide this practical advantage as well as a quasi unlimited choice of molecules adapted to the evolutionary distance that must be investigated. Highly conserved molecular markers and/or gene regions are useful for investigating phylogenetic relationships at higher categorical levels (deep branches of evolutionary history). On the other hand, the hypervariable molecular markers and/or gene regions are useful for elucidating phylogenetic relationships at lower categorical levels (recently diverged branches). Thus, a trade-off must be found between the level of variability of the sequences used and the time interval one wishes to analyse. Generally however, to establish a molecular signature for a given species in order to differentiate it from closely related species from the same taxa with overlapping habitat and very similar morphology means looking at very recent divergence times. Markers of interest for this field of study (speciation and phylogeography) are those that evolve fast (Hewitt, 2001). In animals the mitochondrial (mt) genome has many properties that make it useful for reconstructing recent phylogenetic history (Simon *et al.*, 1994). In vertebrates in particular, these include a high substitution rate (Brown *et al.*, 1982), and a highly conserved gene order that lends itself to the design of PCR primers that work across a wide range of taxa. The mtDNA control region, also called the Displacement Loop (D-Loop) region, is involved in the control of mtDNA replication and RNA transcription. It possesses a hypervariable domain which is believed to be largely selectively neutral, which may account for its very rapid rate of variation. This particular domain of the mtDNA control region is thus a popular sequence for examining population structure and relationships among closely related species. Examples include a wide variety of animal species such as fishes (Lee *et al.*, 1995; Sturmbauer and Meyer, 1993), birds (Randi and Lucchini, 1998) or flies (Brehm *et al.*, 2001). In other taxonomic groups, such as protists (Barta, 2001), it has been found that rRNA sequences such as the Internal Transcribed Spacer (ITS) regions as well as the 18S genes evolve sufficiently fast to characterise a species at the molecular level, and enable the inference of relationships within taxonomic groupings. The situation is less problematic in plants. First, morphological characters are more reliable taxonomic markers and because plants are not mobile, it is possible to sample the same species from the same geographical area. Plants can also be cultivated in laboratory conditions with relative ease, embryos can be frozen and seeds can be stored for decades, thus ensuring that performing new experiments on the same species is possible without requiring a formal molecular signature at the outset.

There exists a rich and rapidly expanding primary literature in the field of molecular systematics, that describe methodologies and discuss the relative merits of individual molecular markers *e.g.* (Graur and Li, 2000; Hillis *et al.*, 1996). Establishing a clear molecular signature for a species that is being targeted for whole genome sequencing represents a fraction of the time, cost and effort of the sequencing itself, and represents a security in case new individuals need to be sampled again. Should this happen during the sequencing project itself, it is also a protection against the risk of mixing different closely related species during the project.

5. Sequencing from a Mixture of Different Haplotypes

The frequency of silent variations that co-exist within the genomes of individuals is variable from species to species. Even within a species, it has been shown that different populations that have remained geographically isolated for thousands of years have accumulated different levels of variations in their genetic pool. For instance the frequency of single nucleotide polymorphisms (SNPs) has been studied in different human populations from Europe, the United States and Africa and showed significant variations (Reich *et al.*, 2001). In recent years SNP discovery has emerged as a field of genomics research on its own (*e.g.* (Altshuler *et al.*, 2000)) motivated by their many applications in multifactorial disease gene hunting. Why is this important for genome sequencing? While such variations may be ardently sought after, they are also potentially a serious source of problems for assembling shotgun sequence reads. Depending on the sequencing strategy, estimating the level of polymorphism in a population may thus be important to guide the choice of DNA samples that will be used to prepare the sequencing templates.

In a BAC to BAC sequencing strategy, as illustrated in the public Human Genome Project (International Human Genome Sequencing Consortium, 2001), individual clones are first selected either based on mapping information or on sequence overlap, and then submitted to shotgun cloning and sequencing. The assembly is thus performed on a DNA sequence that represents a single allelic form. This strategy does not suffer from the presence of SNPs because no polymorphism is expected among the different shotgun reads of a given BAC clone. In this approach SNPs can still be detected if the genome from which the BAC library was constructed is polymorphic, because adjacent BAC inserts may each come from different chromosomes of the same pair. The region of overlap between the two clones will thus reveal polymorphic sequence variations. In a Whole Genome Shotgun (WGS) (Weber and Myers, 1997) assembly however, both alleles of every polymorphic locus in the genome will co-exist. SNPs are noticeable in such sequence assemblies because they appear as high quality base mismatches in the multiple alignment that constitutes each contig. This can be an advantage for SNP discovery (Taillon-Miller *et al.*, 1998) and was for instance one of the reasons for the choice of five donors in the human genome sequencing project carried out by the private company Celera (Venter *et al.*, 2001). But if high quality mismatches can be used by specific algorithms to track SNPs automatically, a high incidence of such variations may also disrupt the assembly process.

Typically, an assembly software such as Phrap (http://www.phrap.org) is designed to reconstruct the order of reads based on sequence overlap, and to produce a consensus sequence that, at each base position, best reflects the alignment of all the reads. The program makes the assumption that sequences from one contig originate from the same locus, with the same haplotype. It follows that Phrap will consider several local occurrences of high quality mismatches (*e.g.* Phred (Ewing and Green, 1998; Ewing *et al.*, 1998) quality above 30) in a given read as a sign that it must be placed in a different consensus sequences, because it must originate from another locus in the genome. Problems arise therefore when an assembly is attempted using reads from different haplotypes: while the assembly software might accept some high quality mismatches due to SNPs in a given consensus, beyond a certain threshold reads will be split in different contigs. This is likely to occur for instance when assembling a WGS dataset from an heterozygous individual that belongs to an outbred population where polymorphism is high. The shotgun reads then contain an equal mixture of two haplotypes, meaning that the assembly will be dependant on the number of SNPs encountered in any given region of overlap between two reads. When sufficient DNA cannot be extracted from a single individual, but instead a pool of several individuals must be used from an outbred population, the situation is much worse.

Extensive studies have not yet been performed for this problem because the application of the WGS strategy on large eukaryote genomes is still limited, hence making it is difficult to estimate the frequency of SNPs that will trigger this effect. The only WGS assembly that has been performed to date on more than one haplotype is the human genome assembly by Celera, which used DNA from five individuals in addition to finished sequences from the public Human Genome Project (Venter *et al.*, 2001). How this actually affected the assembly has not been reported, but the frequency of SNPs in the human population is about one every 1000 bases (Nachman, 2001) suggesting that sequence reads of a few hundred bases will rarely contain more than two SNPs. Other genomes that are or have been sequenced using the WGS approach are either genetic models (Drosophila (Adams *et al.*, 2000), mouse) or agriculturally important species such as rice (Goff *et al.*, 2002; Yu *et al.*, 2002)), and thus already existed as inbred lines.

A number of genomes are now being sequenced following the WGS strategy using DNA samples from several individuals that belong to outbred populations, where the level of polymorphism is likely to be higher than in the human population: these include the mosquito *Anopheles gambiae*, the urochordates *Ciona intestinalis* and *Ciona savignii*, and the pufferfish *Tetraodon nigroviridis*. The outcome of these genome sequencing projects will tell how polymorphism exactly impacts on the final assembly, but it is already clear that standard approaches used for assembling one haplotype will prove difficult to use on these genomes.

If one wants to anticipate potential problems, how is it possible to rapidly estimate the level of polymorphism in a population? It is likely that some genes from this genome, or at least from a close relative, will already be sequenced and available in public databases. The quickest route is then to design primers in introns from these genes and amplify DNA by PCR from a selection of individuals. Sequence comparisons in regions of high quality should reveal SNPs. A decision can then be made as to whether the level of polymorphism is sufficiently low to enable either the selection of one or several individuals for sequencing. If the population is highly polymorphic and one individual does not provide sufficient DNA for the preparation of shotgun libraries, it is worth considering the creation of inbred lines as it might be a better solution than having to make difficult compromises during assembly. This is possible if the mode of reproduction can be controlled in laboratory conditions, if the generation time is not prohibitively long, and if inbreeding depression does not interrupt the process.

6. Analysing Repeated Sequences

6.1 Repeats are Useful Components of a Genome

Repeats come in two broad families, tandem and dispersed, according to their organisation in the genome. Tandem repeats are successive repetitions of a particular motif, ranging in size from one base to several kilobases. They include microsatellites (repeat unit from 1 to 6 bases, but less than a few hundred base pairs overall), minisatellites (repeat unit from 7 to several hundred bases, but less than a few hundred base pairs overall), satellites (arrays of several kilobases, independent of unit size), and the ribosomal RNA genes. Dispersed repeats consist of sequences that have (or had at one point in time) the capacity to replicate themselves and jump from one position to another (transpose), sometimes via an amplification mechanism. These repeats, also called transposable elements, are classified according to their transposition mechanism: retrotransposons, that replicate via an RNA intermediate and hence may produce many new copies from one original element, and DNA transposons, that replicate through a "cut and paste" mechanism (Finnegan, 1992)

As sequencing progresses, identifying and cataloguing repeat sequences in a new genome serves several purposes. First, both tandem and dispersed repeats form important structural components of the genome, and thus may provide markers for the identification of centromeres, telomeres, and other repeat rich structures. In particular, major satellite and rRNA gene clusters form heterochromatic blocks in the genome that are easily recognizable cytogenetically and can serve as useful identifiers when the chromosome formula is difficult to establish. Second, because such sequences occur naturally in multiple copies, any analysis based on sequence comparisons either within the genome or between different genomes will be over-whelmed by large volumes of alignments created by these repeats. In the case of shotgun assemblies of large sections of a genome, and unless the assembly program is designed to deal with these sequences from scratch (*e.g.* ARACHNE (Batzoglou *et al.*, 2002)), repeats will aggregate many different loci in one cluster and prevent the assembly of the sequence. Finally, repeat sequences are important elements of the genome from an evolutionary point of view ((Charlesworth *et al.*, 1994)). For instance transposable elements in particular can influence chromosome evolution by promoting chromosome breakage, deletions, inversions and amplifications ((Lim and Simmons, 1994); (Dimitri *et al.*, 1997); (O'Neill *et al.*, 1998)). The human genome has been extensively studied with respect to transposable element and genome co-evolution. While it is clear that some genes have in fact been derived from such elements (Smit, 1999), there may also be a wider functional association between specific human repeats (Alu sequences) and their host genome (International Human Genome Sequencing Consortium, 2001; Schmid, 1998). Identifying and annotating repeats is therefore important from a practical point of view, but may also be used as an archeological treasure trove for the analysis of genome plasticity and evolution.

6.2 *Identification of Repeats in Shotgun Sequences*

Depending on the type of repeats and the objectives behind their detection, different solutions exist to identify, classify and mask repeats in genomic DNA. The program RepeatMasker (A.F.A. Smit and P. Green, unpublished) was specifically designed to identify both tandem repeats and transposable elements, and is widely used in genome sequencing centres. Its main task is to produce a modified version of the original sequence where repeats are replaced (masked) by generic bases (letter 'N' by default), thus providing a way to avoid such sequences when working with alignment programs that make use of a scoring matrix, such as BLAST or Smith-Waterman. In such programs, the score for aligning a 'N' in front of another nucleotide can be set to zero or to negative values (*e.g.* in the default NUC.4.4 scoring matrix in BLASTN), thus ignoring or penalising masked sequence in the construction of alignments. To identify both tandem and interspersed repeats, RepeatMasker uses cross_match, an efficient implementation of the Smith-Waterman-Gotoh (Gotoh, 1982; Smith and Waterman, 1981) algorithm developed by P. Green. Cross_match will detect microsatellite tandem repeats (mono to pentamers, and a some hexamers, all longer than 20 bp) using a database of all possible repeat motifs. Microsatellite sequences have a finite number of possible motifs (*e.g.* there are four possible dinucleotides only), making it possible to use the same library of microsatellites with any genomic DNA. The interspersed repeat databases screened by RepeatMasker are extracted from the repeat databases (Repbase Update) copyrighted by the Genetic Information Research Institute (G.I.R.I.). However, many interspersed repeats are species specific, and this poses a problem when such repeats must be identified in a new genome not yet referenced in Repbase. In such cases, only interspersed repeats that were present in a common ancestor between the organism under study and a species referenced in Repbase can be found (with a lower sensitivity), although other repeats that share homologies with typical transposable elements genes such as reverse transcriptases or transposases may also be partially identified. It is therefore difficult to divert RepeatMasker from its intended task (identifying and masking known repeats) for the purpose of identifying new repeats, especially transposable elements and satellites, in a new genome.

The fastest route to identifying abundant repeat sequences (tandem and interspersed) is to sequence a random sample of the genome and to attempt to assemble it, for instance using Phrap. The most likely outcome is that repeats will recognise each other and form entangled clusters, thus revealing their presence. This was performed on a set of approximately 10,000 reads during the initial phase of the sequencing of the *Tetraodon nigroviridis* genome (Roest Crollius *et al.*, 2000). This sample represented approximately 3% of the genome, and its assembly produced about 200 contigs of more than one sequence read. The largest contained about 20 sequences, and turned out to be a fragment of the 28S ribosomal RNA gene. Working down the list of contigs in decreasing order of size ultimately enabled the identification of the remaining of the large rRNA units, the centromeric satellite sequence, the 5S rRNA gene, an abundant 10 bp satellite repeat, and fragments of a DNA transposon and of a LINE element. This is therefore a general method to rapidly isolate sequences that are abundant in a given genome, although there are a number of specific additional steps involved to fully characterise each repeat individually.

6.3 Ribosomal RNA Units

The typical eukaryotic rRNA gene array consists of a tandem repetition of a basic unit, separated from the next by an intergenic spacer (IGS). Each unit is transcribed to form a 45S precursor rRNA, but can be further subdivided in three distinct genes separated by "spacer" sequences. The 45S unit starts with a 5' external transcribed spacer (ETS), followed by the 18S, 5.8S and 28S genes separated by two internal transcribed spacers (ITS1 and ITS2), and ending with a 3'ETS. rRNA gene sequences are extremely well conserved from mammals to bacteria, although the number and distribution of the genes and of the repeating units may vary between and within species. In all vertebrates the 5S rRNA gene is also organised in tandem repetitions and generally in separate cluster(s) from those formed by the 18S, 5.8S and 28S genes. In mammals, there are five rRNA gene arrays, totalling several hundred repetitions of the 45S unit. It is thus not surprising that even a small sampling of a genome will already contain several copies of this repeat. Assuming that there are sufficient sequences, the entire unit may be reconstructed as a consensus using assembly programs such as Phrap or Gap4 (Staden, 1996) which should perform well because units are extremely similar to each other, even in the two ITSs. If determining the rRNA sequences is important for any other purpose than masking, it may be faster to sequence selected genomic clones that contain at least one complete copy in order to obtain a high quality result.

Because conservation of the four rRNA genes among eukaryotes is so high, the identification of boundaries between genes and spacers by sequence alignments to known genes is relatively easy. More problematic is the identification of the IGS sequence, that is poorly conserved in evolution and in fact contributes most of the sequence of the repeated unit (30 kb out of 44 kb in human). Again, only the targeted sequencing of a genomic clone containing at least one complete unit will solve this problem. Hybridisation probes derived from such sequences are excellent chromosomal markers in species where the chromosome complement is difficult to delineate, although because of the high degree of conservation, any eukaryote rRNA probe will work equally well.

6.4 Centromeric Satellite Repeats

Centromeres of eukaryotes are often associated with tandem repetitions of a basic repeat unit that show no discernible sequence conservation between species, and no definite sequence specific function has yet been determined for such repeats. However it is clear that in most species several - and sometimes all - chromosomes contain the same satellite sequence, indicating that a mechanism of concerted evolution is operating within populations (Elder and Turner, 1995). Because of the absence of sequence conservation between species, the only lead initially available to identify such repeats is its abundance and its tandemly repeated nature. A popular software to detect tandem repetitions of any size up to 300 bp in genomic DNA is TRF (Tandem Repeat Finder) (Benson, 1999). For instance, to identify the *Tetraodon nigroviridis* 118 bp centromeric repeat (Roest Crollius *et al.*, 2000), TRF was applied to individual sequence reads belonging to a Phrap contig from the initial pilot sequencing phase and suspected of containing tandem arrays. Miropeats is another tool useful to identify any local duplications (tandem, inverted, or otherwise) in genomic DNA (Parsons, 1995). Miropeats has no upper limit in terms of motif size, but it is more intended for a visual inspection of a sequence on a graphical output. TRF in contrast provides more quantitative information on the nature of the repeated motif, such as its average variation in successive units, its size and its consensus sequence.

6.5 Interspersed Repeats

A sequence is rapidly suspected to harbour a transposable element when sequence similarity searches generate matches with for instance reverse-transcriptases, endonucleases (for retrotransposons) or transposases (for DNA transposons). Such genes are part of the panoply of coding sequences found in transposable elements and they retain some degree of sequence conservation, even between classes. Full length retrotransposons with Long Terminal Repeats are then relatively easy to delineate because a simple search for identical sequences a few hundred bases in length, in direct orientation, framing the region where the alignments are located will outline the full sequence at once. A program such as Miropeats is a very convenient tool for this purpose (Parsons, 1995) and can be applied on the assembled sequences of a BAC clone for instance, where there is a chance to find a complete element. However, this ideal situation cannot be expected in whole genome shotgun sequence samples because full length elements are always larger than the average sequence read. In addition, elements that are interspersed in a genome are rarely complete, thus making it difficult to determine the exact limits of what constitutes the repeated unit. Why are the elements often not complete? In fact this largely depends on the time elapsed since a given family of transposable elements was last active in the genome. For instance over half of the transposable elements found in the human genome predate the mammalian radiation, and are thus older than 100 million years (Smit, 1999). In such cases, old transposable elements have the potential to be truncated by small local rearrangements or, more confusingly, to be interrupted by the insertion of a younger element within the old one. Hence beyond the region of similarity in the alignment, which may range in size from a few hundred to a few thousand bases, identifying the exact boundaries of the element using genome shotgun sequences is often difficult and labour intensive. One suggested approach consists of the following steps:

1) Perform a Phrap assembly on the largest set of shotgun sequences that can be handled at one time by the computer resources available. This number will largely depend on the frequency of repeats in the shotgun set, and may require some trials since the computing time does not increase linearly with the number of sequences.

2) Extract consensus sequences from the largest contigs (even if they are incorrectly assembled) and perform a database search against SwissProt to rapidly earmark those clusters that generate alignments with reverse-transcriptases, endonucleases or transposases, and are thus likely to contain transposable elements.

3) For each cluster that identifies such genes, examine the assembly in Consed (Gordon *et al.*, 1998), and remove sequence reads that interrupt the consensus sequence. These are most likely due to early termination of some forms of a given repeat because of truncation, which means that the sequence read probably contains some "unique" sequence.

4) Run Phrap again on the new cleaned-up cluster. The assembly will certainly again be incorrect if the repeat is a LTR retrotransposon or a DNA transposon, because terminal repeats are identical and will have a tendency to pile up at one end of the contig.

5) Assuming both ends of shotgun clones are present in the contig, use this information to identify misassemblies in Consed. These are most likely to be due to terminal repeats that will cause two opposite ends to overlap in the same orientation. Either remove mis-assembled reads from the contig or edit the assembly by manually reorienting the reads in the contig.

This proposed approach may often be a partial solution, mostly depending on the age of the repeat. The longer the period of time the sequence has been inserted in the genome, the more difficult it will be to find a complete copy, and the more divergent will each copy be from each other. Hence this strategy is really intended to identify the sequences that will enable a more efficient masking of the genome prior to sequence similarity searches, and is not intended to outline complete repeats. The latter can only be performed by sequencing complete genomic clones, because then few copies of the repeat are present in the insert and make reconstructing the full sequence much easier, especially with the possibility of performing gap closure and finishing. If the identification of interspersed repeats is an important issue, especially in whole genome shotgun strategies, it would not be considered a waste of time to identify specific genomic clones likely to contain such sequences by hybridisation or end sequencing, and to submit them individually to shotgun sequencing.

Conclusions

Although we have known for decades that genomes of different taxonomic families or geni may vary tremendously in size, architecture and composition, it is nevertheless natural to transpose knowledge acquired on one species to other relatively close species, until we know better. However as more sequencing projects reach completion in closely related prokaryotes, fungi, plants or metazoans, the diversity of genome features becomes more and more obvious even at close evolutionary ranges, and the temptation of inferring properties of one genome from another should be resisted, especially when planning to sequence a new species.

As more genomes are sequenced, molecular phylogeny tools also become more powerful and will probably turn into essential cornerstones of any genome bioinformatics analyses. In particular the prospect of reconstructing in the laboratory the sequence of ancestral genes based on phylogenetic analyses (Chang, 2002) may shed light on the mechanism of new function acquisition. For these reasons placing the species at the correct position in the evolutionary tree is of importance, before the sequencing process starts. Establishing basic parameters of a genome will sometimes be straightforward, such as in many domesticated species, and sometimes arduous. It will always be a compromise between an early start up of the full scale sequencing project versus the speed, cost and benefit of sequencing the genome.

Acknowledgements

I am grateful to Fiona Francis and Gisella Orjeda for helpful comments and criticisms on the manuscript.

References

Adams, M.D., Celniker, S.E., Holt, R.A., Evans, C.A., Gocayne, J.D., Amanatides, P.G., Scherer, S.E., Li, P.W., Hoskins, R.A., Galle, R.F., George, R.A., Lewis, S.E., Richards, S., Ashburner, M., Henderson, S.N., Sutton, G.G., Wortman, J.R., Yandell, M.D., Zhang, Q., Chen, L.X., Brandon, R.C., Rogers, Y.H., Blazej, R.G., Champe, M., Pfeiffer, B.D., Wan, K.H., Doyle, C., Baxter, E.G., Helt, G., Nelson, C.R., Gabor Miklos, G.L., Abril, J.F., Agbayani, A., An, H.J., Andrews-Pfannkoch, C., Baldwin, D., Ballew, R.M., Basu, A., Baxendale, J., Bayraktaroglu, L., Beasley, E.M., Beeson, K.Y., Benos, P.V., Berman, B.P., Bhandari, D., Bolshakov, S., Borkova, D., Botchan, M.R., Bouck, J., Brokstein, P., Brottier, P., Burtis, K.C., Busam, D.A., Butler, H., Cadieu, E., Center, A., Chandra, I., Cherry, J.M., Cawley, S., Dahlke, C., Davenport, L.B., Davies, P., de Pablos, B., Delcher, A., Deng, Z., Mays, A.D., Dew, I., Dietz, S.M., Dodson, K., Doup, L.E., Downes, M., Dugan-Rocha, S., Dunkov, B.C., Dunn, P., Durbin, K.J., Evangelista, C.C., Ferraz, C., Ferriera, S., Fleischmann, W., Fosler, C., Gabrielian, A.E., Garg, N.S., Gelbart, W.M., Glasser, K., Glodek, A., Gong, F., Gorrell, J.H., Gu, Z., Guan, P., Harris, M., Harris, N.L., Harvey, D., Heiman, T.J., Hernandez, J.R., Houck, J., Hostin, D., Houston, K.A., Howland, T.J., Wei, M.H., Ibegwam, C., Jalali, M., Kalush, F., Karpen, G.H., Ke, Z., Kennison, J.A., Ketchum, K.A., Kimmel, B.E., Kodira, C.D., Kraft, C., Kravitz, S., Kulp, D., Lai, Z., Lasko, P., Lei, Y., Levitsky, A.A., Li, J., Li, Z., Liang, Y., Lin, X., Liu, X., Mattei, B., McIntosh, T.C., McLeod, M.P., McPherson, D., Merkulov, G., Milshina, N.V., Mobarry, C., Morris, J., Moshrefi, A., Mount, S.M., Moy, M., Murphy, B., Murphy, L., Muzny, D.M., Nelson, D.L., Nelson, D.R., Nelson, K.A., Nixon, K., Nusskern, D.R., Pacleb, J.M., Palazzolo, M., Pittman, G.S., Pan, S., Pollard, J., Puri, V., Reese, M.G., Reinert, K., Remington, K., Saunders, R.D., Scheeler, F., Shen, H., Shue, B.C., Siden-Kiamos, I., Simpson, M., Skupski, M.P., Smith, T., Spier, E., Spradling, A.C., Stapleton, M., Strong, R., Sun, E., Svirskas, R., Tector, C., Turner, R., Venter, E., Wang, A.H., Wang, X., Wang, Z.Y., Wassarman, D.A., Weinstock, G.M., Weissenbach, J., Williams, S.M., Woodage, T., Worley, K.C., Wu, D.,

Yang, S., Yao, Q.A., Ye, J., Yeh, R.F., Zaveri, J.S., Zhan, M., Zhang, G., Zhao, Q., Zheng, L., Zheng, X.H., Zhong, F.N., Zhong, W., Zhou, X., Zhu, S., Zhu, X., Smith, H.O., Gibbs, R.A., Myers, E.W., Rubin, G.M., and Venter, J.C. 2000. The genome sequence of Drosophila melanogaster. Science. 287: 2185–95.

Altshuler, D., Pollara, V.J., Cowles, C.R., Van Etten, W.J., Baldwin, J., Linton, L. and Lander, E.S. 2000. An SNP map of the human genome generated by reduced representation shotgun sequencing. Nature. 407: 513–516.

Barta, J.R. 2001. Molecular approaches for inferring evolutionary relationships among protistan parasites. Vet. Parasitol. 101: 175–186.

Batzoglou, S., Jaffe, D.B., Stanley, K., Butler, J., Gnerre, S., Mauceli, E., Berger, B., Mesirov, J.P. and Lander, E.S. 2002. ARACHNE: a whole-genome shotgun assembler. Genome Res. 12: 177–189.

Benson, G. 1999. Tandem repeats finder: a program to analyze DNA sequences. Nucleic Acids Res. 27: 573–580.

Brehm, A., Harris, D.J., Hernandez, M., Cabrera, V.M., Larruga, J.M., Pinto, F.M. and Gonzalez, A.M. 2001. Structure and evolution of the mitochondrial DNA complete control region in the Drosophila subobscura subgroup. Insect. Mol. Biol. 10: 573–578.

Brenner, S., Elgar, G., Sandford, R., Macrae, A., Venkatesh, B. and Aparicio, S. 1993. Characterization of the pufferfish (Fugu) genome as a compact model vertebrate genome. Nature. 366: 265–8.

Brodsky, V.Y. and Uryvaeva, I.V. 1985. Genome multiplication in growth and development. Cambridge University Press, Cambridge, UK.

Brown, W.M., Prager, E.M., Wang, A. and Wilson, A.C. 1982. Mitochondrial DNA sequences of primates: tempo and mode of evolution. J. Mol. Evol. 18: 225–239.

Capparelli, R., Cottone, C., D'Apice, L., Viscardi, M., Colantonio, L., Lucretti, S. and Iannelli, D. 1997. DNA content differences in laboratory mouse strains determined by flow cytometry. Cytometry. 29: 261–266.

Carr, J. and Shearer, G., Jr. 1998. Genome size, complexity, and ploidy of the pathogenic fungus Histoplasma capsulatum. J. Bacteriol. 180: 6697–6703.

Chang, B.S., Kazmi, M.A., Sakmar, T.P. 2002. Synthetic gene technology: applications to ancestral gene reconstruction and structure-function studies of receptors. Methods Enzymol. 343:274–294

Charlesworth, B., Sniegowski, P. and Stephan, W. 1994. The evolutionary dynamics of repetitive DNA in eukaryotes. Nature. 371: 215–220.

Crnogorac-Jurcevic, T., Brown, J.R., Lehrach, H. and Schalkwyk, L.C. 1997. Tetraodon fluviatilis, a new puffer fish model for genome studies. Genomics. 41: 177–184.

Daga, R.R., Thode, G. and Amores, A. 1996. Chromosome complement, C-banding, Ag-NOR and replication banding in the zebrafish Danio rerio. Chromosome Res. 4: 29–32.

Dimitri, P., Arca, B., Berghella, L. and Mei, E. 1997. High genetic instability of heterochromatin after transposition of the LINE-like I factor in Drosophila melanogaster. Proc. Natl. Acad. Sci. U S A. 94: 8052–8057.

Elder, J.F., Jr. and Turner, B.J. 1995. Concerted evolution of repetitive DNA sequences in eukaryotes. Q. Rev. Biol. 70: 297–320.

Elgar, G., Clark, M.S., Meek, S., Smith, S., Warner, S., Edwards, Y.J., Bouchireb, N., Cottage, A., Yeo, G.S., Umrania, Y., Williams, G. and Brenner, S. 1999. Generation and analysis of 25 Mb of genomic DNA from the pufferfish Fugu rubripes by sequence scanning. Genome Res. 9: 960–971.

Ewing, B. and Green, P. 1998. Base-calling of automated sequencer traces using phred. II. Error probabilities. Genome Res. 8: 186–194.

Ewing, B., Hillier, L., Wendl, M.C. and Green, P. 1998. Base-calling of automated sequencer traces using phred. I. Accuracy assessment. Genome Res. 8: 175–185.

Finnegan, D.J. 1992. Transposable elements. Curr. Opin. Genet. Dev. 2: 861–7.

Fischer, C., Ozouf-Costaz, C., Roest Crollius, H., Dasilva, C., Jaillon, O., Bouneau, L., Bonillo, C., Weissenbach, J. and Bernot, A. 2000. Karyotype and chromosomal localization of characteristic tandem repeats in the pufferfish *Tetraodon nigroviridis*. Cytogenet. Cell Genet. 88: 50–55.

Gambi, M.C., Ramella, L., Sella, G., Protto, P. and Aldieri, E. 1997. Variation in genome size of benthic polychaete: systemic and ecological relationships. J. of Mar. Biol. Assoc. UK. 77: 1045–1057.

Goff, S.A., Ricke, D., Lan, T.H., Presting, G., Wang, R., Dunn, M., Glazebrook, J., Sessions, A., Oeller, P., Varma, H., Hadley, D., Hutchison, D., Martin, C., Katagiri, F., Lange, B.M., Moughamer, T., Xia, Y., Budworth, P., Zhong, J., Miguel, T., Paszkowski, U., Zhang, S., Colbert, M., Sun, W.L., Chen, L., Cooper, B., Park, S., Wood, T.C., Mao, L., Quail, P., Wing, R., Dean, R., Yu, Y., Zharkikh, A., Shen, R., Sahasrabudhe, S., Thomas, A., Cannings, R., Gutin, A., Pruss, D., Reid, J., Tavtigian, S., Mitchell, J., Eldredge, G., Scholl, T., Miller, R.M.,

Bhatnagar, S., Adey, N., Rubano, T., Tusneem, N., Robinson, R., Feldhaus, J., Macalma, T., Oliphant, A. and Briggs, S. 2002. A draft sequence of the rice genome (Oryza sativa L. ssp. japonica). Science. 296: 92–100.

Goffeau, A., Barrell, B.G., Bussey, H., Davis, R.W., Dujon, B., Feldmann, H., Galibert, F., Hoheisel, J.D., Jacq, C., Johnston, M., Louis, E.J., Mewes, H.W., Murakami, Y., Philippsen, P., Tettelin, H. and Oliver, S.G. 1996. Life with 6000 genes. Science. 274: 546, 563–7.

Gordon, D., Abajian, C. and Green, P. 1998. Consed: a graphical tool for sequence finishing. Genome Res. 8: 195–202.

Gornung, E., Gabrielli, I., Cataudella, S. and Sola, L. 1997. CMA3-banding pattern and fluorescence in situ hybridization with 18S rRNA genes in zebrafish chromosomes. Chromosome Res. 5: 40–46.

Gotoh, O. 1982. An improved algorithm for matching biological sequences. J. Mol. Biol. 162: 705–708.

Graur, D. and Li, W.-H. 2000. Fundamentals of Molecular Evolution. Sinauer Associates, Sunderland, MA, USA.

Gregory, T.R. and Hebert, P.D. 1999. The modulation of DNA content: proximate causes and ultimate consequences. Genome Res. 9: 317–324.

Hewitt, G.M. 2001. Speciation, hybrid zones and phylogeography - or seeing genes in space and time. Mol Ecol. 10: 537–549.

Hillis, D.M., Moritz, C. and Mable, B.K. 1996. Molecular Systematics. Sunderland Associates, Sunderland, MA, USA.

Hinegardner, R. 1972. Cellular DNA content and the evolution of teleostean fishes. The American Naturalist. 106: 621–644.

Howell, W.M. and Black, D.A. 1978. A rapid technique for producing silver-stained nucleolus organizer regions and trypsin-giemsa bands on human chromosomes. Hum. Genet. 43: 53–56.

International Human Genome Sequencing Consortium 2001. Initial sequencing and analysis of the human genome. Nature. 409: 860–921.

Kanmogne, G.D., Bailey, M. and Gibson, W.C. 1997. Wide variation in DNA content among isolates of Trypanosoma brucei ssp. Acta Trop. 63: 75–87.

Lamatsch, D.K., Steinlein, C., Schmid, M. and Schartl, M. 2000. Noninvasive determination of genome size and ploidy level in fishes by flow cytometry: detection of triploid Poecilia formosa. Cytometry. 39: 91–95.

Lee, W.J., Conroy, J., Howell, W.H. and Kocher, T.D. 1995. Structure and evolution of teleost mitochondrial control regions. J. Mol. Evol. 41: 54–66.

Li, W.-H. 1997. Molecular Evolution. Sinauer Associates, Inc., Sunderland, MA, USA.

Lim, J.K. and Simmons, M.J. 1994. Gross chromosome rearrangements mediated by transposable elements in Drosophila melanogaster. Bioessays. 16: 269–275.

Meistrich, M.L., Gohde, W., White, R.A. and Schumann, J. 1978. Resolution of X and Y spermatids by pulse cytophotometry. Nature. 274: 821–823.

Nachman, M.W. 2001. Single nucleotide polymorphisms and recombination rate in humans. Trends Genet. 17: 481–485.

O'Neill, R.J., O'Neill, M.J. and Graves, J.A. 1998. Undermethylation associated with retroelement activation and chromosome remodelling in an interspecific mammalian hybrid. Nature. 393: 68–72.

Parsons, J.D. 1995. Miropeats: graphical DNA sequence comparisons. Comput. Appl. Biosci. 11: 615–619.

Randi, E. and Lucchini, V. 1998. Organization and evolution of the mitochondrial DNA control region in the avian genus Alectoris. J. Mol. Evol. 47: 449–462.

Reich, D.E., Cargill, M., Bolk, S., Ireland, J., Sabeti, P.C., Richter, D.J., Lavery, T., Kouyoumjian, R., Farhadian, S.F., Ward, R. and Lander, E.S. 2001. Linkage disequilibrium in the human genome. Nature. 411: 199–204.

Roest Crollius, H., Jaillon, O., Dasilva, C., Ozouf-Costaz, C., Fizames, C., Fischer, C., Bouneau, L., Billault, A., Quetier, F., Saurin, W., Bernot, A. and Weissenbach, J. 2000. Characterization and repeat analysis of the compact genome of the freshwater pufferfish Tetraodon nigroviridis. Genome Res. 10: 939–949.

Schmid, C.W. 1998. Does SINE evolution preclude Alu function? Nucleic Acids Res. 26: 4541–4550.

Sella, G., Redi, C.A., Ramella, L., Soldi, R. and Premoli, M.C. 1993. Genome size and karyotype in some interstitial polychaete species of the genus Ophryotrocha (Dorvilleidae). Genome. 36: 652–657.

Simon, C., Frati, F., Beckenbach, A., Crespi, B., Liu, H. and Flool, P. 1994. Evolution, weighting, and phylogenetic utility of mitochondrial gene sequences and a compilation of conserved polymerase chain reaction primers. Ann. Entomol. Soc. Am. 87: 651–701.

Smit, A.F. 1999. Interspersed repeats and other mementos of transposable elements in mammalian genomes. Curr. Opin. Genet. Dev. 9: 657–663.

Smith, T.F. and Waterman, M.S. 1981. Identification of common molecular subsequences. J. Mol. Biol. 147: 195–197.

Sola, L. and Gornung, E. 2001. Classical and molecular cytogenetics of the zebrafish, Danio rerio (Cyprinidae, Cypriniformes): an overview. Genetica. 111: 397–412.

Souciet, J., Aigle, M., Artiguenave, F., Blandin, G., Bolotin-Fukuhara, M., Bon, E., Brottier, P., Casaregola, S., de Montigny, J., Dujon, B., Durrens, P., Gaillardin, C., Lepingle, A., Llorente, B., Malpertuy, A., Neuveglise, C., Ozier-Kalogeropoulos, O., Potier, S., Saurin, W., Tekaia, F., Toffano-Nioche, C., Wesolowski-Louvel, M., Wincker, P. and Weissenbach, J. 2000. Genomic exploration of the hemiascomycetous yeasts: 1. A set of yeast species for molecular evolution studies. FEBS Lett. 487: 3–12.

Staden, R. 1996. The Staden sequence analysis package. Mol. Biotechnol. 5: 233–41.

Sturmbauer, C. and Meyer, A. 1993. Mitochondrial phylogeny of the endemic mouthbrooding lineages of cichlid fishes from Lake Tanganyika in eastern Africa. Mol. Biol. Evol. 10: 751–68.

Sumner, A.T. 1972. A simple technique for demonstrating centromeric heterochromatin. Exp. Cell. Res. 75: 304–306.

Sumner, A.T. 1990. Chromosome Banding. Unwin Hyman, London.

Taillon-Miller, P., Gu, Z., Li, Q., Hillier, L. and Kwok, P.Y. 1998. Overlapping genomic sequences: a treasure trove of single-nucleotide polymorphisms. Genome Res. 8: 748–754.

The Arabidopsis Genome Initiative. 2000. Analysis of the genome sequence of the flowering plant Arabidopsis thaliana. Nature. 408: 796–815.

The C. elegans Sequencing Consortium 1998. Genome sequence of the nematode C. elegans: a platform for investigating biology. Science. 282: 2012–2018.

Thomas, C.A. 1971. The genetic organization of chromosomes. Ann. Rev. Genet. 5: 237–256.

Venter, J.C., Adams, M.D., Myers, E.W., Li, P.W., Mural, R.J., Sutton, G.G., Smith, H.O., Yandell, M., Evans, C.A., Holt, R.A., Gocayne, J.D., Amanatides, P., Ballew, R.M., Huson, D.H., Wortman, J.R., Zhang, Q., Kodira, C.D., Zheng, X.H., Chen, L., Skupski, M., Subramanian, G., Thomas, P.D., Zhang, J., Gabor Miklos, G.L., Nelson, C., Broder, S., Clark, A.G., Nadeau, J., McKusick, V.A., Zinder, N., Levine, A.J., Roberts, R.J., Simon, M., Slayman, C., Hunkapiller, M., Bolanos, R., Delcher, A., Dew, I., Fasulo, D., Flanigan, M., Florea, L., Halpern, A., Hannenhalli, S., Kravitz, S., Levy, S., Mobarry, C., Reinert, K., Remington, K., Abu-Threideh, J., Beasley, E., Biddick, K., Bonazzi, V.,

Brandon, R., Cargill, M., Chandramouliswaran, I., Charlab, R., Chaturvedi, K., Deng, Z., Di Francesco, V., Dunn, P., Eilbeck, K., Evangelista, C., Gabrielian, A.E., Gan, W., Ge, W., Gong, F., Gu, Z., Guan, P., Heiman, T.J., Higgins, M.E., Ji, R.R., Ke, Z., Ketchum, K.A., Lai, Z., Lei, Y., Li, Z., Li, J., Liang, Y., Lin, X., Lu, F., Merkulov, G.V., Milshina, N., Moore, H.M., Naik, A.K., Narayan, V.A., Neelam, B., Nusskern, D., Rusch, D.B., Salzberg, S., Shao, W., Shue, B., Sun, J., Wang, Z., Wang, A., Wang, X., Wang, J., Wei, M., Wides, R., Xiao, C., Yan, C., Yao, A., Ye, J., Zhan, M., Zhang, W., Zhang, H., Zhao, Q., Zheng, L., Zhong, F., Zhong, W., Zhu, S., Zhao, S., Gilbert, D., Baumhueter, S., Spier, G., Carter, C., Cravchik, A., Woodage, T., Ali, F., An, H., Awe, A., Baldwin, D., Baden, H., Barnstead, M., Barrow, I., Beeson, K., Busam, D., Carver, A., Center, A., Cheng, M.L., Curry, L., Danaher, S., Davenport, L., Desilets, R., Dietz, S., Dodson, K., Doup, L., Ferriera, S., Garg, N., Gluecksmann, A., Hart, B., Haynes, J., Haynes, C., Heiner, C., Hladun, S., Hostin, D., Houck, J., Howland, T., Ibegwam, C., Johnson, J., Kalush, F., Kline, L., Koduru, S., Love, A., Mann, F., May, D., McCawley, S., McIntosh, T., McMullen, I., Moy, M., Moy, L., Murphy, B., Nelson, K., Pfannkoch, C., Pratts, E., Puri, V., Qureshi, H., Reardon, M., Rodriguez, R., Rogers, Y.H., Romblad, D., Ruhfel, B., Scott, R., Sitter, C., Smallwood, M., Stewart, E., Strong, R., Suh, E., Thomas, R., Tint, N.N., Tse, S., Vech, C., Wang, G., Wetter, J., Williams, S., Williams, M., Windsor, S., Winn-Deen, E., Wolfe, K., Zaveri, J., Zaveri, K., Abril, J.F., Guigo, R., Campbell, M.J., Sjolander, K.V., Karlak, B., Kejariwal, A., Mi, H., Lazareva, B., Hatton, T., Narechania, A., Diemer, K., Muruganujan, A., Guo, N., Sato, S., Bafna, V., Istrail, S., Lippert, R., Schwartz, R., Walenz, B., Yooseph, S., Allen, D., Basu, A., Baxendale, J., Blick, L., Caminha, M., Carnes-Stine, J., Caulk, P., Chiang, Y.H., Coyne, M., Dahlke, C., Mays, A., Dombroski, M., Donnelly, M., Ely, D., Esparham, S., Fosler, C., Gire, H., Glanowski, S., Glasser, K., Glodek, A., Gorokhov, M., Graham, K., Gropman, B., Harris, M., Heil, J., Henderson, S., Hoover, J., Jennings, D., Jordan, C., Jordan, J., Kasha, J., Kagan, L., Kraft, C., Levitsky, A., Lewis, M., Liu, X., Lopez, J., Ma, D., Majoros, W., McDaniel, J., Murphy, S., Newman, M., Nguyen, T., Nguyen, N., Nodell, M., Pan, S., Peck, J., Peterson, M., Rowe, W., Sanders, R., Scott, J., Simpson, M., Smith, T., Sprague, A., Stockwell, T., Turner, R., Venter, E., Wang, M., Wen, M., Wu, D., Wu, M., Xia, A., Zandieh, A., and Zhu, X. 2001. The sequence of the human genome. Science. 291: 1304–51.

Vindelov, L. and Christensen, I.B. 1990. An integrated set of methods for routine flow cytometric DNA analysis. In Darynkiewiez Z. and Crissman, H. (eds.), *Flow Cytometry*. Academic Press, Inc., San Diego. p. 127–137.

Weber, J.L. and Myers, E.W. 1997. Human whole-genome shotgun sequencing. Genome Res. 7: 401–409.

Yu, J., Hu, S., Wang, J., Wong, G.K., Li, S., Liu, B., Deng, Y., Dai, L., Zhou, Y., Zhang, X., Cao, M., Liu, J., Sun, J., Tang, J., Chen, Y., Huang, X., Lin, W., Ye, C., Tong, W., Cong, L., Geng, J., Han, Y., Li, L., Li, W., Hu, G., Li, J., Liu, Z., Qi, Q., Li, T., Wang, X., Lu, H., Wu, T., Zhu, M., Ni, P., Han, H., Dong, W., Ren, X., Feng, X., Cui, P., Li, X., Wang, H., Xu, X., Zhai, W., Xu, Z., Zhang, J., He, S., Xu, J., Zhang, K., Zheng, X., Dong, J., Zeng, W., Tao, L., Ye, J., Tan, J., Chen, X., He, J., Liu, D., Tian, W., Tian, C., Xia, H., Bao, Q., Li, G., Gao, H., Cao, T., Zhao, W., Li, P., Chen, W., Zhang, Y., Hu, J., Liu, S., Yang, J., Zhang, G., Xiong, Y., Li, Z., Mao, L., Zhou, C., Zhu, Z., Chen, R., Hao, B., Zheng, W., Chen, S., Guo, W., Tao, M., Zhu, L., Yuan, L. and Yang, H. 2002. A draft sequence of the rice genome (Oryza sativa L. ssp. indica). Science. 296: 79–92.

From: *Genome Mapping and Sequencing*
© 2003 Horizon Scientific Press, Wymondham, UK

2

Radiation Hybrid Mapping - A Whole Chromosome Approach

Panos Deloukas

Abstract

Radiation hybrid mapping is a method that combines aspects of both genetic and physical mapping and has proven to be key in the rapid construction of whole genome maps. A panel of radiation hybrid cell lines, each retaining a different portion of the donor genome in the background of a recipient cell, is used to score the presence or absence of sequence tagged sites (STSs). The frequency with which markers co-segregate corresponds to their proximity to one another. Statistical analysis of the similarity of the retention patterns allows definition of marker order and intermarker distances. This chapter aims to provide hands on experience in using this mapping methodology and extends into the logistics of building a radiation hybrid map to support the assembly of bacterial clone maps spanning whole chromosomes.

1. Introduction

Construction of both genetic and physical maps at increasing levels of resolution has been one of the first milestones in the course of the Human Genome Project (HGP). As for any other organism, the ultimate map of the human genome is its complete nucleotide sequence. The decision to proceed and sequence the human genome on a clone by clone basis using existing technology and through a well co-ordinated and systematic approach was taken in the mid nineties by the international community. Several bacterial genomes have been or are currently sequenced using direct shotgun sequence analysis of the whole genome, an approach that requires no prior mapping. However, it has been demonstrated that sequencing of larger genomes relies heavily on the ability to build clone maps. Genome projects such as the yeast *Saccharomyces cerevisiae* which was completed in 1996, the nematode *Caenorhabditis elegans* completed in 1998, the fruit fly *Drosophila Melanogaster* completed in 1998, and the plant Arabidopsis Thaliana completed in 2000 had detailed bacterial clone maps available well in advance of large scale genomic sequencing (Waterston and Sulston, 1995; Goffeau *et al.*, 1996; Mozo *et al.*, 1999; Hoskins *et al.*, 2000). In contrast, such maps were not available up front for most of the human chromosomes. In addition, most of the available long range YAC-based maps (Chumakov *et al.*, 1995), although a valuable resource, were either incomplete and/or had problems caused by chimeric or deleted clones. A turning point in the construction of long range maps was the introduction of whole genome radiation hybrid mapping (Walter *et al.*, 1994).

Whole genome radiation hybrid (WGRH) mapping is a method amenable to automation and high-throughput and provides a means to localise any sequence tagged site (STS) to a defined map position in the genome, by use of the polymerase chain reaction (PCR). For example, gene maps are constructed using STSs derived from expressed sequence tags, ESTs (Schuler *et al.*, 1996; Hukriede *et al.*, 1999; Hudson *et al.*, 2001). WGRH maps have become the primary tool in integrating linkage and physical maps as well as in anchoring and ordering bacterial, large-insert, clone contigs used in sequencing complex genomes (Bentley *et al.*, 2001; Olivier *et al.*, 2001). Compared to other long range mapping reagents, *e.g.* YACs, radiation hybrids have the advantage of providing complete genome-coverage and overcome the type of problems that chimerism and rearrangements often cause in mapping.

2. Theory and Strategies

2.1 Radiation Hybrid Mapping

Radiation hybrid (RH) mapping is a method based on studies by Goss and Harris (Goss and Harris, 1975, 1977) in which they fused lethally irradiated human diploid lymphoblast cells with Chinese hamster cells and selected hybrids on the basis of the retention of an X-linked marker. The order and distance of a number of loci on the X chromosome was inferred on the basis of their absence rates in a panel of such hybrids and relative to the locus under selection. A decade later, (Cox *et al.*, 1990) used a mono-chromosomal hamster-human somatic cell hybrid as the donor and found that the selectable marker could lie on a hamster chromosome of the irradiated donor genome. In contrast to the findings made by Goss and Harris it was shown that the hybrids retained human DNA fragments at high frequencies although no direct selection was applied. The basic assumption made, was that the further apart any two loci are on the human chromosome in the donor hybrid, the more likely that the irradiation will result in breaking them apart and hence the two loci will not co-segregate in the same radiation hybrids. Thus, scoring the presence or absence of a marker in each hybrid of the panel provides a retention pattern and the frequency with which markers co-segregate corresponds to their proximity to one another, which allows determination of their order. The method combines aspects of both genetic and physical mapping. In contrast to recombination events, the frequency of breaks induced by the irradiation seems to be linearly related to physical distance without cold- or hot-spots of breakage along the chromosome. (Walter *et al.*, 1994) re-examined the original work of Goss and Harris and succeeded in constructing whole genome radiation hybrid (WGRH) panels using a diploid human cell line as the donor. WGRH panels became a standard tool in map construction of complex genomes and high resolution maps from a variety of species have been reported in the past few years (Hudson *et al.*, 1995; Stewart *et al.*, 1997; Deloukas *et al.*, 1998; Watanabe *et al.*, 1999; Avner *et al.*, 2001; Breen *et al.*, 2001).

2.2 Considerations in the Choice of a WGRH Panel

The type of donor and recipient cell line as well as the dose of radiation to be used in the construction of a radiation hybrid (RH) panel will depend mainly on the goals of the mapping project to be undertaken. The source of donor cells can be either a monochromosomal hybrid cell line if the aim of the project is to generate a map of a single chromosome, or a diploid male cell line for projects aiming at the construction of whole genome maps. In either case, between 80-100 hybrids are required for the RH panel. Thus, the use of monochromosomal hybrids as donors in whole genome projects will require over 1900 hybrids, which is impractical.

Both the average size of DNA fragments of the irradiated donor cells and the frequency with which they are retained in the hybrids determine the overall resolution of an RH panel. Typically, the higher the dose of radiation the more breaks are induced in the donor DNA, which results in smaller fragments, and lower retention of the fragments in the hybrids. Note that hybrids suitable for WGRH panels should have a retention frequency between 12 and 50% and such panels constructed with a diploid male cell line as the donor have on average a lower resolution for the sex chromosomes than for the autosomes. The construction of WGRH panels is beyond the scope of this chapter; detailed protocols can be found in Stewart and Cox (Stewart and Cox, 1997) as well as in several of the papers cited in Table 1. The higher the resolution of an RH panel, the more markers need to be assayed upfront before any statistically significant linkage can be observed along a whole chromosome. So, when starting a mapping project the selection of which resolution panel to use should be made in light of the size and nature of the available pool of markers. For example, the distribution of gene-based markers in mammalian genomes is uneven and an RH map based solely on such markers and constructed with a too high resolution WGRH panel may have gaps in gene-poor regions. WGRH panels constructed with a radiation dose of 3000 rads seem to provide the long-range continuity required for first generation maps and have been successfull in a variety of species.

Radiation hybrid cell lines are usually unstable and, when re-grown, often tend to loose some of the retained donor fragments. In some instances they even appear to gain fragments; for example when a subpopulation of cells

Table 1. Characteristics of commonly used radiation hybrid panels.

Donor Genome	Recipient	Panel Name	Number of RH clones	Radiation dose	Retention	Resolution	Conversion	Reference
				(rads)	(%)	(kb)	cR / Mb	
Human	Hamster	Genebridge4	83	3,000	32	1000	4	(Gyapay et al.,
Human	Hamster	Stanford G3	93	10,000	18	500	33	(Stewart et al.,
Human	Hamster	Stanford TNG	90	50,000	16	100	250	(Olivier et al.,
Mouse	Hamster	T31	93	3,000	20-25	350	10	(McCarthy et a
Rat	Hamster	T55v3	96	3,000	27-29	410	9.5	(Watanabe et a
Zebra fish	Hamster	T51	94	3,000	20-25	350	16.4	(Geisler et al.,
Zebra fish	Hamster	LN54	93	4-5,000	22	500	6.8	(Hukriede et a
Bovine	Hamster	Womack 5000	90	5,000	28	na	na	(Womack et al
Bovine	Hamster	TM112	94	3,000	28			
Canine	Hamster	RHDF$_{5000}$	126	5,000	20-25	630	na	(Priat et al., 19
Porcine	Hamster	ImpRH	118	6-7,000	27	na	na	(Yerle et al., 1
Equine	Hamster	TM99	90	3,000	28	na	na	(Kiguwa et al.,
Baboon	Hamster	T25	93	3,000	20-25	na	na	

na: not available

takes over. Because of that it is imperative to grow each hybrid in a single batch and extract enough DNA to last for the entire mapping project.

2.3 RH Mapping in the Context of Constructing a Sequence-Ready Map of a Whole Chromosome

Broadly speaking, the mapping strategy adopted by the HGP to decode the human genome consisted of: (a) generating adequate resources of DNA markers, (b) using long-range reagents to order the markers, (c) screening genomic libraries with the mapped markers to isolate overlapping clones and build contigs and (d) closing the gaps between the contigs by means of chromosome walking techniques. The Sanger Centre, now the Wellcome Trust Sanger Institute, in its bid to map and sequence one third of the human genome adopted RH mapping as the long range component of this strategy. RH maps were to serve as long range scaffolds for anchoring and often orienting contigs of large-insert bacterial clones (*e.g.* P1-artificial chromosomes (PACs) and bacterial artificial chromosomes (BACs)). Given that the average insert size of such clones is in the range of 100-200 kb, we considered that RH maps with a density of 15 markers per Mb, should be adequate to support the construction of sequence-ready maps. All markers of the RH map can then be used to isolate bacterial clones and assemble contigs on the basis of landmark content and restriction fingerprint analysis (see Chapter 7). This approach was applied in projects aiming to produce sequence-ready maps of human chromosomes 1, 6, 9, 10, 13, and 20. The chromosome 20 project (http://www.sanger.ac.uk/HGP/Chr20/) will be used as an example.

3. Building a Human, Chromosome Specific, STS Collection

As described above, the first step in our strategy is to construct an RH map with a density of at least 15 markers per Mb. For chromosome 20, which at the start of the project was reported to have an estimated length of 72 Mb, construction of such a map required 1080 STSs. We choose to start by making full use of the then publicly available resources and imported all established STS markers reported to map on chromosome 20 from the Genome Database (GDB; http://www.gdb.org), Généthon (http://www.genethon.fr), the Whitehead Institute (WI; http://www-genome.wi.mit.edu/ftp/distribution/human_STS_releases/), the Stanford Human Genome Center (SHGC; ftp://shgc.stanford.edu/pub/hgmc/), the Cooperative Human Linkage Centre (CHLC; http://lpgws.nci.nih.gov/html-chlc/chlc Markers.html), and the Radiation Hybrid data base (RHdb; ftp://ftp.ebi.ac.uk/pub/databases/RHdb/) via anonymous ftp. Inevitably, one expects a high degree of redundancy among such sets of markers. It is well advised that all corresponding DNA sequences (whenever available) are subjected to standard DNA similarity analysis (*e.g.* BLAST) to assure mapping of a non redundant set of markers. STSs for which only the sequences of the two corresponding primers are available, can be analysed using the electronic PCR program (Schuler, 1997). Once a non redundant set of STS markers is defined, an automated pipe-line could be set up to compare any new DNA sequences, considered for STS development, against the whole collection. At the start of the project, the number of publicly available STS markers was far below the desired figure of 15 per Mb. In addition, a high proportion of all markers was derived from expressed sequence tags (EST) which meant that gene poor regions of the chromosome were most probably underrepresented in the STS collection. The approach we took to deal with both problems at once was to construct a small insert library of flow-sorted chromosome 20 DNA and develop STSs from single pass sequence reads of a random set of clones. This approach is applicable to all human chromosomes, however, the use of a suitable somatic cell hybrid line is probably required for human chromosomes 9-12, which cannot be separated currently by flow-sorting.

3.1 Construction of Small Insert Libraries Using Flow-Sorted DNA - Sequencing and Primer Design

Flow-sorting (Ormerod, 1994) allows the isolation of pure samples of individual chromosomes of human and other organisms. Flow-sorted human chromosomes have been used extensively in mapping as a DNA source for construction of chromosome specific libraries (Ross and Langford, 1997). As mentioned above, our aim is to use such libraries as a source for developing genomic STS-based markers scattered along the chromosome. Protocol 1 describes a method for constructing small insert *Hind*III libraries in pBluescript II (SK+) using flow-sorted DNA. For the chromosome 20 project we used an EBV-transformed lymphoblastoid cell line and isolated DNA corresponding to approximately 250,000 chromosome copies at a purity >95%. Standard blue-white selection of recombinant clones can be used with the *E. coli* XL-1 blue strain (Protocol 1). The constructed library, referred to as SC20pF, gave approximately 85% recombinant clones. The transformation efficiency was estimated to be 1.6 x 10^7 recombinants per µg of insert DNA.

The number of recombinant clones that needs to be sequenced will depend on the length of the chromosome, the targeted marker density, and the number of STS markers available from other sources. Of the 1513 recombinant clones from the SC20pF library we randomly picked for sequencing, 1259 (83.2%) gave sequence reads suitable for primer design. Despite the blue-white selection, around 10% of the sequenced clones had no insert mainly due to the use of a colony picking robot. Primer design was carried out using the *PRIMER 0.5* software (http://www-genome. wi.mit.edu/ftp/distribution/software/primer.0.5/manual.asc). We typically set the product length between 80 and 250 bp. After masking known human DNA repeats we obtained primer pairs for 618 sequences (49.1%).

4. Long Range Mapping Using the Whole Genome Radiation Hybrid Panel Genebridge 4 (GB4)

4.1 Development of STS Markers Suitable for RH Mapping

In general, STS markers for use with a human-hamster RH panel should amplify by PCR a human sequence of specific length and should not co-amplify a hamster-derived sequence of the same size. So, to develop appropriate PCR conditions it is usually enough to test primer pairs using human and hamster genomic DNA. For the chromosome 20 project, we also screen markers for amplification of DNA isolated from a mouse cell line carrying human chromosome 20 (GM13140; Coriell Cell Repositories). It is advisable to have such a control in order to eliminate at an early stage markers that are not on the targeted chromosome. Coriell Cell Repositories which is a source for monochromosomal somatic cell hybrid lines of all human chromosomes, provide conditions on how to grow and maintain the GM13140 cell line. DNA isolation can be performed using standard protocols (Sambrook *et al.*, 1989).

Undertaking a whole chromosome project requires the processing of several hundreds of STSs. To streamline this process for chromosome 20, we test primer pairs simultaneously at three annealing temperatures (typically 55, 60, and 65 °C) using four control DNAs: human (male placental; Sigma), hamster (Chinese Hamster Ovary; Lofstrand Labs Ltd), mouse (BALB/C genomic; Clontech), and chromosome 20 hybrid. PCR reactions are carried out in 96-well plates as described in Protocol 2 by simply adjusting the volumes in steps 1 and 2. Although this approach has a higher reagent cost per marker, it is suitable for high throughput as less than 10% of primer pairs need to be tested at additional annealing temperatures to determine success or failure. Over 80% of all primer pairs tested at this stage were successful. It is worth mentioning that this fraction does not include markers which constantly give a low yield of PCR amplification when assayed with the human control DNA. We have observed that such markers fail when assayed against the RH panel.

In parallel with testing primer pairs for optimal PCR conditions for RH mapping one can generate probes to be used in applications such as screening of genomic libraries by hybridisation (Chapter 6). For example, in the chromosome 20 mapping project for each marker that gave a clean positive signal with both the human and the chromosome 20 hybrid DNA and was negative for the mouse DNA control, the obtained PCR product (human control DNA) was excised from the agarose gel and stored in a 0.5 ml tube containing 100 µl of water at 4 °C.

4.2 Assaying STSs Using the GB4 RH Panel by PCR

Many different PCR protocols have been described in the literature and most of them are likely to be suitable for assaying STSs against RH panels. The one given below, Protocol 2, has been optimised to work on DNA templates of variable purity (*e.g.* DNA from YAC pools) and aims at boosting target amplification (high Mg and dNTP concentration). Typically, most STSs show robust amplification even at higher annealing temperatures (*i.e* ≥65°C), a useful feature when attempting to reduce background. The use of certain equipment and reagents in this protocol aims to minimise manipulation of samples post PCR. For example, thermocyclers with heated lids eliminate the need of adding a layer of mineral oil to the samples and the use of sucrose and cresol red in the PCR reaction allows samples to be loaded directly on agarose gels. Sucrose also seems to reduce non-specific amplification. Although such modifications are essential for automating certain steps of the process , conventional thermocyclers (*i.e.* without heated lids) can be used.

STS markers that have successfully passed the development step are assayed against the GB4 panel as described in Protocol 2. It is highly recommend that all assays are done in duplicate. In our laboratory, markers which give discrepant results for more than three hybrids in the duplicate assay, are re-assayed and if unsuccessful they are excluded from further analysis. The use of duplicate assays aids in the identification of technical failures and/or non robust assays whereas the elimination of highly discrepant results contributes to the accuracy of the map.

4.3 Automation

Both the development of STS markers suitable for RH mapping and their subsequent typing against the GB4 panel are PCR-based, a technique amenable to automation. Automation is key to high throughput; we set up a semi-automated pipeline using a number of robotic devices that are commercially available or have been designed in house. The reader should consider the information given below as a guideline; a plethora of companies nowadays offer more or less sophisticated devices and integrated platforms for automating PCR reactions.

In the first step of our process, primer pairs are screened against four control DNAs. We automated reaction assembly using a Genosis 100 (Tecan) for which we designed adapters to hold the v-shaped reservoirs of the Biomek 1000 (Beckman) robot in order to minimise reagent loss. The second step requires the handling of the GB4 panel; we use the Hydra^{TM} 96 (Robbins Scientific) which is a 96-channel dispenser, for all dilution steps and aliquoting of the DNA into 96-well thermocycler plates. When the same unit is used to aliquot different panels it is recommend that the user performs an acid wash followed by thorough rinsing with H_2O in between operations to avoid cross contamination. The master mix (Protocol 2) is added to the plates carrying the GB4 panel using a Biomek 1000 (Beckman) equipped with an 8-channel pipetting device (MP200).

The loading of the samples post-PCR on to the agarose gels is done using a Flexys robot (Genomic Solutions PLC) equipped with an 8-channel pipetting device which was developed in house. The instrument is calibrated with the upper left well of the gel. Gel trays were designed to hold six combs and were manufactured to high specification to ensure reproducibility. In addition, we have designed metal combs which hook on one side of the gel tray but can move freely from the other side (pouring hot agarose causes the tray first to expand and then to contract). To prevent any shrinking, gels are cast into trays without the gates (taped) which allows the agarose to set into the upper and bottom groove and thus attach to the tray. The gels are kept moistened by wrapping them with Saranwrap.

Following electrophoresis, gel images are acquired using a high resolution solid-state camera (KODAK MEGAPLUS, model 1.4i) attached to a Macintosh computer equipped with a Neotech IG24 image grabber card. The camera and the UV transilluminator are accommodated in a dark cabinet, designed in-house. At the front of the cabinet there is a metal frame which opens and closes like a drawer and allows the gel tray first to be slotted in a fixed position relative to the four circular LEDs of the frame and then moved above the transilluminator. We use the Gel Gem / Gel Print software (ME Electronics Ltd) in conjunction with a Sony UP890CE thermoprinter to create image files of the gels and obtain hard copies. We then carry out image analysis using the Band Analysis software (ME

Electronics Ltd). This software can call and analyse simultaneously the two image files representing a duplicate assay (*i.e.* an STS typed twice against the GB4 panel). For each marker the output of the individual assays and the consensus are stored together in a report file. Report files are transferred overnight to a UNIX environment and read into our central database, rhace.

4.4 Software, Data Storage and Data Analysis

Scoring the presence or absence of a human target sequence in each hybrid DNA of an RH panel provides a pattern of retention (or vector) for each marker. RH results are thus encoded as a vector of values where 1 (one) indicates that the marker was retained in a hybrid, 0 (zero) indicates that it was not, and 2 (two) indicates an ambiguity in the duplicate typing or that the hybrid was not assayed. Constructing a map requires computational analysis of the RH vectors of the assayed markers. Two markers are considered to be linked if they have vectors of statistically significant similarity (defined by a LOD score), and a measure of their separation is obtained from analysis of the degree of difference between the two vectors. Markers which have identical vectors cannot be resolved by the RH panel used and are referred as totally linked. Vector similarity is thus used to estimate both order and distance between markers. The unit of map distance is the cR (centiRay) and represents 1% probability of breakage between two markers for a given X-ray dosage (*i.e.* the one used to construct the RH panel). Therefore the correlation between cR units and physical distance in bp will differ from one panel to the other and can only be determined by extrapolation.

The most commonly used software packages for construction of RH maps are MultiMap (Matise *et al.*, 1994; http://compgen.rutgers.edu/ multimap/), RHMAP (Lunetta *et al.*, 1996; http://www.sph.umich.edu/ /statgen/boehnke/rhmap.html), RHMAPPER (Slonim *et al.*, 1997; http://www.genome.wi.mit.edu/ftp/pub/software/rhmapper); and SAMapper (Stewart *et al.*, 1997; http://shgc.stanford.edu). More recently developed packages such as RHO (Ben-Dor *et al.*, 2000; http://www.cs.technion.ac. il/Labs/cbl/CGI/rh-wizard.pl) and CONCORDE (Agarwala *et al.*, 2000; http://www.math.princeton.edu/tsp/concorde.html) as well as a new method for the rapid construction of reliable RH frameworks (Bo *et al.*, 2002) claim improved marker ordering and reduced computing time. A general page with links to RH software tools is available at http:// compgen.rutgers.edu/rhmap/ #programs/.

RHMAP is the most widely used package and seems to be an excellent choice for small scale mapping projects. In our experience, RHMAP is less well suited for large scale projects compared to MultiMap and RHMAPPER. We developed a set of extensions, referred to as Z-extensions, for RHMAPPER version 1.1 (Soderlund *et al.*, 1998) and we currently use Z-RHMAPPER for construction of RH maps. The Z-extensions are available at ftp.sanger.ac.uk/pub/zmapper and the users manual at http://www.sanger. ac.uk/Software/Z/ Protocol 3 describes the map construction process with Z-RHMAPPER.

Any large scale mapping project requires a powerful and flexible type of database that allows both the storage and display of a variety of mapping data. At the Sanger Institute all mapping projects make use of ACeDB (A *C. elegans* Data Base; http://www.acedb.org) as their central data repository. It is a flexible data base as it allows 'models' (configurable data files) that can be specifically tailored to the needs of different mapping projects and is powerful not only in storing mapping information but also displaying it in a graphical fashion. Some key features of the current model used in human mapping projects that relate to this chapter are (a) the use of grid displays for representing individual clones of radiation hybrid panels and genomic libraries. A simple click can associate landmark data to an object in the grid; (b) the use of graphical displays for STS and clone based maps; (c) the graphical representation of features in DNA sequences (*e.g.* primers, repeat elements, regions with homology to coding sequences etc.). The chromosome 20 ACeDB, known as 20ace, is available at ftp://ftp.sanger.ac.uk/pub/human/chr20/. In addition, the data in 20ace can be accessed via a WWW browser using Webace.

5. The Chromosome 20 RH Map Project

The project started by importing from publicly available sources all markers known to map on chromosome 20. After filtering for markers representing the same locus we compiled a non-redundant set of 1035 STSs composed of:

Genetic Markers :   207
Genomic (random):   336
EST-derived: 492

This marker pool was combined with 1137 additional STSs that were generated from two in house efforts. The human gene map project contributed 485 additional EST-derived markers and the sequencing of a small-insert library prepared with flow-sorted chromosome 20 DNA contributed 652 random genomic markers. The total marker pool consisted of 2172 STSs and of those 1713 were suitable for RH mapping. Markers dropped out during the PCR optimization step for the following reasons:

Multiple human products (141)
No human product (70)
Low yield of human product - weak amplification (72)
Co-amplification of a rodent product of similar size (85)
No amplification of the chromosome 20 control DNA (116)

The last reason implies that the 116 markers do not represent chromosome 20 loci. Of those, 43 were derived from clone sequences of the flow-sorted library indicating that contamination was 6.6%. The remaining 73, however, were markers reported to map on chromosome 20. We have tested 39 of those using Coriell's monochromosomal Mapping panel 2 and assigned all of them to a chromosome other than 20.

The 1713 markers were typed in duplicate across the GB4 panel and for 1691 of them (98.7%) we obtained a consensus vector with ≤3 ambiguities. We typed 85 of the 92 hybrids of the GB4 panel. The hybrids omitted are 4A5, 4O10, 4U3, 4R12, 4BB10, 4B2 and 4B9. The vectors of 1585 STSs were imported into Z-RHMAPPER and used to assemble a framework supported at a LOD score of 2.5 by an interactive approach (Protocol 3). The resulting framework is composed of 40 STSs and was used to place 802 additional markers. Together with the 651 totally linked markers the map harbors 1493 STSs and is available at http://www.sanger.ac.uk/cgi-bin/rhtop?chr=20/. The total map length is 406 cR_{3000}. Given a length estimate of 62.7 Mb for chromosome 20, 1 Mb is on average 6.475 cR_{3000}. The average marker density on the map is 23.8 STSs per Mb. The resolution of the GB 4 RH panel and thus of this map is in the range of 500 kb. The relatively even distribution and the high density of markers across the map has proved adequate to support the reliable construction of bacterial clone contigs spanning the whole chromosome (Deloukas et al., 2001). In addition, the map with most of the known genes and EST clusters present, constitutes an integrated transcript map resource of the chromosome.

6. Protocols

Protocol 1. Construction Of Small Insert Libraries Using Flow-Sorted DNA

Equipment and Reagents

- 1 µg/µl pBluescript II (SK+) (Stratagene)
- 10 x NEB2 buffer (New England Biolabs)
- *Hind*III (20 U/µl; New England Biolabs)
- 0.02 U/µl calf intestinal alkaline phosphatase (Boehringer) freshly diluted from 1 U/µl stock in H_2O
- 10 x TBE: 108 g Tris, 55 g boric acid, 9.3 g Na_2EDTA, made to 1 litre
- Agarose (electrophoresis grade; Gibco-BRL)
- Ethidium bromide 10 mg/ml (Sigma)
- 150 mM nitrilotriacetic acid (NTA)-trisodium salt (Sigma)
- TE buffer: 10 mM Tris-HCl, 1 mM EDTA pH 8.0
- Phenol/chloroform 1:1 v/v; chloroform
- 5 M NaCl
- Ethanol (100% and 70%)
- T4 DNA ligase (400 U/µl; New England Biolabs)
- 10 x ligase buffer (New England Biolabs)
- ~4 ng/µl flow-sorted DNA (prepared according to Ross and Langford (1997). Typically, start with 250,000 copies of the selected human chromosome.
- *E. coli* XL1-blue electroporation competent cells (Strategene)
- LB medium (sterile)
- Sterile Pulser cuvettes; 0.1 cm electrode gap (Biorad)
- Pulse Controller and Gene Pulser (Biorad)

Method

1. To a 0.5 ml tube add 3 µl pBluescript II (SK+), 5 µl 10 x NEB2 buffer, 1 µl *Hind*III, 41 µl H_2O. Incubate the tube for 3 h at 37 °C.
2. Analyse a 1 µl sample (use undigested pBluescript II (SK+) as control) on a 0.7% agarose gel in 1 x TBE buffer containing 0.4 µg/ml ethidium bromide. Photograph gel under a UV transilluminator. Check that the digestion was completed. The expected size is 2,961 bp.
3. To the remainder of the reaction, add 2 µl of 0.02 U/µl calf intestinal alkaline phosphatase. Incubate the tube for 30 min at 37 °C.

4. Add 5.8 μl of 150 mM NTA-trisodium salt to the tube and incubate it for 25 min at 68 °C. Add 44 μl of TE buffer to the tube.

5. Extract the DNA with 100 μl phenol/chloroform. Transfer the aqueous phase to a 1 ml tube. Back extract the organic phase with 100 μl TE, combine the aqueous phases and extract with 200 μl chloroform. Transfer the aqueous phase to a 1.5 ml tube. Back extract the organic phase with an equal volume of TE and combine the aqueous phases.

6. Add to the tube 5M NaCl to a final concentration of 0.2 M and then 2 vol. of absolute ethanol. Precipitate the DNA at -20 °C overnight.

7. Pellet the DNA in a microcentrifuge at 14000 rpm for 10 min at room temperature. Wash the pellet with 1 ml of 70% ethanol, then re-spin the tube for 2 min. Air-dry the DNA pellet briefly and re dissolve it in 60 μl TE buffer (approximately 50 ng/μl).

Note: Although optional it is recommended at that point to test an aliquot of the vector preparation for self ligation in the presence and absence of polynucleotide kinase (Boehringer). Set up two ligation reactions each containing 0.5 μl vector, 1 μl 10 x ligase buffer, 0.5 μl of T4 DNA ligase, and 7.5 μl H_2O. Add 0.5 polynucleotide kinase to one reaction and 0.5 μl of H_2O to the other, incubate both reactions for 1 h at room temperature. Analyse both samples on a 0.7% agarose gel. The sample without polynucleotide kinase should show no sign of ligation, whereas the treated sample should show several ligation products and very little sign of the 2.9 kb fragment.

8. To a 0.5 ml tube add 8 μl (~30 ng) flow-sorted DNA, 1 μl 10 x NEB2 buffer, and 1 μl *Hind*III. Incubate the tube for 3 h at 37 °C.

9. Heat the tube for 20 min at 65 °C to inactivate *Hind*III.

10. In a 0.5 ml tube, ligate 5 μl (~15 ng) of *Hind*III digested flow-sorted DNA (step 9) to 1 μl of vector DNA (50 ng; from step 7) in a 10 μl reaction containing 1 μl of 10 x ligase buffer, 0.5 μl of T4 DNA ligase, and 2.5 μl H_2O. Incubate the tube for 16 h at 16 °C.

11. Heat the tube for 10 min at 65 °C to inactivate the T4 DNA ligase. Add 10 μl of TE to the tube. At this point the ligation mix can be stored at -70 °C.

12. Electroporate 40 μl E. coli XL1-blue electrocompetent cells with 1 μl of the ligation mix (1800 V; 200 ohms; 25 μF). Immediately, add 1 ml of LB medium to the cells and mix gently once.

13. Transfer the cells to a sterile tube and incubate the tube for 45 min at 37 °C with shaking at 250 rpm. Add glycerol to a final

concentration of 15% (v/v) and freeze the cells on dry-ice. Store cells at -70°C.

Protocol 2. Typing STSs Against the GB4 RH Panel by PCR

Equipment and Reagents

- $T_{0.1}E$: 10 mM Tris-HCl, 0.1 mM EDTA pH 8.0
- dNTP mix: 5 mM each dATP, dCTP, dGTP, dTTP in H_2O (Pharmacia)
- PCR buffer (10x): 500 mM Tris, 67 mM $MgCl_2$, 22 g/l $(NH_4)_2SO_4$ in $T_{0.1}E$
- Sucrose solution: 28% w/v sucrose, 0.2 g/l cresol red (sodium salt; Sigma) in H_2O
- Primer pair mix: 100 ng/µl each in H_2O
- Taq DNA polymerase, 5 U/µl ('Amplitaq', Perkin-Elmer)
- GB4 RH panel (96 samples): 93 radiation hybrid DNAs plus human and hamster genomic control DNA, all at 3 ng/µl (Research Genetics). Add a negative control such as H_2O or $T_{0.1}E$

Note: Aliquot concentrated DNA panels and store at -20 °C; avoid repeated thawing-freezing. Dilute DNA with the following buffer: 2:1 v / v H_2O / $T_{0.1}E$; 4.4 mg/l cresol red. Adjust the pH to 8.5 using NaOH. Diluted DNA can be stored at 4 °C for several days.

- β-mercaptoethanol (BDH Laboratory Supplies)
- Ethidium bromide 10 mg/ml (Sigma)
- Agarose (electrophoresis grade; Gibco-BRL)
- 10 x TBE (Protocol 1)
- Thermocycler (MJ Research, INC. PTC-225 DNA Engine Tetrad)
- 96-well thermocycler plates (Costar Thermowell™ 6511)
- OmniSeal TD Mats (Hybaid; HB-MT-SRS-5)
- 100 bp ladder (Pharmacia)
- Agarose gel electrophoresis apparatus
- 300 nm UV transilluminator and gel photography system
- UV Stratalinker 2400 (Strategene)

Method

1. Assemble the following master mix:
   - PCR buffer (10x):   2   µl

- dNTP mix: 1.5 µl
- Sucrose solution: 5.4 µl
- β-mercaptoethanol: 0.1 µl
- primer mix: 0.75 µl
- Amplitaq: 0.25 µl

Volumes are given per single reaction and should be scaled up accordingly; prepare mix with approximately 10% excess when using multichannel pipettes for dispensing.

Note: For testing primer pairs for optimal PCR conditions the master mix is assembled with the appropriate control DNA template instead of the primer mix. Primer pairs to be tested are then added in step 2.

2. Dispense 10 µl of each of the 96 samples of the RH panel into a 96-well thermocycler plate and add 10 µl of the master mix to each well. Fit a mat on top of the plate. Tap gently the plate on the bench to ensure that all the reagents are at the bottom of the well.

Note: Any DNA panel such as DNA isolated from pools of PAC clones (see Chapter 6), can be used at this point.

Note: Reuse mats up to 20 times but UV-irradiate them between runs to avoid cross contamination; use the Stratalinker with the "Energy" mode set at 240 mJoules.

3. Place the plate in the thermocycler, close the heated lid, and cycle as follows:
   - 94 °C 5 min
   followed by 35 cycles of:
   - 93 °C 30 sec
   - T_{an} 50 sec 30 cycles (The optimal annealing temperature of each marker)
   - 72 °C 50 sec
   followed by:
   - 72 °C 5 min   1 cycle
   followed by:
   - 15 °C (indefinite)

4. Remove the mat and analyse 15 µl samples on a 2.5% agarose gel in 1 x TBE buffer containing 0.4 µg/ml ethidium bromide (9 V/cm, 20 min).

5. Place the gel on a 300 nm UV transilluminator and photograph.

Protocol 3. Map Construction with Z-RHMAPPER

The process of map construction with Z-RHMAPPER can be divided in to four steps:
- (a) Set a project in Z-RHMAPPER and load RH vectors into the database.
- (b) Create groups of totally linked markers.
- (c) Identify markers to build a framework.
- (d) Place all other markers relative to the framework.

(a) Set a project and import all marker data into the database

A number of parameters need to be specified up front. We normally set the two parameters which provide estimates of laboratory error, 'ALPHA' (false negative) and 'BETA' (false positive), to 0.001 but recommend using higher values when analysing pooled results (*i.e.* generated in different laboratories). The 'RETENTION_FREQUENCY' is set to a value that depends on the characteristics of the RH panel used. For the GB4 panel we set it to 0.4. Marker data should be in a tab delimited format as shown in the example below with marker name first and RH vector last, whereas multiple fields of information can be specified in between.

Marker	Chr	Sequence	RH vector
10751	20	Z52150	10110010101011010011110011̇0.........0001100111010010010010

For a whole genome project it is best to set a different project for each chromosome. When no information is available on the chromosomal origin of the used markers the alternative approach is to set first a single project, import all the markers and then use the 'link_groups' command at a relevant TWO_POINT_CUTOFF (LOD score) value to identify markers that most probably map to the same chromosome. The value will depend on the panel used; for example a LOD score of 10 is suitable for the GB4 panel. It is recommended to eliminate markers with very low and very high retention frequencies from the analysis. Cut off values will depend on the average retention frequency of the used RH panel. For example, we do not include markers assayed on the GB4 panel which has an average retention frequency of 32%, with a retention frequency below 10% or above 60%. This approach will often result in two groups of markers for metacentric chromosomes, one for each chromosome arm. Markers that map to multiple chromosomes will appear as members of multiple groups and should be eliminated from the analysis when they appear in more than two.

(b) Create groups of totally linked markers

The 'Ztl' command is used to create automatically groups of totally linked (TL) markers. It identifies first all markers with no ambiguities in the RH vector, referred to as canonical. Each canonical marker with a unique vector is then made the representative of a TL group. Markers with a vector identical to that of the TL group's representative, referred to as buried, become members of this group. Markers which have ambiguities in the RH vector and are not part of a TL group are termed orphans. If at a later point a marker becomes part of the framework map (see below) it will be preferentially used as the representative of its TL group. The canonical markers are used as input to the 'Zgrow' and 'Zplace' commands described below.

(c) Construction of a framework map

This is the most crucial step of the whole process. The objective is to identify a set of evenly spaced markers along the chromosome whose order is significantly better, typically at 1000:1 odds, than the next best. The command 'assemble_framework1' is used to assemble automatically a framework but running time can increase dramatically with groups of more than 200 markers. In the absence of any other mapping information the best candidates to be evaluated first for framework construction are canonical markers representing large groups of totally linked markers. In this step we run 'assemble_framework1' with the 'FRAMETHRESH' parameter set to LOD 3. Such frameworks can then be extended using the 'Zgrow' command and all the remaining TL group representatives. Typically, we lower the 'FRAMETHRESH' parameter to LOD 2.5 in this step. Additional markers can then be tested in subsequent rounds. The addition of markers at the two ends of the framework during the final rounds of this step should be carefully checked to avoid erroneous expansion of the overall map length. An independent mapping method such as fluorescent in situ hybridisation (FISH) can be used to assess both the integrity of the end markers of the framework but also their proximity to the true telomeric ends of the analysed chromosome.

When independent mapping information is available, a more interactive approach can be taken. For example, a set of well spaced markers can be selected from the genetic linkage map of an organism. We followed this approach in the construction of the chromosome 20 and all other human chromosome specific RH maps. A set of microsatellite markers ordered with high odds on the Généthon human genetic linkage map (Dib *et al.*, 1996) can be selected across the human chromosome (except the Y chromosome) and typed against the RH panel. The order of markers on the genetic map can then be evaluated as a hypothesis in Z-RHMAPPER. We use the 'Zrip' command set to 3 (maximum is 5) to verify the quality of candidate orders. It works by sliding a window across an order, finding all permutations of markers within that window, and remembering the best order within that window. Our approach is to start with markers at one end of the chromosome. Once the first three most telomeric markers are defined, a framework can be assembled in a step wise fashion by testing one marker at a time and asking that the candidate order is supported at a LOD score greater than 2.5 over the next best order. It is recommended that the generated best orders are also checked using the 'Zevaluate' command which gives estimates of cR distances between markers. For the GB4 panel, running Z-RHMAPPER under the conditions described above, a value of 5 cR_{3000} corresponds on average to 1 Mb. Following this approach we have assembled a framework of 30 microsatellite markers for chromosome 20 which was then extended by the 'Zgrow' command to include a total of 40 markers.

(d) Place all other markers relative to the framework

A comprehensive RH map can be assembled using the 'Zplace' command which executes the 'create_placement_map' command to place non-buried markers on the framework. Each marker is tried sequentially against each interval (bin) of the framework and the LOD score of the corresponding order is stored. The marker is placed in the bin with the highest LOD score and the log difference to the next best bin is reported. The order in which markers appear within a framework bin is mainly based on the cumulative distance of each marker from the top framework marker defining the bin. The operator can utilize interactively some of the commands mentioned above to refine the order of markers within the bin. In our experience, this operation, which is very time consuming, does not result in significant changes. The map is automatically saved in a file called fw.map. We use the default values of LOD 5 and 15 cR for the 'PLACEMENT_LINKAGE' and 'PLACEMENT_TOO_FAR' parameters, respectively. The buried markers are attached to the map using the 'Zprint_map' command. The map output looks like the example given below which is a section of the chromosome 20 RH map between framework markers D20S97 and D20S882.

D20S97	0.89	F
PRNP	1.84	P0.00
D20S742	0.00	P0.84
D20S1095	0.05	P0.84
D20S500	0.46	P0.86
D20S835	0.00	P1.41
D20S1130	0.00	P1.41
stSG25078	0.00	P1.36
stSG25115	0.00	P1.44
stSG25587	0.06	P1.43
D20S849	0.94	P1.32 stSG5017
D20S895	1.63	P1.13
D20S95	0.00	P0.01
stSG2589	0.10	P0.01
D20S882	0.00	F D20S816 stSG12840 stSG15193 D20S805 SGC33687

The first and second column in the map table report the marker name and the distance, in cR_{3000}, to the next marker, respectively. The markers that form the framework are indicated by an F in the third column whereas those placed relative to the framework by a P. For the placement markers, a LOD score value is also reported. As an example P1.32 means that the probability with which D20S849 maps into this framework interval is 20.89 times higher than that of D20S849 mapping into the adjacent interval towards the telomere. Placed markers which are located close to a framework marker will have very low P values as they can map to the bins either side of the framework marker with very similar probabilities. The buried markers of a TL group are shown in the fourth column (*e.g.* D20S849 and stSG5017 belong to the same TL group but only D20S849 was used to calculate the map).

It is good practice to specify always the running conditions of the map construction algorithm as the estimated cR distances between markers are influenced by certain parameters. For example, in Z-RHMAPPER increasing the value for the 'ALPHA' (false negative) and 'BETA' (false positive) parameters results in slightly smaller distances between markers. As mentioned above, the correlation between cR units and physical distance in bp will differ from one panel to the other and can only be determined by extrapolation. For example, the often-quoted size estimate of human chromosome 20 is 72 Mb (Morton, 1991) whereas sequencing this chromosome gave a revised size estimate of 62.7 Mb (Deloukas *et al.*, 2001). The total length of our chromosome 20 RH map is 406 cR_{3000} meaning that $1cR_{3000}$ corresponds on average to 177.4 or 154.4 kb depending on the physical size estimate used. Furthermore, marker retention is not always uniform across RH panels and has an impact on the estimated physical distance between markers. For example, the very high retention of centromeric fragments causes an overestimation of the cR distance between markers flanking the centromere. Correcting for this effect, 1 cR_{3000} corresponds on average to 193 kb in the euchromatic part of chromosome 20. Very high retention frequency should also be expected for markers in the region surrounding the selectable marker used in the construction of an RH panel (*e.g.* in the case of GB4, the thymidine kinase (TK) gene on the q arm of human chromosome 17). As mentioned earlier, the mapping resolution of an RH panel depends on both the average size of DNA fragments of the donor cells and the frequency with which they are retained in the hybrids. Map resolution will decrease within such regions of high retention. Low retention has a similar effect on map resolution. It is therefore best to construct a chromosome X map using an RH panel made with an XX donor cell.

7. Radiation Hybrid Mapping Resources (panels, maps, servers)

Whole genome radiation hybrid panels have been constructed for a number of vertebrate species including human, baboon, cat, cow, dog, horse, mouse, pig, rat, and zebra fish. The most comprehensive collection of commercially available WGRH panels is maintained by Research Genetics (http://www.resgen.com/intro/mapping.php3). Table 1 summarizes the characteristics of some of the most widely used WGRH panels whereas a comprehensive list of nearly all available WGRH panels is available at http://compgen.rutgers.edu/rhmap.

An advantage of RH mapping is that once a dense map is established and made available on the WWW any researcher can map a marker relative to this map. The hard way is to re-compute the map including one's own data. This task requires the availability of all raw data, that is the RH vectors. The main public repository of such data is the Radiation Hybrid database (RHdb; http://www.ebi.ac.uk:80/RHdb/) maintained by the European Bioinformatics Institute at Hinxton UK. An alternative is to use servers, which accept the RH vector(s) of a researcher and compute the most probable position of the corresponding marker(s) relative to the available map, usually the framework. Several centres have set up such servers and for the human genome are available at the Stanford Human Genome Center (G3 and TNG panels; http://shgc.stanford.edu), The Wellcome Trust Sanger Institute (GB4; http://www.sanger.ac.uk/HGP/Rhmap) and the White Head Institute for Biomedical Research (GB4; http://www-genome.wi.mit.edu/cgi-bin/contig/phys_map). A tip to every user of these servers: make sure that the position of each hybrid in "your" vector(s) is identical to the position the server expects for this hybrid; do not enter results for hybrids not used by that centre nor for control DNAs. The actual map displays are also made available by the above centres or can be found in databases such as the National Center for Biotechnology Information (NCBI; http://www.ncbi.nlm.nih.gov), RHdb and the Genome Data Base (GDB; http://www.gdb.org). For example, NCBI is hosting the integrated human GeneMap99 (GB4 and G3; http://www.ncbi.nlm.nih.gov/genemap/). Integration of RH, genetic and other physical maps is a quite straightforward procedure as long as there are enough anchor points, markers, that are common to all of them. The user of an integrated map needs to have a good understanding of the resolution and limitations of each map and corresponding methodology, respectively (Deloukas, 2001). The overall resolution of the integrated map is that of the lowest resolution component.

RH mapping has become the method of choice for the rapid construction of STS-based maps of complex genomes and beside the human the most advanced so far are the mouse, rat and zebrafish maps. NCBI maintains a summary page (http://www.ncbi.nlm.nih.gov/Genomes/index.html) with links to RH maps and servers for these genomes. For mouse, the most comprehensive resource is hosted at The Jackson Laboratory (http://www.jax.org/resources/documents/cmdata/rhmap/RHIntro.html). It acts as a repository of data generated with the T31 WGRH panel and provides both an integrated RH map display and a corresponding server. The University of Iowa is hosting a rat gene map (http://ratest.uiowa.edu/mapping-new/) and server (http://huge.eng.uiowa.edu/cgi-bin/place-marker.pl). Other sites maintaining data on the rat genome are: Otsuka Pharmaceutical Co, Japan (http://ratmap.ims.u.tokyo.ac.jp/menu/RF.html) and the The Rat Genome Database (RatMap; http://ratmap.gen.gu.se/) at the Göteborg University, Sweden. Zebrafish RH maps and servers are available at http://zfrhmaps.tch.harvard.edu/ZonRHmapper/mapService.htm (T51 panel) and http://dir.nichd.nih.gov/lmg/lmgdevb.htm (LN54 panel). Sites for the bovine, canine and porcine genomes can be found at http://bos.cvm.tamu.edu/bovgbase.html/ (Womack-5000 panel), http://www-recomgen.univ-rennes1.fr/doggy.html (RHDF$_{5000}$ panel) and http://imprh.toulouse.inra.fr/ (ImpRH panel), respectively. Maps of the baboon, equine and cat genomes are being developed and the reader is pointed to the links available at http://compgen.rutgers.edu/rhmap. The Roslin Institute is hosting a database, Arkdb (http://www.thearkdb.org/), for the pig, horse, cat, chicken, sheep and other genomes.

Acknowledgements

The author wishes to thank Dr Mark Ross for providing protocols for the construction of small insert libraries from flow-sorted DNA and the members of the Radiation Hybrid and Chromosome 20 mapping groups who assisted in the compilation of the protocols.

References

Agarwala, R., Applegate, D.L., Maglott, D., Schuler, G.D., and Schaffer, A.A. 2000. A fast and scalable radiation hybrid map construction and integration strategy. Genome Res. 10: 350–64.

Avner, P., Bruls, T., Poras, I., Eley, L., Gas, S., Ruiz, P., Wiles, M.V., Sousa-Nunes, R., Kettleborough, R., Rana, A., Morissette, J., Bentley, L., Goldsworthy, M., Haynes, A., Herbert, E., Southam, L., Lehrach, H., Weissenbach, J., Manenti, G., Rodriguez-Tome, P., Beddington, R., Dunwoodie, S., and Cox, R.D. 2001. A radiation hybrid transcript map of the mouse genome. Nat. Genet. 29: 194–200.

Ben-Dor, A., Chor, B., and Pelleg, D. 2000. RHO—radiation hybrid ordering. Genome Res. 10: 365–78.

Bentley, D.R., Deloukas, P., Dunham, A., French, L., Gregory, S.G., Humphray, S.J., Mungall, A.J., Ross, M.T., Carter, N.P., Dunham, I., Scott, C.E., Ashcroft, K.J., Atkinson, A.L., Aubin, K., Beare, D.M., Bethel, G., Brady, N., Brook, J.C., Burford, D.C., Burrill, W.D., Burrows, C., Butler, A.P., Carder, C., Catanese, J.J., Clee, C.M., Clegg, S.M., Cobley, V., Coffey, A.J., Cole, C.G., Collins, J.E., Conquer, J.S., Cooper, R.A., Culley, K.M., Dawson, E., Dearden, F.L., Durbin, R.M., de Jong, P.J., Dhami, P.D., Earthrowl, M.E., Edwards, C.A., Evans, R.S., Gillson, C.J., Ghori, J., Green, L., Gwilliam, R., Halls, K.S., Hammond, S., Harper, G.L., Heathcott, R.W., Holden, J.L., Holloway, E., Hopkins, B.L., Howard, P.J., Howell, G.R., Huckle, E.J., Hughes, J., Hunt, P.J., Hunt, S.E., Izmajlowicz, M., Jones, C.A., Joseph, S.S., Laird, G., Langford, C.F., Lehvaslaiho, M.H., Leversha, M.A., McCann, O.T., McDonald, L.M., McDowall, J., Maslen, G.L., Mistry, D., Moschonas, N.K., Neocleous, V., Pearson, D.M., Phillips, K.J., Porter, K.M., Prathalingam, S.R., Ramsey, Y.H., Ranby, S.A., Rice, C.M., Rogers, J., Rogers, L.J., Sarafidou, T., Scott, D.J., Sharp, G.J., Shaw-Smith, C.J., Smink, L.J., Soderlund, C., Sotheran, E.C., Steingruber, H.E., Sulston, J.E., Taylor, A., Taylor, R.G., Thorpe, A.A., Tinsley, E., Warry, G.L., Whittaker, A., Whittaker, P., Williams, S.H., Wilmer, T.E., Wooster, R., and Wright, C.L. 2001. The physical maps for sequencing human chromosomes 1, 6, 9, 10, 13, 20 and X. Nature. 409: 942–3.

Bo, T.H., Jonassen, I., Eidhammer, I., and Helgesen, C. 2002. A fast top-down method for constructing reliable radiation hybrid frameworks. Bioinformatics. 18: 11-8.

Breen, M., Jouquand, S., Renier, C., Mellersh, C.S., Hitte, C., Holmes, N.G., Cheron, A., Suter, N., Vignaux, F., Bristow, A.E., Priat, C., McCann, E., Andre, C., Boundy, S., Gitsham, P., Thomas, R., Bridge, W.L., Spriggs, H.F., Ryder, E.J., Curson, A., Sampson, J., Ostrander, E.A., Binns, M.M., and Galibert, F. 2001. Chromosome-specific single-locus FISH probes allow anchorage of an 1800-marker integrated radiation-hybrid/linkage map of the domestic dog genome to all chromosomes. Genome Res. 11: 1784–95.

Chumakov, I.M., Rigault, P., Le Gall, I., Bellanne-Chantelot, C., Billault, A., Guillou, S., Soularue, P., Guasconi, G., Poullier, E., Gros, I., *et al.* 1995. A YAC contig map of the human genome. Nature. 377: 175-297.

Cox, D.R., Burmeister, M., Price, E.R., Kim, S., and Myers, R.M. 1990. Radiation hybrid mapping: a somatic cell genetic method for constructing high-resolution maps of mammalian chromosomes. Science. 250: 245–50.

Deloukas, P. 2001. Map integration. From a genetic map to a physical gene map and ultimately to the sequence map. Methods Mol. Biol. 175: 129–42.

Deloukas, P., Matthews, L.H., Ashurst, J., Burton, J., Gilbert, J.G., Jones, M., Stavrides, G., Almeida, J.P., Babbage, A.K., Bagguley, C.L., Bailey, J., Barlow, K.F., Bates, K.N., Beard, L.M., Beare, D.M., Beasley, O.P., Bird, C.P., Blakey, S.E., Bridgeman, A.M., Brown, A.J., Buck, D., Burrill, W., Butler, A.P., Carder, C., Carter, N.P., Chapman, J.C., Clamp, M., Clark, G., Clark, L.N., Clark, S.Y., Clee, C.M., Clegg, S., Cobley, V.E., Collier, R.E., Connor, R., Corby, N.R., Coulson, A., Coville, G.J., Deadman, R., Dhami, P., Dunn, M., Ellington, A.G., Frankland, J.A., Fraser, A., French, L., Garner, P., Grafham, D.V., Griffiths, C., Griffiths, M.N., Gwilliam, R., Hall, R.E., Hammond, S., Harley, J.L., Heath, P.D., Ho, S., Holden, J.L., Howden, P.J., Huckle, E., Hunt, A.R., Hunt, S.E., Jekosch, K., Johnson, C.M., Johnson, D., Kay, M.P., Kimberley, A.M., King, A., Knights, A., Laird, G.K., Lawlor, S., Lehvaslaiho, M.H., Leversha, M., Lloyd, C., Lloyd, D.M., Lovell, J.D., Marsh, V.L., Martin, S.L., McConnachie, L.J., McLay, K., McMurray, A.A., Milne, S., Mistry, D., Moore, M.J., Mullikin, J.C., Nickerson, T., Oliver, K., Parker, A., Patel, R., Pearce, T.A., Peck, A.I., Phillimore, B.J., Prathalingam, S.R., Plumb, R.W., Ramsay, H., Rice, C.M., Ross, M.T., Scott, C.E., Sehra, H.K., Shownkeen, R., Sims, S., Skuce, C.D., Smith, M.L., Soderlund, C., Steward, C.A., Sulston, J.E., Swann, M., Sycamore, N., Taylor, R., Tee, L., Thomas, D.W., Thorpe, A., Tracey, A., Tromans, A.C., Vaudin, M.,

Wall, M., Wallis, J.M., Whitehead, S.L., Whittaker, P., Willey, D.L., Williams, L., Williams, S.A., Wilming, L., Wray, P.W., Hubbard, T., Durbin, R.M., Bentley, D.R., Beck, S., and Rogers, J. 2001. The DNA sequence and comparative analysis of human chromosome 20. Nature. 414: 865–71.

Deloukas, P., Schuler, G.D., Gyapay, G., Beasley, E.M., Soderlund, C., Rodriguez-Tome, P., Hui, L., Matise, T.C., McKusick, K.B., Beckmann, J.S., Bentolila, S., Bihoreau, M., Birren, B.B., Browne, J., Butler, A., Castle, A.B., Chiannilkulchai, N., Clee, C., Day, P.J., Dehejia, A., Dibling, T., Drouot, N., Duprat, S., Fizames, C., Bentley, D.R., *et al.* 1998. A physical map of 30,000 human genes. Science. 282: 744–6.

Dib, C., Faure, S., Fizames, C., Samson, D., Drouot, N., Vignal, A., Millasseau, P., Marc, S., Hazan, J., Seboun, E., Lathrop, M., Gyapay, G., Morissette, J., and Weissenbach, J. 1996. A comprehensive genetic map of the human genome based on 5,264 microsatellites. Nature. 380: 152–4.

Geisler, R., Rauch, G.J., Baier, H., van Bebber, F., Brobeta, L., Dekens, M.P., Finger, K., Fricke, C., Gates, M.A., Geiger, H., Geiger-Rudolph, S., Gilmour, D., Glaser, S., Gnugge, L., Habeck, H., Hingst, K., Holley, S., Keenan, J., Kirn, A., Knaut, H., Lashkari, D., Maderspacher, F., Martyn, U., Neuhauss, S., Haffter, P., *et al.* 1999. A radiation hybrid map of the zebrafish genome. Nat. Genet. 23: 86–9.

Goffeau, A., Barrell, B.G., Bussey, H., Davis, R.W., Dujon, B., Feldmann, H., Galibert, F., Hoheisel, J.D., Jacq, C., Johnston, M., Louis, E.J., Mewes, H.W., Murakami, Y., Philippsen, P., Tettelin, H., and Oliver, S.G. 1996. Life with 6000 genes. Science. 274: 546, 563–7.

Goss, S.J., and Harris, H. 1975. New method for mapping genes in human chromosomes. Nature. 255: 680–4.

Goss, S.J., and Harris, H. 1977. Gene transfer by means of cell fusion I. Statistical mapping of the human X-chromosome by analysis of radiation-induced gene segregation. J. Cell Sci. 25: 17–37.

Gyapay, G., Schmitt, K., Fizames, C., Jones, H., Vega-Czarny, N., Spillett, D., Muselet, D., Prud'Homme, J.F., Dib, C., Auffray, C., Morissette, J., Weissenbach, J., and Goodfellow, P.N. 1996. A radiation hybrid map of the human genome. Hum. Mol. Genet. 5: 339–46.

Hoskins, R.A., Nelson, C.R., Berman, B.P., Laverty, T.R., George, R.A., Ciesiolka, L., Naeemuddin, M., Arenson, A.D., Durbin, J., David, R.G., Tabor, P.E., Bailey, M.R., DeShazo, D.R., Catanese, J., Mammoser, A., Osoegawa, K., de Jong, P.J., Celniker, S.E., Gibbs, R.A., Rubin, G.M., and Scherer, S.E. 2000. A BAC-based physical map of the major autosomes of Drosophila melanogaster. Science. 287: 2271–4.

Hudson, T.J., Church, D.M., Greenaway, S., Nguyen, H., Cook, A., Steen,

R.G., Van Etten, W.J., Castle, A.B., Strivens, M.A., Trickett, P., Heuston, C., Davison, C., Southwell, A., Hardisty, R., Varela-Carver, A., Haynes, A.R., Rodriguez-Tome, P., Doi, H., Ko, M.S., Pontius, J., Schriml, L., Wagner, L., Maglott, D., Brown, S.D., Lander, E.S., Schuler, G., and Denny, P. 2001. A radiation hybrid map of mouse genes. Nat. Genet. 29: 201–5.

Hudson, T.J., Stein, L.D., Gerety, S.S., Ma, J., Castle, A.B., Silva, J., Slonim, D.K., Baptista, R., Kruglyak, L., Xu, S.H., and *et al.* 1995. An STS-based map of the human genome. Science. 270: 1945–54.

Hukriede, N., Fisher, D., Epstein, J., Joly, L., Tellis, P., Zhou, Y., Barbazuk, B., Cox, K., Fenton-Noriega, L., Hersey, C., Miles, J., Sheng, X., Song, A., Waterman, R., Johnson, S.L., Dawid, I.B., Chevrette, M., Zon, L.I., McPherson, J., and Ekker, M. 2001. The LN54 radiation hybrid map of zebrafish expressed sequences. Genome Res. 11: 2127–32.

Hukriede, N.A., Joly, L., Tsang, M., Miles, J., Tellis, P., Epstein, J.A., Barbazuk, W.B., Li, F.N., Paw, B., Postlethwait, J.H., Hudson, T.J., Zon, L.I., McPherson, J.D., Chevrette, M., Dawid, I.B., Johnson, S.L., and Ekker, M. 1999. Radiation hybrid mapping of the zebrafish genome. Proc. Natl. Acad. Sci. USA. 96: 9745–50.

Kiguwa, S.L., Hextall, P., Smith, A.L., Critcher, R., Swinburne, J., Millon, L., Binns, M.M., Goodfellow, P.N., McCarthy, L.C., Farr, C.J., and Oakenfull, E.A. 2000. A horse whole-genome-radiation hybrid panel: chromosome 1 and 10 preliminary maps. Mamm. Genome. 11: 803–5.

Lunetta, K.L., Boehnke, M., Lange, K., and Cox, D.R. 1996. Selected locus and multiple panel models for radiation hybrid mapping. Am. J. Hum. Genet. 59: 717–25.

Matise, T.C., Perlin, M., and Chakravarti, A. 1994. Automated construction of genetic linkage maps using an expert system (MultiMap): a human genome linkage map. Nat. Genet. 6: 384–90.

McCarthy, L.C., Terrett, J., Davis, M.E., Knights, C.J., Smith, A.L., Critcher, R., Schmitt, K., Hudson, J., Spurr, N.K., and Goodfellow, P.N. 1997. A first-generation whole genome-radiation hybrid map spanning the mouse genome. Genome Res. 7: 1153–61.

Morton, N.E. 1991. Parameters of the human genome. Proc. Natl. Acad. Sci. USA. 88: 7474–6.

Mozo, T., Dewar, K., Dunn, P., Ecker, J.R., Fischer, S., Kloska, S., Lehrach, H., Marra, M., Martienssen, R., Meier-Ewert, S., and Altmann, T. 1999. A complete BAC-based physical map of the Arabidopsis thaliana genome. Nat. Genet. 22: 271–5.

Olivier, M., Aggarwal, A., Allen, J., Almendras, A.A., Bajorek, E.S., Beasley, E.M., Brady, S.D., Bushard, J.M., Bustos, V.I., Chu, A., Chung, T.R., De Witte, A., Denys, M.E., Dominguez, R., Fang, N.Y., Foster, B.D., Freudenberg, R.W., Hadley, D., Hamilton, L.R., Jeffrey, T.J., Kelly,

L., Lazzeroni, L., Levy, M.R., Lewis, S.C., Liu, X., Lopez, F.J., Louie, B., Marquis, J.P., Martinez, R.A., Matsuura, M.K., Misherghi, N.S., Norton, J.A., Olshen, A., Perkins, S.M., Perou, A.J., Piercy, C., Piercy, M., Qin, F., Reif, T., Sheppard, K., Shokoohi, V., Smick, G.A., Sun, W.L., Stewart, E.A., Fernando, J., Tejeda, Tran, N.M., Trejo, T., Vo, N.T., Yan, S.C., Zierten, D.L., Zhao, S., Sachidanandam, R., Trask, B.J., Myers, R.M., and Cox, D.R. 2001. A high-resolution radiation hybrid map of the human genome draft sequence. Science. 291: 1298–302.

Ormerod, M.G., Ed. 1994. Flow Cytometry: A Practical Approach. Oxford University Press, Oxford.

Priat, C., Hitte, C., Vignaux, F., Renier, C., Jiang, Z., Jouquand, S., Cheron, A., Andre, C., and Galibert, F. 1998. A whole-genome radiation hybrid map of the dog genome. Genomics. 54: 361–78.

Ross, M.T., and Langford, C.F. 1997. The use of flow-sorted chromosomes in genome mapping. In: Genome mapping: a practical approach. P.H. Dear ed. IRL Press, Oxford. p. 165–184.

Sambrook, J., Fritsch, E.F., and Maniatis, T. 1989. Molecular Cloning: A Laboratory Manual. Cold Spring Harbor Laboratory Press, Cold Spring Harbor, New York.

Schuler, G.D. 1997. Sequence mapping by electronic PCR. Genome Res. 7: 541–50.

Schuler, G.D., Boguski, M.S., Stewart, E.A., Stein, L.D., Gyapay, G., Rice, K., White, R.E., Rodriguez-Tome, P., Aggarwal, A., Bajorek, E., Bentolila, S., Birren, B.B., Butler, A., Castle, A.B., Chiannilkulchai, N., Chu, A., Clee, C., Cowles, S., Day, P.J., Dibling, T., Drouot, N., Dunham, I., Duprat, S., East, C., Hudson, T.J., *et al.* 1996. A gene map of the human genome. Science. 274: 540–6.

Slonim, D., Kruglyak, L., Stein, L., and Lander, E. 1997. Building human genome maps with radiation hybrids. J. Comput. Biol. 4: 487–504.

Soderlund, C., Lau, T., and Deloukas, P. 1998. Z extensions to the RHMAPPER package. Bioinformatics. 14: 538–9.

Stewart, E.A., and Cox, D.R. 1997. Radiation hybrid mapping. In: Genome mapping: a practical approach. P.H. Dear ed. IRL Press, Oxford. p. 73–93.

Stewart, E.A., McKusick, K.B., Aggarwal, A., Bajorek, E., Brady, S., Chu, A., Fang, N., Hadley, D., Harris, M., Hussain, S., Lee, R., Maratukulam, A., O'Connor, K., Perkins, S., Piercy, M., Qin, F., Reif, T., Sanders, C., She, X., Sun, W.L., Tabar, P., Voyticky, S., Cowles, S., Fan, J.B., Cox, D.R., *et al.* 1997. An STS-based radiation hybrid map of the human genome. Genome Res. 7: 422–33.

Walter, M.A., Spillett, D.J., Thomas, P., Weissenbach, J., and Goodfellow, P.N. 1994. A method for constructing radiation hybrid maps of whole genomes. Nat. Genet. 7: 22–8.

Watanabe, T.K., Bihoreau, M.T., McCarthy, L.C., Kiguwa, S.L., Hishigaki, H., Tsuji, A., Browne, J., Yamasaki, Y., Mizoguchi-Miyakita, A., Oga, K., Ono, T., Okuno, S., Kanemoto, N., Takahashi, E., Tomita, K., Hayashi, H., Adachi, M., Webber, C., Davis, M., Kiel, S., Knights, C., Smith, A., Critcher, R., Miller, J., James, M.R., *et al.* 1999. A radiation hybrid map of the rat genome containing 5,255 markers. Nat. Genet. 22: 27–36.

Waterston, R., and Sulston, J. 1995. The genome of Caenorhabditis elegans. Proc. Natl. Acad. Sci. USA. 92: 10836–40.

Womack, J.E., Johnson, J.S., Owens, E.K., Rexroad, C.E., 3rd, Schlapfer, J., and Yang, Y.P. 1997. A whole-genome radiation hybrid panel for bovine gene mapping. Mamm. Genome. 8: 854–6.

Yerle, M., Pinton, P., Robic, A., Alfonso, A., Palvadeau, Y., Delcros, C., Hawken, R., Alexander, L., Beattie, C., Schook, L., Milan, D., and Gellin, J. 1998. Construction of a whole-genome radiation hybrid panel for high-resolution gene mapping in pigs. Cytogenet. Cell Genet. 82: 182–8.

From: *Genome Mapping and Sequencing*
© 2003 Horizon Scientific Press, Wymondham, UK

3

Construction of Large-Insert Bacterial Clone Libraries

Jeffrey Garnes, Marguerite Ciancio and Andreas Gnirke

Abstract

This chapter provides an introduction to the theory and practice of genomic library construction in fosmid and BAC vectors. It contains a protocol for preparing agarose-embedded high-molecular-mass genomic DNA from materials that require grinding and homogenization to access the DNA. It describes the preparation of fosmid and BAC-cloning vector, partial digestion of the genomic DNA, preparative pulsed-field gel electrophoresis and guides the reader step-by-step through the entire process of fosmid and BAC-library construction.

1. Introduction

Most genome projects adopt the time-honored "divide-and-conquer" strategy whereby the genomic DNA is first broken down into smaller pieces and sampled in the form of a library of cloned DNA fragments. Individual clones are analyzed by restriction fingerprinting (Coulson *et al.*, 1986; Olson *et al.*, 1986; Marra *et al.*, 1997; Marra *et al.*, 1999), STS content (Olson *et al.*, 1989; Green and Olson, 1990; Green and Green, 1991; Foote *et al.*, 1992; Hudson *et al.*, 1995) or, more recently, direct sequencing (Fleischmann *et al.*, 1995; Myers *et al.*, 2000; Venter *et al.*, 2001). The pieces are then re-assembled to build a composite clone-based map of the underlying genome.

As is true for any jigsaw puzzle, the larger the individual pieces the fewer are needed, and the easier it is to reconstruct the entire genome. Yeast artificial chromosomes (YACs; Burke *et al.*, 1987), with a cloning capacity exceeding 1 Mb, are unrivalled in terms of sheer clone size and hence their ability to provide long-range contiguous genome coverage. On the other hand, YAC cloning is perhaps the most difficult cloning system, and, once constructed, YAC libraries are plagued by the relatively high proportion of chimeric and unstable clones. Moreover, there are no simple methods to isolate even modest quantities of pure YAC DNA, thereby precluding the use of YACs as direct sequencing templates. On the other end of the spectrum are small-insert clones in plasmid vectors or M13 phage. These clones provide ease of DNA preparation and compatibility with high-throughput automated sequencing at the expense of long-range continuity. Contigs of small-insert clones are therefore comparatively small, leaving genome maps highly fragmented.

This chapter describes how to make fosmids (Kim *et al.*, 1992) and bacterial artificial chromosomes (BACs; Shizuya *et al.*, 1992), large-insert plasmids which provide ease of plasmid-DNA purification and a respectable and useful insert size, about halfway between conventional plasmids and YACs. BACs in particular are key components of virtually all contemporary strategies to map and sequence complex genomes. In the classic clone-by-clone strategy, BACs serve as mapped intermediate substrates for the generation of small-insert libraries for sequencing and provide a facile path to finished sequence. In whole-genome-shotgun strategies, BACs provide scaffolds that link, organize and order the otherwise highly fragmented and unordered sequence assemblies derived from small-insert clones.

Fosmids are similar to cosmids (Collins and Hohn, 1978) in that cos-mediated *in vitro* packaging in bacteriophage lambda heads is used to deliver 38-52 kb molecules into the *E. coli* host where they are propagated as circular plasmids. The only difference is that fosmids have a backbone derived from the single-copy F-plasmid of *E. coli* whereas cosmids are high-copy-number replicons and hence more prone to deletions and rearrangements by recombination.

BACs, too, have an F-factor derived single-copy plasmid backbone. However, BACs are transferred into the *E. coli* host by means of electroporation rather than via lambda packaging and transduction. While electroporation overcomes the strict upper size limit of lambda packaging it has a markedly lower cloning efficiency for large DNA molecules. Moreover, while lambda packaging excludes clones below a minimum size threshold, quite the contrary holds for electroporation in that there is a strong bias towards clones harboring small or no inserts. Consequently, in the absence of biological size exclusion, the success of large-insert BAC-library construction depends critically on efficient physical means to exclude small clonable fragments prior to ligation and electroporation.

Fosmid cloning is fast, efficient, robust and forgiving in terms of quantity and quality of the source DNA. It takes only about two weeks to construct a fosmid library, including the DNA preparation. In contrast, constructing a large-insert BAC library is a serious undertaking that can easily take three to six months - with no guarantee for success. Indeed, few laboratories worldwide have reduced large-insert BAC cloning to a routine exercise.

We strongly recommend constructing a fosmid library as a prelude to BAC cloning. Fosmid cloning is an excellent practice round for investigators who are not yet familiar with the intricacies of large-insert cloning. Moreover, sequencing and analyzing a small number of gene-specific and some randomly selected fosmids can provide valuable insights in gene structure and genome organization such as presence and distribution of repeat elements and other potential pitfalls for future genome mapping and sequence assembly early on. Finally, a fosmid library is valuable in its own right, a stable resource that does not necessarily become obsolete once a larger-insert BAC library becomes available. Computer simulations, common sense and actual experience with whole-genome shotgun sequencing all suggest that it is crucial to have several clone libraries of widely different but narrow size distributions. With cloned inserts between 40 and 45 kb, fosmids fall between small-insert plasmids (3-5 kb) and BACs (150 kb). The compact, pre-defined size distribution of fosmid libraries means *a priori* knowledge of the spacing between clone-end sequences and hence statistical power for mapping and sequence-assembly algorithms.

2. Library Statistics

The three most fundamental statistical parameters of a library are (1) the haploid genome size, (2) the size of the library, *i.e.*, the number of recombinant clones and (3) the average size of the cloned inserts. Let us consider a 1-Gb genome that has been sampled in a library comprising 10^6 clones with an average insert size of 10^4 bp. The total number of basepairs cloned in this library is $10^6 \times 10^4 = 10^{10}$, ten times the haploid genome size or ten genome equivalents. The number of genome equivalents in a library is also called its redundancy or depth d. A library that contains ten genome equivalents is sometimes referred to as a 10X or 10–hit library. The depth ten indicates the average number of clones for any given locus present in the library.

The probability (P) of finding n clones by screening an ideal d-deep library with a short single-copy probe can be calculated from the Poisson equation $P(n) = (d^n/n!) \times e^{-d}$. The probability of finding one or more clones with this probe is $1 - P(0) = 1 - e^{-d}$. The chance of finding at least one clone in a 1X library is therefore about 63%; increasing the library size ten-fold renders finding a clone almost certain (P=99.995%) – at least in theory.

It is important to keep in mind that these calculations are based on the assumptions that the initial fragmentation was completely random and that there is no bias at any cloning step. In practice, either assumption can only be approximated. Fragmentation by partial digestion with one restriction enzyme will depend on the distribution of restriction sites along the chromosomes. Any region that is devoid of a restriction site will be absent whereas regions with clusters of restriction sites will lead to fragments that may fall below the lower size cut of the library. Perhaps the best approximation to random fragmentation is hydrodynamic shearing. However, even a perfectly random fragmentation will rarely give rise to an ideal library. Many of the downstream steps such as recovery of DNA fragments from an agarose gel, transformation and maintenance in the microbial host will be somewhat influenced by parameters such as GC content or melting temperature, let alone biological activities of the cloned sequences themselves. Moreover, most if not all real-world genomes contain regions that are refractory to cloning and are therefore underrepresented or even absent in a real library. If possible one should therefore construct a larger library than suggested by the Poisson equation above.

The choice of cloning system, insert-size range and distribution, and number of clones necessary depends on numerous factors including the goal of the genome project, the role of the library within the genome project, the method for determining overlap among the clones, the density of STSs or probes in the genome to be studied, timing issues, freezer capacity and budget. Nonetheless, details aside, as a general rule, a library for comprehensive genome analysis should provide at least 10-fold, better 20-fold statistical genome coverage.

3. Special Instrumentation

The protocols in this chapter require access to a pulsed-field gel apparatus (*e.g.*, CHEF DRIII, Biorad #170-3700). The electroporation settings in the protocol for BAC libraries are for a specific electroporation device (Invitrogen CellPorator #11609-013 with Voltage booster #11612-017).

For large-scale clone handling including picking, arraying, replication and screening of libraries we recommend using a robotic colony picker and liquid handler (*e.g.*, QBot or QpixII from Genetix PLC, UK). Finally, storage of large arrayed libraries (and of multiple copies thereof) requires sufficient −80°C freezer capacity as well as a plate-labeling system and corresponding database.

4. Protocols

Our protocols for the construction of fosmid and BAC libraries use partial digestion with the four-basepair-cutter *Mbo*I to generate clonable DNA fragments and include a preparative pulsed-field gel run to size-select the *Mbo*I fragments prior to cloning into the *Bam*HI site of the cloning vector.

Constructing large-insert genomic libraries from partial restriction digests requires a method of DNA preparation that is on one hand gentle and avoids excessive double-strand breakage by hydrodynamic shearing and, on the other hand, is thorough enough to eliminate even trace amounts of endogenous nuclease or of inhibitors of the restriction enzyme. As a rule, the bulk of the starting DNA should be at least five times the size of the desired cloning range. For BAC libraries it is important to protect the DNA from shear forces by embedding the source material in solid agarose prior to complete lysis of cells or nuclei. Although fosmid libraries can be prepared from DNA prepared in liquid form we prefer the agarose format for these libraries as well. General guidelines, tips and step-by-step protocols for preparing, handling and analyzing high-molecular-mass DNA have been published elsewhere (Riethman *et al.*, 1997).

4.1 Preparation of Agarose-Embedded Genomic DNA

It is impossible to write a universal protocol that works for any imaginable organism. Detailed protocols for isolating high-molecular-mass DNA from soft mammalian tissues, sperm, cultured cells, yeast and gram-negative bacteria can be found in Riethman *et al.* (1997). Additional protocols have been compiled by Birren and Lai (1993). The protocol below has worked well for "hard" cases, organisms that cannot be embedded in agarose as provided by nature, including chitin-armored beetles and cellulose-clad plants. It involves grinding of frozen material with mortar and pestle, homogenization and cell lysis in a glass tissue grinder, embedding of the crude nuclear fraction in agarose followed by complete lysis and extraction of remaining soluble cell constituents by a series of washes in a solution containing a high concentration of EDTA and a strong detergent. For other, softer or looser material that can be easily teased apart and dispersed to a near-homogenous suspension, steps 1–7 can be skipped. Some materials, for example a suspension of L1 larvae from C. elegans, require more than EDTA and detergent to gain access to the genomic DNA. For these materials one can treat the agarose block with proteinase K during steps 12-14 (see Riethman *et al.* (1997) for a detailed protocol).

The protocol provides enough agarose blocks to carry out the fosmid-library protocol as described in 4.2.3 twice (although it is rarely necessary to repeat it). Most BAC libraries will require a scaled-up protocol. For example, we have used up to 6 g of frozen beetles without changing the volumes in steps 1-7, but resuspended and embedded the final pellet in a larger volume (steps 8 and 9). The concentration of the DNA in the agarose block will obviously depend on the amount of DNA in the starting material, the efficiency of tissue grinding and homogenization and on how much material, cells and nuclei gets lost during filtration and centrifugation. None of these parameters are necessarily known beforehand. For fosmid libraries we prefer a "blind" approach, that is to proceed with the fosmid library protocol without prior assessment of the agarose-embedded DNA. BAC-library construction is much more sensitive to the quality and quantity of DNA in the agarose blocks and may require optimizing the concentration of crude nuclei in the agarose block using an analytical partial digest similar to steps 1-10 of protocol 4.2.3 as an assay for quality and quantity of the agarose-embedded DNA. If necessary, small DNA fragments can be electrophoresed out of the agarose blocks immediately prior to the partial digestion (Osoegawa *et al.*, 1998).

Protocol 4.1. Preparation of Agarose-Embedded High-Molecular-Mass DNA

Equipment and Reagents

- Frozen starting material (1-2 g)
- Porcelain mortar (~50-100 ml) and pestle
- Liquid nitrogen and dry ice
- DNA Homogenization Buffers DHB_{200} and DHB_{500} and DHB_0 (30 mM Tris-HCl, pH 8.0, 100 mM NaCl, 10 mM EDTA, 200 or 500 or 0 mM sucrose, respectively)
- 40-ml Dounce homogenizer (Kontes tissue grinder; VWR #KT885300-040)
- 45°C waterbath

- 2% Seaplaque GTG low-melting-point agarose (BioWhittaker Molecular Applications #50111) in DHB_0 (melted and held at 45°C)
- Coarse (150-200 μm) polypropylene or nylon mesh (*e.g.*, Spectra #148557 or #145556)
- Fine (40 μm) nylon mesh (*e.g.*, Spectra #145585)
- Small (~8 cm diameter) funnel
- Plug molds (we use 400-μl (25x8x2 mm) re-usable molds from the Washington University machine shop (kreitlerj@msnotes.wustl.edu); 100-μl disposable plug molds are available from BioRad #1703706)
- 6-well tissue-culture plates
- LDS solution (1% lauryl sulfate, lithium salt, Sigma #L-4632, 10 mM Tris-HCl, pH 8.0, 100 mM EDTA)
- 0.2 X NDS solution (1X NDS is 0.5 M EDTA, 10 mM Tris, 1% N-laurylsarcosine Sigma #L-9150, pH adjusted to 9.5 with NaOH)
- 0.5 M EDTA, pH 8.0
- TE, pH 8.0

Method

1. Place tube containing 1-2 g frozen material in liquid nitrogen. Fill Dounce homogenizer with 25 ml ice-cold DHB_{200} and put on ice. Pour liquid nitrogen into mortar with pestle until both are cold enough to hold the liquid nitrogen for about 2 minutes. Fill mortar once more and wait until the liquid nitrogen has almost evaporated.
2. Immediately transfer frozen material into pre-cooled mortar and grind the material to a fine powder.
3. With an autoclaved spatula transfer the frozen powder to the Dounce homogenizer tube containing 25 ml DHB_{200}. Use the spatula to disperse the powder in the liquid. Insert glass pestle "A" and homogenize the material by carefully moving the pestle up and down (at least five times) until it begins to run smoothly taking some care not to splash the liquid. Continue the homogenization by at least five strokes with glass pestle "B".
4. Split the liquid in two 15-ml conical tubes. Centrifuge for 1 min at 1,000 rpm in a clinical swing-out centrifuge to remove remaining debris. Immediately decant the supernatant into a 50-ml tube and put on ice.
5. Fold a round piece of fine nylon gauze (150 mm diameter) to a conical filter. Put a similarly folded coarse nylon gauze on top of it. Use a paper clip or a gloved hand to hold the sandwich together while filtering the crude homogenate through the gauze using a

funnel to collect the filtrate in a 50-ml tube on ice. This step may take up to 15 min, depending on the consistency of the homogenate. If the flow halts, save the unfiltered liquid and replace the clogged top (coarse) gauze with a fresh one.

6. Transfer the filtrate to a 30-ml Corex centrifuge tube and centrifuge for 10 min at 6,500 rpm (7,000 x g) in a Sorvall HB-6 rotor (or equivalent) at 4°C. Carefully decant the supernatant. With some materials (*e.g.*, fat caterpillars) there will be a lipid layer floating on top of the supernatant. Use a 1-ml pipettor to resuspend the pellet of crude nuclei with 1 ml ice-cold DHB_{200} taking some care to avoid carry-over of residual lipid.

Note: It is important to work quickly and to resuspend the pellet completely. Nuclei are somewhat sticky and can easily clump together. Keeping the material on ice seems to help keeping the nuclei in suspension.

7. Transfer the 1 ml suspension to a fresh 30-ml Corex tube and bring the volume up to 10 ml with ice-cold DHB_{200}. Carefully underlay the suspension with 10 ml ice-cold DHB_{500} buffer (using either a 10-ml syringe with needle or a 10-ml pipette) and centrifuge for 10 min at 7,500 rpm (9,000 x g) in a Sorvall HB-6 rotor at 4°C. Carefully decant the supernatant.

8. Resuspend the pellet in 1 ml ice-cold DHB_{200} as above and repeat step 7.

9. Use a 1-ml pipettor to resuspend the pellet with 0.4 ml ice-cold DHB_{500}, measure the volume and bring it up to 0.8 ml with ice-cold DHB_{500}.

Note: The concentration of nuclei will depend on the nature of the starting material. To determine the concentration one can remove an aliquot and count cells and nuclei under the light microscope using a hemacytometer. However it is often impossible to tell the difference between nuclei, nucleated cells and other debris. In these cases one can stain the DNA with DAPI and count nuclei under the fluorescence microscope. This assay will give an accurate count and indication of the expected DNA concentration. The DNA concentration (in μg per ml agarose block) can be calculated as follows:

DNA concentration (μg/ml) = (number of nuclei/ml x haploid genome size (in bp) $\times$ ploidy $\times$ 660 $\times$ 10^6) $\div$ (6 $\times$ 10^{23}).

10. Place the glass tube containing the resuspended nuclei in a 45°C waterbath for exactly 2 min, add 1 volume (0.8 ml) 2% SeaPlaque GTG agarose in DHB_0 held at 45°C and mix by pipetting up and

down. Hold the tube in a beaker filled with 37-45°C water while quickly dispensing the agarose mixture into four 400-μl slots of a plug mold that has been taped at the bottom with autoclave tape. Put the plug mold for 10 min at 4°C.

11. Remove the tape and extrude the solidified agarose blocks into a 6-well plate, 2 blocks per well, using a disposable 200-μl pipette tip or a bent glass Pasteur pipette.

12. Add 5 ml LDS to each well, free any agarose blocks that stick to each other or to the plastic surface and incubate for 1 h at 37°C on an orbital shaker set to about 100 rpm.

13. Carefully remove the used LDS and replace it with 5 ml fresh LDS. Incubate overnight at 37°C with gentle agitation.

14. Repeat step 13.

15. Replace LDS with 5 ml 0.2X NDS. Incubate for 1 h at 37°C with gentle agitation.

Note: LDS precipitates at 4°C. Do not store the agarose blocks at 4°C before the LDS has been sufficiently diluted by at least two washes with 0.2X NDS.

16. Replace 0.2X NDS with 5 ml fresh 0.2X NDS and continue incubation at 37°C with shaking overnight.

17. Replace 0.2X NDS with 5 ml 0.5 M EDTA, pH 8. Gently shake the plate for 1 h at room temp. Equilibrate the agarose blocks in 0.5 M EDTA by five additional changes of 0.5 M EDTA, each time letting the blocks equilibrate with the solution for about 1 h.

18. After the last addition of 0.5 M EDTA, pH 8, place three layers of Parafilm between the 6-well plate and its lid to prevent evaporation. Store the agarose blocks at 4 °C under 0.5 M EDTA.

Note: Color and transparency of the agarose blocks depend on the presence of insoluble material in the crude nuclei. Appearance is of secondary importance. We have constructed BAC and fosmid libraries from agarose blocks that were opaque and dark brown. It is however critical that the DNA is high-molecular-mass, accessible and digestable with restriction enzyme and not contaminated with endogenous nuclease. All of these properties will become evident at the partial-digestion step (step 10 in protocol 4.2.3, for example).

4.2 Construction of Fosmid Libraries

The following protocols describe the construction of fosmid libraries from size-selected *Mbo*I partial digests in the original fosmid vector pFOS1

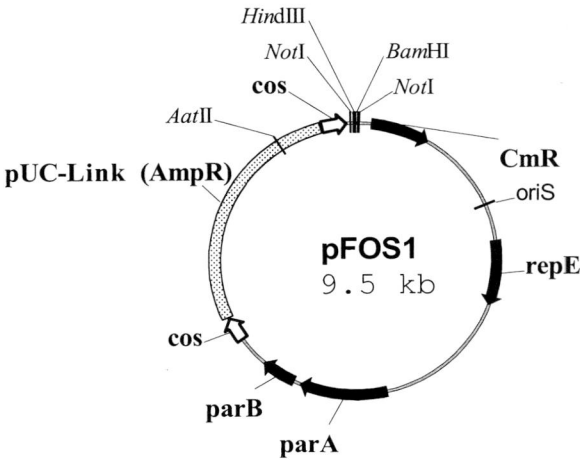

Figure 1. Map of pFOS1. The high-copy number pMB1 replicon of the pUC link between the two cos sites allows isolation of sufficient quantities of pFOS1 plasmid. To prepare cloning vector, pFOS1 plasmid is linearized at the *Aat*II site followed by complete dephosphorylation to prevent concatemerization of the vector during ligation. The *Bam*HI cloning site is opened by complete digestion with *Bam*HI. Genomic *Mbo*I fragments that have been size-selected to 40–45 kb are ligated to an excess of pFOS1 cloning arms. Molecules that contain two cos sites, neither less than 38 kb nor more than 52 kb apart and in the right orientation, are packaged into bacteriophage lambda heads *in vitro*. The packaging machinery removes the pUC link sequences at the ends of the ligation products. After circularization inside the *E. coli* host, the fosmid is under the control of the F-factor-derived functions oriS and repE which mediate unidirectional replication and parA and parB which control and maintain the copy number of the fosmid at one or two copies per cell. The lambda packaging extract used in the protocol below preferentially packages molecules between 47 and 52 kb, about 7 kb of which are vector. A large proportion of cloned fosmid inserts are therefore between 40 and 45 kb in size.

(Kim *et al.*, 1992). A stab of pFOS1 plasmid in *E. coli* strain POP2136 can be requested from Dr. U.-J. Kim, California Institute of Technology (ung@its.caltech.edu). pFOS1-containing *E. coli* POP2136 should be grown at 30°C in LB under double-selection with 50 mg/L Amp and 25 mg/L Chloramphenicol (Cam). To prepare large quantities (>100 μg) of pFOS1 plasmid we recommend growing a 500-ml LB culture at 30°C and using a commercial resin-based maxiprep kit (*e.g.*, the large-construct kit from Qiagen which includes a digestion with exonuclease III to minimize the contamination with (linear) *E. coli* genomic DNA) followed by standard CsCl banding (p. 1.42 in Sambrook *et al.*, (1989)).

Protocol 4.2.1 Preparation of pFOS1 Vector Arms

Equipment and reagents

- CsCl-purified pFOS1 plasmid (50 μg)
- 20 u/μl *Aat*II (NEB #117L)
- 10X NEB4 buffer
- 20 u/μl calf intestine alkaline phosphatase (CIP; Roche #1097075)
- 10X CIP buffer (Roche)
- 20 u/μl *Bam*HI (NEB #136S)
- 10X *Bam*HI buffer (NEB)
- 100X BSA (NEB)

Method

1. To a 1.5-ml tube add 50 μg pFOS1 plasmid, 50 μl 10X NEB4, and H$_2$O to a final volume of 490 μl. Remove 3 μl and save for step 2. To the remainder add 10 μl 20 u/μl *Aat*II. Incubate for 2 h at 37°C.
2. Remove 3 μl and analyze along with the pre-digestion sample on a 0.7% agarose gel. Linearized pFOS1 should run as a single band at 9.5 kb.
3. Extract the reaction once with 500 μl phenol/chloroform. To the aqueous phase add 1/20 volume 5 M NaCl and 2.1 volumes Ethanol. Precipitate the DNA for 2 h at –20°C. Centrifuge for 15 min at 14,000 rpm in a microcentrifuge. Wash the pellet with 1 ml 70% Ethanol and respin the tube for 2 min at room temperature. Aspirate supernatant, spin briefly and remove any remaining liquid. Air dry the pellet and resuspend in 50 μl TE, pH 8.0.
4. To 50 μl linearized pFOS1 add 400 μl H$_2$O, 50 μl 10X CIP buffer and 1 μl 20 u/μl CIP. Incubate for 1 h at 37°C.
5. Add 1 μl 20 u/μl CIP and incubate for 1 h at 55°C.
6. Add 50 μl 0.5 M EDTA, pH 8, and inactivate CIP for 10 min at 75°C.
7. Extract with 550 μl phenol/chloroform. To the aqueous phase add 1/20 volume 5 M NaCl and 2.1 volumes Ethanol. Precipitate the DNA for 2 h at –20°C. Centrifuge for 15 min at 14,000 rpm in a microcentrifuge. Wash the pellet with 1 ml 70% Ethanol and respin the tube for 2 min at room temperature. Aspirate supernatant, spin briefly and remove any remaining liquid. Air dry the pellet and resuspend in 50 μl TE, pH 8.0.

8. To 50 μl dephosphorylated pFOS1 add 390 μl H$_2$O, 50 μl 10X *Bam*HI buffer and 5 μl 100X BSA. Remove 3 μl and save for next step. To the remainder add 5 μl 20 u/μl *Bam*HI (~2 u *Bam*HI per μg pFOS1). Incubate for 1 h at 37°C.

9. Analyze 3 μl of the reaction along with the pre-digestion aliquot on a 0.7% agarose gel. The vector arms should run as two bands at 7 kb and 2.5 kb, respectively.

10. Heat inactivate the enzyme for 20 min at 65°C.

11. Extract once with 500 μl phenol/chloroform. To the aqueous phase add 1/20 volume 5 M NaCl and 2.1 volumes Ethanol. Precipitate the DNA for 2 h at –20°C. Centrifuge for 15 min at 14,000 rpm in a microcentrifuge. Wash the pellet with 1 ml 70% Ethanol and respin the tube for 2 min at room temperature. Aspirate supernatant, spin briefly and remove any remaining liquid. Air dry the pellet and resuspend in 50 μl TE, pH 8.0.

12. Determine the DNA concentration (by A_{260} or fluorometry with Hoechst 33258). Expect about 75-80% recovery (~40μg). Dilute the vector arms in TE, pH 8 to a final concentration of 100 ng/μl and store at –20°C.

Protocol 4.2.2 Preparation of Plating Bacteria for Fosmid Libraries

Equipment and Reagents

- *E. coli* strain XL1-Blue MR (Stratagene #200300)
- 10-cm LB agar plates
- LB broth
- 1 M MgSO$_4$
- 10 mM MgSO$_4$
- 20% maltose
- LB broth containing 10 mM MgSO$_4$ and 0.2% maltose
- 10 mM MgSO$_4$ containing 20% (v/v) glycerol

Method

1. Inoculate 5 ml LB containing 10 mM MgSO$_4$ and 0.2% maltose with a single colony of *E. coli* strain XL1-Blue MR grown on an LB plate. Incubate overnight with shaking.

2. Bring 250 ml LB containing 10 mM MgSO$_4$ and 0.2% maltose to 37°C and inoculate with 2.5 ml overnight culture. Shake at 200 rpm

at 37°C. After 2 h start to follow the growth by measuring the A_{600} until the culture has reached 1.0 A_{600} (which takes about 4 h).

3. Put on ice and spin for 10 min at 1,000 x g in a 250 ml conical tube (alternatively use five 50-ml tubes) in a swing-out rotor in a tabletop centrifuge at 4°C.

4. Decant supernatant. Add 5 ml ice-cold 10 mM $MgSO_4$ and pipette up and down to resuspend the pellet. Fill up to 125 ml with ice-cold 10 mM $MgSO_4$ and centrifuge as in the previous step.

5. Decant supernatant. Add 5 ml ice-cold 10 mM $MgSO_4$ containing 20% (v/v) glycerol, pipette up and down to resuspend the pellet. Bring volume to 25 ml with ice-cold 10 mM $MgSO_4$ containing 20% (v/v) glycerol and freeze in 100-μl and 500-μl aliquots in screw-cap cryovials at −80°C.

Protocol 4.2.3 Partial *Mbo*I Digestion and Size Selection to 40-45 kb

Equipment and reagents

- Pulsed-field gel apparatus (BioRad #170-3700) with wide (21×14 cm) gel casting tray.
- Molecular-biology grade agarose (*e.g.*, SeaKem GTG agarose; Biowhittaker Molecular Applications #50074)
- "High-molecular-weight" DNA marker (Invitrogen #15618-010) 1:5 diluted in TE, pH 8, plus Ficoll 400, BPB and XC
- "low-range" PFG marker (NEB #350)
- Two 400-μl blocks of agarose-embedded DNA (see 4.1)
- 50 mg/ml acetylated BSA (Invitrogen #15561-020)
- 10X *Mbo*I buffer w/o Mg^{++} (500 mM Tris-HCl, pH 7.9, 1 M NaCl, 10 mM DTT)
- 1X *Mbo*I buffer w/o Mg^{++} (500 μg/ml acetylated BSA, 50 mM Tris-HCl, pH 7.9, 100 mM NaCl, 1 mM DTT)
- 1 M $MgCl_2$
- 5 u/μl *Mbo*I (NEB #147S)
- Disposable plastic spatulas (Sarstedt #81.970)
- Ready-to-use dialysis tubing (Spectrum #132110)
- TE, pH 8 with 100 μg/ml acetylated BSA
- 20 mg/ml glycogen (Roche #901303)
- lambda-DNA concentration standards (1, 3, 10, 30 and 100 ng/μl in TE, pH 8; dilutions made from Invitrogen #25250-010)

Method

1. Dialyze two 500-μl blocks of agarose-embedded DNA in 50 ml TE, pH 8 for 30 min at room temperature with occasional agitation. Repeat dialysis three times, the last time overnight at 4°C.

2. Remove liquid and add 5 ml 1X *Mbo*I buffer w/o Mg^{++}. Keep on ice for 1 h with occasional gentle swirling.

3. Repeat step 2.

4. In the meantime, add 1 μl 5 u/μl *Mbo*I to 199 μl 1X *Mbo*I buffer w/o Mg^{++}, mix well, and prepare a 5-fold serial dilution in 1X *Mbo*I buffer w/o Mg^{++} in 1.5-ml microcentrifuge tubes on ice as follows:

Tube # (n)	1	2	3	4	5	6	7	8
Buffer (μl)	199	160	160	160	160	160	160	160
5 u/μl *Mbo*I (μl)	1	–	–	–	–	–	–	–
content of tube # n–1 (μl)	–	40	40	40	40	40	40	–
*Mbo*I (u in 160 μl)	4	0.8	0.16	0.03	0.006	0.001	0.0003	–

5. After transferring 40 μl of tube 6 to tube 7, mix well, remove 40 μl from tube 7 and discard. Tube 8 is a mock digestion, *i.e.*, contains reaction buffer but no *Mbo*I. Keep all tubes on ice.

6. Carefully place the two equilibrated agarose blocks inside the lid of a petri dish. With a glass microscope coverslip cut each agarose block perpendicular to the long axis into 5 pieces of about 80 μl each. Using a disposable plastic spatula and a pipette tip put two "aliquots" in a microfuge tube labeled "0" and store on ice until the preparative electrophoresis (step 9). Put one slice each into tubes 1-8. Mix gently and leave on ice for 1 h. During this time the *Mbo*I restriction enzyme will start diffusing into the block. However, in the absence of Mg^{++} it will not digest the embedded DNA.

7. Add 2.5 μl 1 M MgCl$_2$ to each tube and mix gently. Put tubes back on ice for another 30 min before transferring the tubes to a 37°C water bath. Incubate for 2 h.

8. Add 1 ml 50 mM EDTA, mix gently and put on ice. After 30 min, carefully remove liquid and replace with 1 ml TE, pH 8.

9. In the meantime tape the teeth of a gel comb with autoclave tape to form two slots (each about 6 cm wide) plus at least three conventional wells on either side and between the slots. Melt 180 ml 1% regular molecular-biology grade agarose in 0.5X TBE and pour a 21×14 cm gel. Save about 5 ml of the melted agarose and keep at 65°C.

10. Place one agarose slice from tube 0 on a clean piece of Parafilm and cut it in two unequal pieces (about 10 µl and 70 µl). Load the smaller "aliquot" upright in the left wide slot and the remainder in the right slot. Do the same with slices 8 through 1. (By working in this order one does not need to change the coverslip every time). Do not leave a gap between the larger pieces from tubes 1-7. Load a thin slice of "low-range" PFG marker in a center lane and two slices in two outside lanes. Seal all "plug" containing slots and wells with melted agarose. Place the gel in a pulsed-field gel box filled with 2 L fresh 0.5X TBE. Heat the "high-molecular-weight" gel marker for 5 min at 65°C and load 20 µl in the middle and 20 µl each into two outside wells. Set the electric field angle to 120° and the switching time to a linear ramp from 1 to 5 s and run the gel at 6 V/cm for 13 h at 14°C.

11. Make a cut between the preparative (containing the larger plugs) and the analytical (size markers and the small plugs) portions of the gel and stain only the latter two pieces with Ethidium bromide. View on a UV transilluminator covered with plastic wrap and make two small notches to mark the position of the two largest bands of the "high-molecular-weight" marker (38.4 and 48.5 kb). Use the Ethidium bromide-stained analytical lanes to determine which reactions produced digestion products between 40 and 50 kb. Reassemble the gel and excise the 40-45-kb region from the corresponding unstained preparative lanes. Stain and reassemble the gel and take a picture to have a permanent record (see Figure 2).

12. Place a strip of dialysis tubing for 30 min in a container filled with H_2O. Insert the agarose slice, close one end of the tubing and add 750 µl TE, pH 8, containing 100 µg/ml acetylated BSA. Squeeze out any air bubbles. Close the other end and place dialysis bag in a gel box with the long axis of the slice perpendicular to the electrical field. Fill the gel box with 0.5X TBE and electroelute for 2 h at 80V at 4°C. Reverse the polarity for 1 min and recover all liquid from within the tubing. Extract twice with an equal volume phenol/chloroform and once with chloroform, each time shaking the tube by hand and centrifuging for 5 min in a microcentrifuge. To the aqueous phase (usually 400 µl) add 1/20 volume 5 M NaCl, 0.5 µl 20 mg/µl glycogen and 2.1 volumes Ethanol, mix and put for 2 h at −20°C.

13. Spin for 30 min at 14,000 rpm in a microcentrifuge at 4°C. Carefully decant the supernatant and wash pellet once with 1 ml 70% Ethanol. Respin for 5 min at room temperature. Carefully

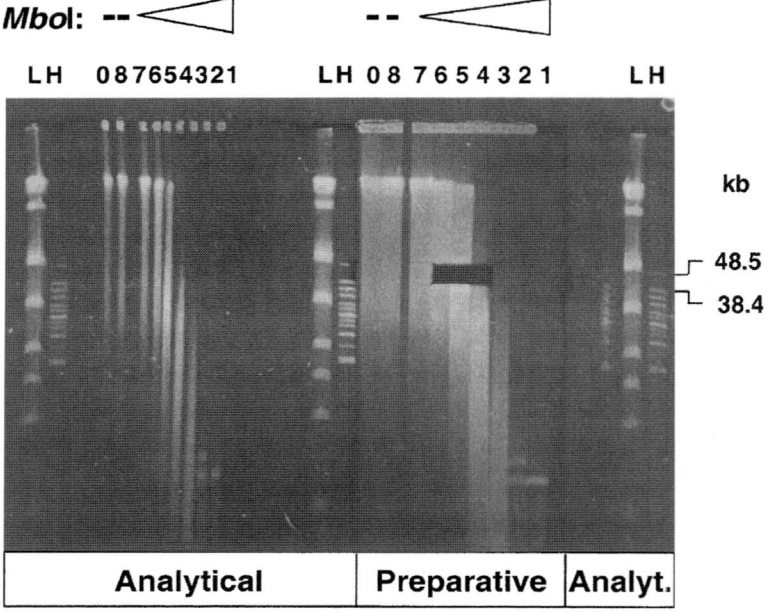

Analytical | **Preparative** | **Analyt.**

Figure 2. Partial *Mbo*I digestion and size selection to 40-45 kb. Lanes marked with L contain low-range PFG marker. H denotes "high-molecular weight" size marker. "Analytical" and "preparative" portions of the gel are indicated. Samples 0 (control), 8 (mock digestion) and 7-1 (increasing amounts of *Mbo*I) were treated exactly as described in steps 1-8 of protocol 4.2.3. The DNA preparation used in this particular experiment already contained DNA fragments of less than 100 kb, perhaps caused by mechanical shearing during grinding and homogenization of the starting tissue or by endogenous nuclease (lane 0). However, there is no increase in fragments <100 kb in the mock digestion (lane 8), indicating the absence of nuclease activity in the final DNA preparation. Reactions 4-6 produced *Mbo*I fragments in the size range required for fosmid cloning. The DNA in lane 6 is slightly under-digested, whereas reaction 4 is slightly over-digested. Fragments from all three lanes were excised and used for cloning to allow representation of both *Mbo*I-rich and *Mbo*I-poor regions of the genome.

aspirate supernatant, spin briefly and remove any remaining liquid. Air dry the pellet by leaving the tube open for about 10 min. Add 10 μl TE, pH 8 and let the pellet re-hydrate for a about 30 min on ice before resuspending it completely by slowly pipetting up and down.

14. Heat a series of lambda-DNA concentration standards (1, 3, 10, 30 and 100 ng/μl) for 5 min at 65°C. Mix 1 μl of each quantitation standard as well as 1 μl of the size-selected *Mbo*I fragments with 5 μl TE, pH 8 and 1 μl 6X gel-loading buffer and run for 15-30 min at 50 V on a 0.8% miniagarose gel. Estimate the concentration of

the *Mbo*I fragments relative to the lambda standards. Typical concentrations range from 5 to 50 ng/μl. Store at 4°C until use.

Protocol 4.2.4 Fosmid Ligation and *In Vitro* Lambda Packaging

Equipment and Reagents

- pFOS1 vector arms (100 ng/μl)
- 40-45-kb *Mbo*I fragments (for example 30 ng/μl)
- 400 u/μl T4 DNA ligase (NEB #202S)
- 10X ligase buffer (NEB)
- Gigapack XL lambda packaging extract (Stratagene #200209)
- SM buffer (p. A.7 in Sambrook *et al.*, (1989))
- Cell-culture-grade DMSO (Sigma #D-2650)
- Plating bacteria (see protocol 4.2.2)
- LB broth (prewarmed to 37°C)
- 10-cm LB-Cam (15 mg/L) agar plates
- 96-well format plasmid-isolation kit (*e.g.*, Qiagen R.E.A.L prep #26173)
- 10 u/μl *Not*I (NEB #189L)
- 10X NEB3 buffer (NEB)
- 100X BSA (NEB)
- 20 u/μl HindIII (NEB #104L)
- 10X NEB2 buffer (NEB)

Method

1. To a 0.5-ml tube add H_2O (*e.g.*, 3.2 μl), 1 μl 10X ligase buffer, 2 μl 100 ng/μl pFOS1 vector arms, 100 ng size-selected *Mbo*I fragments (*e.g.*, 3.3 μl) and 0.5 μl 400 u/μl T4 DNA ligase. Set up another reaction with TE, pH 8 instead of the *Mbo*I fragments. Incubate both reactions overnight at 16°C.
2. Remove one tube of packaging extract from dry ice. Hold between two fingers until content starts to thaw. Immediately add 3.3 μl of the control ligation without *Mbo*I fragments, gently stir with pipette tip and put in microtube rack. Repeat this procedure three times with 3 × 3.3 μl of the ligation with *Mbo*I fragments. Leave packaging reactions at room temperature for 90 min.
3. Add 500 μl SM and mix by inverting. Spin briefly and pool the three packaged ligations in one 2.2 -ml tube. Use 2 × 2.5 μl from

both the minus-insert control and from the pooled fosmid library to determine the titer (step 4). To the remaining ~1.5 ml of the library add 113 μl DMSO (7% f.c.), mix well, and store in 250-μl aliquots in screw-cap cryovials at -80°C.

4. To titer the library, thaw 100 μl plating bacteria (see 4.2.2) and dilute 1:5 with 400 μl 10 mM $MgSO_4$. To a 1.5-ml tube add 25 μl diluted plating bacteria and 2.5 μl library. Set up duplicate reactions for the library plus two for the minus-insert control plus two with plain SM buffer. Mix gently and leave at room temperature for 30 min.

5. Add 200 μl prewarmed LB and incubate for 45 min at 37°C, inverting the tubes every 15 min.

6. Put on ice and plate 2 × 100 μl on 10-cm LB-Cam (15 mg/L) plates. Incubate overnight at 37°C.

7. Count colonies and calculate the titer. Typical titers range from 50 to 500 colony-forming units (cfu) per μl frozen packaged library. The titer of the minus-insert control should be less than 0.1% of that.

8. Pick up to 96 transformants (sample across the entire size range of colonies) and grow minicultures for 18 h at 37°C in 1.2 ml 2X YT-Cam (15 mg/L) broth in a 2-ml deep-well block. Use a commercial plasmid kit and isolate fosmid DNA according to the manufacturer's instructions. Resuspend the fosmid DNA in 25 μl TE, pH 8 and set up analytical digests with *Not*I (step 9) and a suitable six-basepair-cutter such as HindIII (step 10)

9. To a 1.5-ml tube add 836 μl H_2O, 220 μl 10X NEB3 buffer, 22 μl 100X BSA and 22 μl 10 u/μl *Not*I. Mix and dispense 96 × 10 μl into a PCR tray. Add 5 μl fosmid DNA, mix and incubate for 3 h at 37°C and 15 min at 65°C in a PCR machine. Add 3 μl 6X gel-loading buffer and run on two 1% 0.5X TBE pulsed-field gels for 16 h at 120°, 6V/cm, 0.5-2 s and 14°C.

10. To a 1.5-ml tube add 858 μl H_2O, 220 μl 10X NEB2 buffer, 22 μl 20 u/μl HindIII. Mix and dispense 96 × 10 μl into a PCR tray. Add 5 μl fosmid DNA, mix and incubate for 3 h at 37°C and 15 min at 65°C in a PCR machine. Add 3 μl 6X gel-loading buffer and analyze on two 0.8% agarose gels.

Protocol 4.2.5 Large-scale Transduction and Plating of Fosmid Libraries

Equipment and Reagents

- Fosmid library (25,000 cfu in <0.5 ml SM; see 4.2.4)

- Plating bacteria (see 4.2.2)
- LB (prewarmed to 37°C)
- 80% glycerol (optional)
- 22×22 cm square LB-Cam (15 mg/L) agar plates (dried)

Method

1. To a 50-ml conical tube add 2 ml 10 mM $MgSO_4$, 500 μl thawed plating bacteria and 25,000 cfu fosmid library in up to 0.5 ml SM. Mix gently and leave at room temperature for 30 min.
2. Add 8 ml prewarmed LB and shake for 45 min at 200 rpm at 37°C. Put on ice.
3. Optional: add 2 ml 80% glycerol and freeze the transduction mix in 1-ml aliquots in screw-cap cryovials in a styrofoam container at −80°C. In our hands, freezing in liquid nitrogen or dry-ice/ethanol can result in a significant (>50%) loss of titer.
4. Plate on ten 22×22 cm square LB-Cam (15 mg/L) agarose plates using glass beads to distribute the liquid evenly on the surface of the plate. Discard the beads and then let the plate air dry for about 10 min. Incubate overnight at 37°C.

4.3 Construction of BAC Libraries

The fundamental problem of BAC cloning is the strong bias of electroporation for small plasmids and the poor overall cloning efficiency for large inserts. Since there is no practical method for size fractionation of large circular molecules, small clonable fragments have to be excluded prior to ligation. The size selection is done by preparative pulsed-field gel electrophoresis. Pulsed-field gels are easily overloaded, leading to mobility shifts relative to size markers and poor separation. Moreover, DNA molecules at high concentrations have a tendency to aggregate somewhat, resulting in the retardation of small molecules and contamination of seemingly large size fractions with small clonable DNA fragments. On the other hand, the concentration of large-insert fragments has to be high enough to recover intact molecules at sufficient concentration for the subsequent ligation. In other words, one has to find the right balance of DNA concentration and running conditions to achieve efficient separation while avoiding dilution of the precious fragments. All these purification steps including the handling of large DNA molecules in solution after electroelution from the gel become increasingly harder as the insert size increases.

To make matters worse, the cloning efficiency for fragments greater than 150 kb in size is extremely low, requiring extraordinary precautions during the preparation of the cloning vector to keep the background of non-recombinant clones below 5%. Most BAC vectors include a mechanism to screen or select against recircularized empty vector. pBeloBacII (Kim *et al.*, 1996) and its derivatives pCUGI (Luo *et al.*, 2001) and pIndigoBAC (Birren *et al.*, 1997) allow blue/white selection through alpha complementation of the lacZ gene. pBACe.3.6 (Frengen *et al.*, 1999) and its derivatives allow positive selection through insertional inactivation of a SacBII gene cassette and selection for SacBII-deficient transformants on sucrose-containing plates. However neither mechanism provides a safeguard against empty clones that have deleted or otherwise functionally inactivated gene cassettes around the cloning site. It is therefore important to carefully titrate each enzymatic reaction so as to avoid even trace amounts of unwanted side products caused by star activity or exonuclease and to prepare sufficient quantities of quality-controlled BAC cloning vector at the beginning of the project.

In a nutshell, the novice BAC cloner will commonly face pitfalls on two fronts, the preparation of insert DNA *and* the cloning vector. In the absence of a positive control for either one there is no easy way to troubleshoot experiments.

The protocols below are based on the pioneering work and methods published by Pieter de Jong and colleagues (Osoegawa *et al.*, 1998). We clone size-selected *Mbo*I fragments between the *Bam*HI sites of pTARBAC1 (Zeng *et al.*, 2001), a derivative of pBACe3.6 (Frengen *et al.*, 1999) that includes a yeast centromere and selectable marker for re-cloning of the corresponding genome segment from other individuals or strains by homologous recombination in yeast.

A stab of pTARBAC1 plasmid in *E. coli* DH10B can be requested from Dr. Pieter de Jong, Children's Hospital Oakland Research Institute (pdejong @chori.org). pTARBAC1-containing *E. coli* should be grown at 37°C in LB under double-selection with 50 mg/L ampicillin (Amp) and 25 mg/L chloramphenicol (Cam). To prepare large quantities (>50 μg) of pTARBAC1 plasmid we recommend growing a 500-ml LB culture at 37°C and using a commercial resin-based maxiprep kit (*e.g.*, the large-construct kit from Qiagen which includes a digestion with exonuclease III to minimize the contamination with (linear) *E. coli* genomic DNA) followed by standard CsCl banding (p. 1.42 in Sambrook *et al.*, 1989).

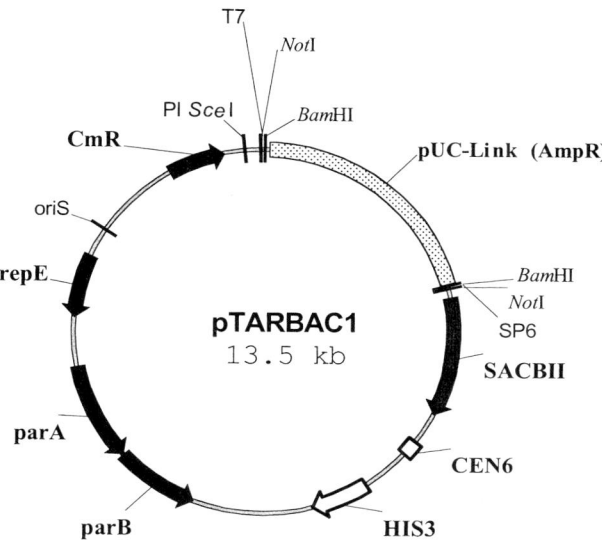

Figure 3: Map of pTARBAC1. pTARBAC1 is a derivative of pBACe3.6 featuring a yeast centromere (CEN6) and selectable marker (HIS3). The pUC link in the multiple cloning site contains a high-copy-number origin allowing the isolation of sufficient quantities of pTARBAC1 plasmid. For preparation of the BAC vector for cloning, the pUC link is excised by a carefully controlled complete *Bam*HI digestion followed by careful dephosphorylation of the *Bam*HI cloning site. Since it is difficult to achieve complete dephosphorylation without damaging the cloning site, a bulk ligation of the vector preparation is performed to convert residual molecules carrying a 5'-phosphate to linear multimers or circles. The non-ligated linear 10.7-kb monomer is purified on a pulsed-field gel and tested by a kinase-dependent ligation assay for its cloning competence. Genomic *Mbo*I fragments that have been size-selected to 150-200 kb are ligated to a quality-controlled vector preparation. After electroporation of the *E. coli* host, circular plasmids are under the control of the F-factor derived functions oriS, repE, parA and parB and maintained at 1-2 copies per cell. The cloned *Mbo*I fragment separates the SacBII gene from its promoter. The resulting SacBII-deficient E. coli cells grow in the presence of sucrose whereas non-recombinant, SacBII-expressing cells do not.

Protocol 4.3.1 Preparation of pTARBAC1-cloning Vector

Equipment and Reagents

- Pulsed-field gel apparatus (BioRad #170-3700) with wide (21×14 cm) gel casting tray.
- Molecular-biology grade agarose (*e.g.*, SeaKem GTG agarose; Biowhittaker Molecular Applications #50074)

- CsCl-purified pTARBAC1 plasmid (>50 μg)
- 20 u/μl *Bam*HI (NEB #136S)
- 10X *Bam*HI buffer (NEB)
- 100X BSA (NEB)
- 2 u/μl *Bam*HI in 1X *Bam*HI buffer with 1X BSA (fresh dilution)
- 1 u/μl calf intestine alkaline phosphatase (CIP; Roche #713023)
- 10X CIP buffer (Roche)
- 0.5 M EDTA, pH 8
- 1u/μl T4 DNA ligase (Invitrogen #15224-025)
- 5X T4 DNA ligase buffer (Invitrogen)
- 400 u/μl T4 DNA ligase (NEB #202S)
- 10X T4 DNA ligase buffer (NEB)
- "High-molecular-weight" DNA marker (Invitrogen #15618-010) 1:5 diluted in TE, pH 8, plus Ficoll 400, BPB and XC
- 10 u/μl T4 polynucleotide kinase (NEB #201S)

Method

1. To a 1.5-ml tube add 1.4 μg pTARBAC1 plasmid, 28 μl 10X BamHI buffer, 2.8 μl 100X BSA and H_2O to a final volume of 280 μl. Dispense 1 × 36 μl, 1 × 34 μl and 9 × 20 μl into eleven 1.5-ml tubes. Add 4 μl of freshly diluted 2 u/μl *Bam*HI to 36 μl cocktail. Prepare four serial 1:2 dilutions by transferring 20 μl from this tube to 20 μl cocktail etc. Prepare a similar series from a 6+34 μl dilution and put on ice. Each 20 μl reaction contains 100 ng DNA and between 0.25 and 6 u *Bam*HI. Incubate all reactions plus a minus-enzyme control for 1 h at 37°C.

2. Analyze each sample on a 0.7% agarose gel and determine the minimum enzyme concentration necessary to achieve a complete digest (*i.e.*, the 2.8 kb pUC link and the 10.7 kb BAC cloning vector). Expect complete digestion at and above 1 u *Bam*HI per reaction. Multiply the units *Bam*HI with 500 to set up the preparative digest (step 3).

3. To a 1.5-ml tube add 50 μg pTARBAC1, 50 μl 10X *Bam*HI buffer, 5 μl 100X BSA, the amount of *Bam*HI as determined in the previous step (*e.g.*, 25 μl 20 u/μl *Bam*HI) and H_2O to a final volume of 500 μl. Incubate for 1 h at 37°C.

4. Put reaction on ice and analyze 1 μl along with 100 ng uncut pTARBAC1 on a 0.7% agarose gel.

5. Extract complete digest with 500 μl phenol/chloroform. To the aqueous phase add 1/20 volume 5 M NaCl and 2.1 volumes

Ethanol. Precipitate the DNA for 2 h at –20°C. Centrifuge for 15 min at 14,000 rpm in a microcentrifuge. Wash the pellet with 1 ml 70% Ethanol and respin the tube for 2 min at room temperature. Aspirate supernatant, spin briefly and remove any remaining liquid. Air dry the pellet and resuspend in 50 μl TE, pH 8.0.

6. To a 1.5-ml tube add 223.5 μl H_2O, 1.5 μl (~1.5 μg) BamHI-cut pTARBAC1 and 25 μl 10X CIP buffer. Dispense 1 × 48 μl and 7 × 25 μl into eight 1.5-ml tubes. Add 2 μl 1 u/μl CIP to 48 μl cocktail and prepare seven 1:2 serial dilutions by transferring 25 μl from this tube to 25 μl cocktail etc. Each 25 μl reaction contains 150 ng DNA and between 0.002 and 1 u CIP. Incubate all 8 reactions plus a minus-enzyme control for 1 h at 37°C.

7. Add 2 μl 0.5 M EDTA and heat inactivate the enzyme for 10 min at 65°C. Add 73 μl TE, pH 8 and extract with 100 μl phenol/chloroform. To the aqueous phase add 1/20 volume 5 M NaCl and 2.1 volumes Ethanol. Precipitate the DNA for 2 h at –20°C. Centrifuge for 15 min at 14,000 rpm in a microcentrifuge. Wash the pellet with 1 ml 70% Ethanol and respin the tube for 2 min at room temperature. Aspirate supernatant, spin briefly and remove any remaining liquid. Air dry the pellet and resuspend in 10 μl TE, pH 8.0. Analyze 1 μl each on a 0.7% agarose gel.

8. To the remainder add 2.4 μl 5X T4 DNA ligase buffer and 0.6 μl 1 u/μl T4 DNA ligase and incubate for 3 h at 16°C.

9. Analyze each reaction on a 0.7% agarose gel to determine the lowest concentration of CIP that prevents ligation. Large concatemers run above, circular ligation products below the respective linear fragments of 2.8 and 10.7 kb in size. Expect complete dephosphorylation at and above 0.125 u CIP per reaction. Multiply the units CIP with 300 to set up the preparative dephosphorylation (step 10).

10. To a 1.5-ml tube add 45 μl BamHI-cut pTARBAC1 (step 3), 100 μl 10X CIP buffer, the amount of CIP as determined in the previous step (e.g. 37.5 μl 1 u/μl CIP) and H_2O to 1 ml final volume. Incubate for 1 h at 37°C.

11. Add 100 μl 0.5 M EDTA, pH 8 and heat inactivate the enzyme for 10 min at 75°C. Split into two tubes and extract each with an equal volume phenol/chloroform. To the aqueous phase add 1/20 volume 5 M NaCl and 2.1 volumes Ethanol. Precipitate the DNA for 2 h at –20°C. Centrifuge for 15 min at 14,000 rpm in a microcentrifuge. Wash the pellets with 1 ml 70% Ethanol and respin the tube for 2 min at room temperature. Aspirate supernatant, spin briefly and remove any remaining liquid. Air dry the pellets and resuspend in

a total of 30 μl TE, pH 8.0. Determine DNA concentration (by A_{260} or fluorometry with Hoechst 33258). Expect about 35 μg. Adjust the DNA concentration to 100 ng/μl.

12. To a 1.5-ml tube add 25 μl dephosphorylated *Bam*HI-cut pTARBAC1 (25 μg), 100 μl 5X T4 DNA ligase buffer, 365 μl H_2O and 10 μl 1 u/μl T4 DNA ligase. Incubate for 1 h at 16°C. Extract with 500 μl phenol/chloroform. To the aqueous phase add 1/20 volume 5 M NaCl and 2.1 volumes Ethanol. Precipitate the DNA for 2 h at –20°C. Centrifuge for 15 min at 14,000 rpm in a microcentrifuge. Wash the pellet with 1 ml 70% Ethanol and respin the tube for 2 min at room temperature. Aspirate supernatant, spin briefly and remove any remaining liquid. Air dry the pellet, resuspend in 750 μl TE, pH 8.0.

13. Melt 2 × 180 ml 1% regular molecular-biology grade agarose in 0.5X TBE and pour two 21×14 cm gels using a preparative comb to form a 10-cm-wide slot with analytical wells on either side. Place gels in two pulsed-field gel boxes filled with fresh 0.5X TBE.

14. To 750 μl ligase-treated BAC vector (step 12) add 150 μl 6X gel-loading buffer, load 450 μl into each preparative slot and run along with preheated (5 min at 65°C) "high-molecular-weight" marker for 16 h at 6 V/cm, 120° angle, 3 s switching time and 14°C.

15. Make two cuts at the left and right edges of the preparative lane and stain the two flanking marker-containing lanes with Ethidium bromide. View on a UV transilluminator covered with plastic wrap and cut two small notches to mark the position of the linear 10.7-kb vector band. Reassemble the gel and excise the linear vector band from each of the unstained preparative lanes. Stain and reassemble the gels and take a picture to have a permanent record.

16. Place two strips of dialysis tubing for 30 min in a container filled with H_2O. Insert the agarose slices, close one end of the tubings and add just enough TE, pH 8, containing 100 μg/ml BSA, to cover the agarose slice. Squeeze out any air bubbles. Close the other end and place the dialysis bags in a gel box with the long axis of the slice perpendicular to the electrical field. Fill the gel box with 0.5X TBE and electroelute for 2h at 80V at 4°C. Reverse the polarity for 1 min and recover all liquid from within the tubings. Extract twice with an equal volume phenol/chloroform and once with chloroform. To the aqueous phase add 1/20 volume 5 M NaCl and 2.1 volumes Ethanol, mix and precipitate for 2 h at –20°C.

17. Spin for 30 min at 14,000 rpm in a microcentrifuge at 4°C. Carefully decant the supernatant and wash pellet once with 1 ml

70% Ethanol. Respin for 5 min at room temperature. Carefully aspirate supernatant, spin briefly and remove any remaining liquid. Resuspend the air-dried pellet in 100 μl. Determine the concentration by Hoechst 33258 fluorometry. Expect about 10 μg. Dilute to 25 ng/μl with TE, pH 8.

18. To a 0.5-ml tube add 5 μl 10 ng/μl BAC-cloning vector, 1 μl 10X T4 DNA ligase buffer, 0.5 μl 10 u/μl T4 polynucleotide kinase, 0.5 μl 400 u/μl T4 DNA ligase and H$_2$O to a final volume of 10 μl. Set up a second reaction without kinase and a third one without kinase and without ligase. Incubate for 1 h at 16°C.

19. Add 1.5 μl 6X gel-loading buffer and run on a 0.4% agarose gel. Only the complete reaction with prior phosphorylation should give rise to visible ligation products. The other two samples should look identical and show only the linear 10.7-kb vector band.

Protocol 4.3.2 Partial *Mbo*I Digestion and Size Selection to 150-200 kb

Equipment and Reagents

- Pulsed-field gel apparatus (BioRad #170-3700) with wide (21×14 cm) gel casting tray.
- Molecular-biology grade agarose (*e.g.*, SeaKem GTG agarose; Biowhittaker Molecular Applications #50074)
- "low-range" PFG marker (NEB #350)
- Yeast PFG marker (NEB #345)
- Four to ten 400-μl blocks of agarose-embedded DNA (see 4.1)
- 50 mg/ml acetylated BSA (Invitrogen #15561-020)
- 10X *Mbo*I buffer w/o Mg^{++} (500 mM Tris-HCl, pH 7.9, 1 M NaCl, 10 mM DTT)
- 1X *Mbo*I buffer w/o Mg^{++} (500 μg/ml acetylated BSA, 50 mM Tris-HCl, pH 7.9, 100 mM NaCl, 1 mM DTT)
- 1 M MgCl$_2$
- 5 u/μl *Mbo*I (NEB #147S)
- Disposable plastic spatulas (Sarstedt #81.970)
- Ready-to-use dialysis tubing (Spectrum #132110)
- TE, pH 8 with 100 μg/ml BSA
- 20 mg/ml glycogen (Roche #901303)
- 15, 12, 9, 6, 3, 1.5 and 0.75 mM MgCl$_2$
- 1X NDS (0.5 M EDTA, 10 mM Tris, 1% N-laurylsarcosine Sigma #L-9150, pH adjusted to 9.5 with NaOH)

- 0.2 mg/ml proteinase K (Roche #745723) in 1X NDS (freshly prepared)
- 0.1 M PMSF (Roche #1359061) in isopropanol
- 0.5X TBE (autoclaved)
- wide-bore 200-μl and 1-ml pipette tips
- lambda-DNA concentration standards (0.5, 1.0, 1.5, 2, 3, 5 and 10 ng/μl in TE, pH 8; dilutions made from Invitrogen #25250-010)

Method

1. Dialyze one 400-μl block of agarose-embedded DNA in 50 ml TE, pH 8 for 30 min at room temperature with occasional gentle swirling. Repeat dialysis five times, the last time overnight at 4°C.
2. Remove liquid and add 5 ml 1X *Mbo*I buffer w/o Mg^{++}. Keep on ice for 1 h with occasional gentle swirling.
3. Repeat step 2.
4. Carefully place the equilibrated agarose block inside the lid of a petri dish. Use a glass microscope coverslip to cut the agarose block first along the long axis in two halves and then perpendicular to the first cut into 10 pieces, ~40 μl each.
5. Put each slice in a 1.5-ml tube. Store one "aliquot" in a tube labeled "C" at 4°C until step 9. Mix 2 μl 5 u/μl *Mbo*I with 18 μl 1X *Mbo*I buffer w/o Mg^{++}. Mix 4 μl of the diluted enzyme with 2 ml 1X *Mbo*I buffer w/o Mg^{++} and add 100 μl of this enzyme dilution (0.001 u/μl) to "plugs" 1-8. The ninth receives 100 μl 1X *Mbo*I buffer w/o Mg^{++} without enzyme.
6. Mix gently and leave overnight at 4°C. During this time the *Mbo*I restriction enzyme will start diffusing into the block. However, in the absence of Mg^{++} it will not digest the embedded DNA.
7. Add 10 μl 15 mM MgCl$_2$ to tube 1 and 9, mix and put on ice. Add 10 μl of 12, 9, 6, 3, 1.5 and 0.75 mM MgCl$_2$ to tube 2 to 7, respectively. The final MgCl$_2$ concentration ranges from 1 mM (tubes 1 and 9) to 0.1 mM (tube 7). Tube 8 receives 10 μl H$_2$O instead. Mix and keep on ice for 2 h before starting the reaction in a 37°C waterbath. Incubate for 1 h at 37°C.
8. Add 1 ml 50 mM EDTA, mix gently and put on ice. After 30 min, carefully remove liquid and replace with 1 ml TE, pH 8.
9. Run all samples including plug "C" (step 5) along with yeast and low-range PFG markers on a 1% agarose gel in 0.5X TBE. Set the electric field angle to 120° and the switching time to a linear ramp from 0.1 to 40 s and run the gel at 6 V/cm for 16 h at 14°C.

10. Stain the gel with Ethidium bromide. The bulk of the DNA in lanes C (untreated control), 8 (*Mbo*I, no Mg^{++}) and 9 (no *Mbo*I, 1 mM Mg^{++}) should either stay in the well or run at the limiting mobility of the gel. Increased background smearing in lane 9 relative to lane C indicates endogenous nuclease activity in the starting DNA preparation. Lanes 1 through 8 should give a typical partial-digestion pattern ranging from a near complete digestion (lane 1) to perhaps a light background smear (lane 8) no more intense than that in lane C. The partially digested DNA should run in almost straight smears (see Figure 4A). Excessive "bulges" (see Figure 4B) indicate that

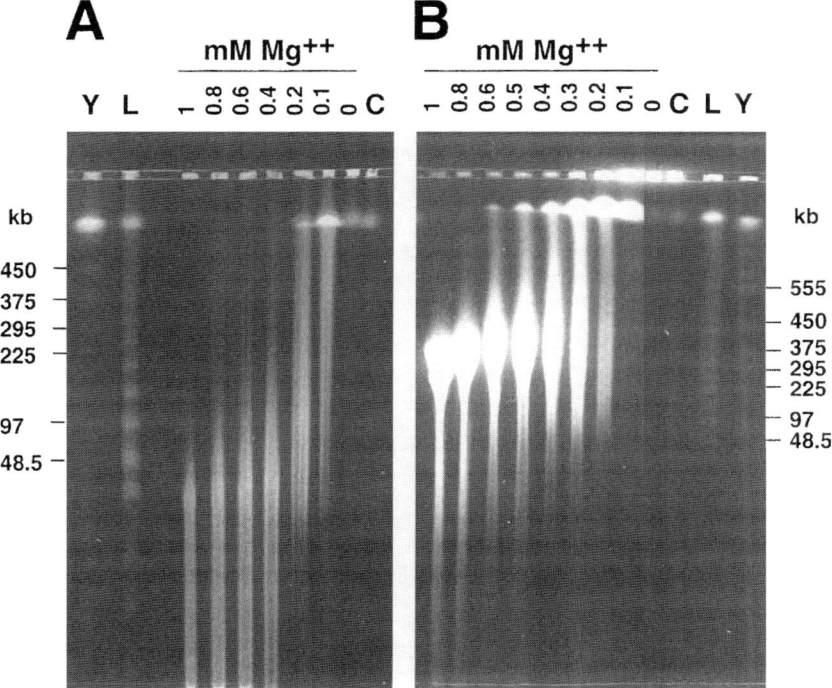

Figure 4. Examples of analytical partial *Mbo*I digestions. The numbers above the lanes indicate the final MgCl$_2$ concentration in the reactions. C indicates the non-incubated control plug. The test for residual nuclease activity (incubation in the presence of MgCl$_2$ but without *Mbo*I; reaction 9 in protocol 4.3.2, step 5) had been performed beforehand and is not shown. For the DNA on gel A, 0.2 mM MgCl$_2$ was chosen for the preparative digest. Gel B was overloaded resulting in bulging lanes and a significant mobility shift (retardation) relative to the size markers. Nonetheless, preparative partial digests of the same DNA at 0.3-0.5 mM MgCl$_2$ yielded a BAC library with a narrow insert-size distribution of (140 ± 10) kb. The sizes of a subset of bands in the yeast (Y), lambda-ladder or low-range PFG marker lanes (L) are indicated.

the concentration of agarose-embedded DNA is too high. Overloading the gel will lead to an overestimation of the true size of the DNA fragments and to the contamination of large DNA fragments with co-migrating smaller molecules. Determine which condition produced the most partial-digest fragments in the 150-250 kb size range. Expect the optimum $MgCl_2$ concentration between 0.2 and 0.6 mM.

Note: If necessary, repeat the analytical digest to fine-tune the conditions. If one is not limited by starting material, it is advisable to set up three slightly different scaled-up preparative digests in the following step, bracketing the optimal Mg^{++} concentration as determined in the analytical partial digest.

11. For each preparative partial digest, repeat steps 1-4 using three 400-μl blocks instead of one.

12. Mix 2 μl 5 u/μl *Mbo*I with 18 μl 1X *Mbo*I buffer w/o Mg^{++}. In a 15-ml tube mix 6 μl of the diluted enzyme with 3 ml 1X *Mbo*I buffer w/o Mg^{++} and put on ice. Carefully add all thirty 40-μl plugs, mix gently and put on ice for about 1 h. Mix at least once more and keep overnight at 4°C.

13. Add 300 μl of the $MgCl_2$ dilution that produced the best partial digest as determined in step 10. Mix gently and keep on ice for 2 h. Mix once more and incubate for 1 h at 37°C.

14. Add 500 μl 0.5 M EDTA, pH 8, mix gently and put on ice.

15. Carefully replace liquid with 5 ml 0.2 mg/ml proteinase K in 1X NDS. Mix gently and incubate for 3 h at 50°C.

16. Replace liquid with 10 ml TE, pH 8. Mix gently and dialyze at room temperature for 30 min with occasional agitation. Repeat once.

17. Dilute 0.3 ml 0.1 M PMSF in 30 ml TE, pH 8 and dialyze the plugs in this solution 3 × 30 min at room temperature. Dialyze once against plain TE, pH 8.

18. Melt 180 ml 1% regular molecular-biology grade agarose in 0.5X TBE and pour a 21×14 cm gel using a preparative comb to form a wide slot with at least 3 analytical wells on either side. Position the comb exactly 2.5 cm from the top of the gel. Load the plugs into the slot, leaving no gap between them. Put a thin slice of low-range and yeast PFG markers on either side. Seal with melted agarose. Place gel in the opposite orientation (*i.e.* loaded samples towards the + electrode of the electrical field) in a pulsed-field gel box

filled with fresh 0.5X TBE. Set the electric field angle to 120° and the switching time to 15 s and run the gel at 5 V/cm for 10-12 h at 14°C. The purpose of this run is to elute fragments smaller than 150 kb into the electrophoresis buffer.

Note: It is important to optimize the exact run time beforehand using low-range and yeast PFG markers. Chose the shortest run time that leads to elution of the 150-kb "rung" of the lambda ladder. High concentration of agarose-embedded genomic DNA may lead to a mobility shift relative to the size markers requiring slightly (10-20%) longer run times to lose 150-kb fragments from the more concentrated partially digested DNA sample.

19. Replace electrophoresis buffer with fresh 0.5X TBE, reorient the gel such that the loaded samples point towards the – electrode and run gel under the exact same conditions as before (10-12 h at 5 V/cm, 120°, 15 s, 14°C). The purpose of this run is to run fragments not lost in the previous step back into the well.

20. Load two slices of low-range PFG marker into the outside analytical wells on either side and continue the electrophoresis for 16 h at 6 V/cm, a linear switching ramp from 0.1-40 s at 14°C. The purpose of this run is to size-fractionate the partial-digest fragments for cloning.

21. Make two cuts at the left and right edges of the preparative lane and stain the two flanking gel pieces with Ethidium bromide. View on a UV transilluminator covered with plastic wrap and mark the positions of the ~100-kb and ~300-kb size markers. Use the yeast marker lane to identify the "rungs" of the lambda ladder. The 150-kb band should not be visible in the marker lane loaded at the beginning of the reverse gel run indicating that fragments smaller than ~150-kb did run off the gel. Reassemble the gel, excise the 100-300 kb region of the unstained preparative lane, making size cuts about every half cm. Stain and reassemble the remainder of the gel and take a picture to have a permanent record (see Figure 5).

22. Take a ~2-mm "biopsy" from the mid section of each slice and load along with low-range and yeast PFG markers on a 1% agarose gel. Run the gel at 6 V/cm, 120°, 0.1-40s and 14°C. Stain the gel with Ethidium bromide and determine which of the gel slices contain DNA fragments in the size range suitable for cloning.

Note: For short-term storage (<2 weeks) keep the gel slices in TE, pH 8 at 4°C. For long-term storage keep the gel slices in 0.5 M EDTA at

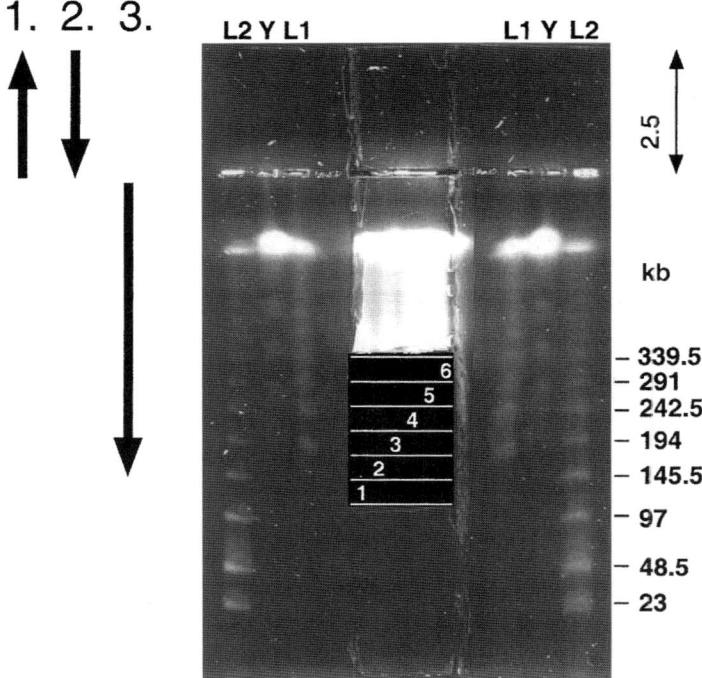

Figure 5. Example of a preparative pulsed-field gel for BAC-library construction. Agarose plugs containing partially digested genomic DNA were loaded on the gel and flanked by yeast (Y) and low-range PFG markers (L1). Molecules smaller than ~150 kb in size were eluted by running the DNA towards the top of the gel. After running the remaining DNA back to the origin, fresh low-range PFG marker was loaded in the two outside lanes (L2). After the third segment of the gel run, only the marker lanes were stained with Ethidium bromide. The lower "rungs" of the ladder in lane L1, including the 145-kb band, had been run off during the first run. Horizontal strips were excised from the central unstained portion of the gel. The two lowest strips did not contain much DNA. DNA eluted from the third slice was successfully cloned, resulting in a BAC library with an average insert size of approximately 190 kb. The next larger fraction contained DNA fragments exceeding 200 kb and produced mostly empty clones.

4°C. Reduce the EDTA concentration by dialyzing at least 5 times against a TE, pH 8 before electroelution.

23. For each gel slice containing DNA fragments between 150 and 250 kb, place one strip of dialysis tubing for 30 min in a container filled with H₂O. Insert agarose slice, close one end of the tubing and add just enough autoclaved 0.5X TBE to cover the agarose slice. Squeeze out any air bubbles. Close the other end and place the

dialysis bag in a gel box with the long axis of the slice perpendicular to the electrical field. Fill the gel box with 0.5X TBE and electroelute for 3 h at 80V at 4°C. Reverse the polarity for 1 min and transfer the dialysis bag to a beaker containing 1-2 L TE, pH 8 and a stir bar. Dialyze overnight at 4°C.

24. Use a wide-bore pipette tip to recover the liquid from the bag. Store the electroeluted DNA in a 1.5-ml tube at 4°C. Remove 1 μl with a regular pipette tip, add 5 μl TE, pH 8 and 1 μl 6X gel-loading buffer and analyze along with 1 μl pre-heated (5 min at 65°C) lambda-concentration standards on a 0.8% miniagarose gel for 15-30 min at 50 V. Estimate the DNA concentration relative to the lambda standards. Expect a concentration between 1 and 5 ng/μl. Samples below 1 ng/μl usually clone poorly.

Protocol 4.3.3 BAC Ligation and Electroporation

Equipment and Reagents

- Pulsed-field gel apparatus (BioRad #170-3700) with wide (21×14 cm) gel casting tray.
- "low-range" PFG marker (NEB #350)
- 25 ng/μl pTARBAC1 cloning vector (see 4.3.1)
- ≥1 ng/μl size-selected *Mbo*I fragments (see 4.3.2)
- 1u/μl T4 DNA ligase (Invitrogen #15224-025)
- 5X T4 DNA ligase buffer (Invitrogen)
- 30% (w/v) PEG 8000 (USB #19959) in 0.5X TE, pH 8
- Dialysis membrane filters (25 mm diameter; Millipore #VWM S0025)
- Electrocompetent *E. coli* DH10B (Invitrogen #18290-015)
- Electroporator equipped with voltage booster (Invitrogen #11609-013 plus #11612-017)
- Electroporation cuvettes (0.15 cm electrode gap; Invitrogen #11608-031)
- SOC broth (Invitrogen #15544-034)
- 10-cm and 22×22 cm LB-Cam (15 mg/L)/5% (w/v) sucrose agar plates
- 2X YT Cam (15 mg/L) broth
- 10 mg/ml proteinase K (Roche #745723) in H_2O
- 0.1 M PMSF (Roche #1359061) in isopropanol

- 96-well format plasmid-isolation kit (*e.g.*, Qiagen R.E.A.L prep #26173)
- 10 u/μl *Not*I (NEB #189L)
- 10X NEB3 buffer (NEB)
- 100X BSA (NEB)
- Optional: PI-SceI (NEB #696L)

Method

1. To a 1.5-ml tube add 10 μl 5X T4 DNA ligase buffer, 50 ng ~150-kb *Mbo*I fragments (in less than 35 μl), 4 μl 25 ng/μl pTARBAC1 vector, 1 μl 1 u/μl T4 DNA ligase and H$_2$O to 50 μl final volume. Carefully mix using a wide-bore pipette tip. Set up a minus-insert control using TE, pH 8, instead of the *Mbo*I fragments. Incubate for 2 h at 16°C.

Note: The concentration of insert DNA in the ligation should be around 1 ng/μl. The molar ratio of vector to insert should be between 5 and 10 (7.5 in this example). Use slightly more vector for smaller inserts and slightly less vector for larger inserts. T4 DNA ligase from NEB (400 u/μl) can be substituted for the ligase from Invitrogen (1 Weiss u/μl). However we recommend using the Invitrogen buffer for either enzyme.

2. Fill a 10-cm petri dish with ~30 ml 30% (w/v) PEG 8000 in 0.5X TE, pH 8 and float two pencil-labeled filter membranes on the surface, shiny side up. Carefully, using a wide-bore pipette tip, transfer the ligation mixes to the floating filters. Dialyze for 30 min to 1 h at 4°C or until the volume drops below 30 μl. Use a pipette tip cut off at an angle to retrieve as much of each drop as possible. Store at 4°C.

3. Thaw 100 μl electrocompetent *E. coli* DH10B on ice and dispense 4 × 20 μl in pre-cooled 1.5-ml tubes on ice. With a wide-bore pipette tip add 2 × 2 μl ligation mix and 2 × 2 μl vector-only control. Put on ice. With a wide-bore tip pipette the DNA/DH10B mixture between the electrodes of a pre-cooled 0.15 cm electroporation cell. Set the voltage booster to 4,000 Ω, the electroporator to 330 μF, low Ω and fast charge rate. Charge the instrument with 330 V, switch to "arm" and trigger as soon as the voltage display on the electroporator has dropped to 315. The actual voltage displayed on the voltage booster will be about 1.9 kV (13 kV/cm). Retrieve the hanging drop with a regular pipette tip and transfer immediately to a

10-ml tube containing 0.5 ml SOC broth prewarmed to 37°C. Shake at 200 rpm for 1 h at 37°C, plate 2 × 200 μl on LB-Cam/Sucrose plates and incubate overnight at 37°C. Expect between 50 and 500 colonies per plate. The vector-only control should give less than 5% of that.

4. Pick up to 96 transformants and grow minicultures for 18 h at 37°C in 1.2 ml 2X YT-Cam (15 mg/L) broth in a 2-ml deep-well block. Use a commercial plasmid kit and isolate BAC DNA according to the manufacturer's instructions. Resuspend the BAC DNA in 25 μl TE, pH 8.

5. To a 1.5-ml tube add 836 μl H₂O, 220 μl 10X NEB3 buffer, 22 μl 100X BSA and 22 μl 10 u/μl NotI. Mix and dispense 96 × 10 μl into a PCR tray. Add 5 μl BAC DNA, mix and incubate for 3 h at 37°C and 15 min at 65°C in a PCR machine. Add 3 μl 6X gel-loading buffer and run on two 1% 0.5X TBE pulsed-field gels for 16 h at 120°, 6V/cm, 1-15 s and 14°C.

6. Stain the gel with Ethidium bromide. Determine the percentage recombinant and non-recombinant clones as well as the average insert size of recombinant clones calculating the insert size from the sum of the NotI fragments other than the 10.7-kb vector band. If the average insert size is too small, try a larger size fraction (protocol 4.3.2, step 22). If non-recombinant clones give rise to rearranged vector bands less than 10.7 kb in size go back to step 1 and use a slightly different (usually smaller) ratio of BAC vector to insert. If the results are satisfactory calculate the number of 50-μl ligation reactions necessary to produce the library and proceed to the next step.

Note: Non-vertebrate DNA often gives too many NotI fragments per BAC for this analysis. Such BACs can be linearized by cutting the PI-SceI site in the pTARBAC1 vector with the corresponding "homing" endonuclease PI-SceI to measure their size on a pulsed-field gel.

7. In 1.5-ml tubes set up the required number of identical optimized 50-μl ligation reactions (see step 1) and incubate for 2 h at 16°C.

8. Add 2 μl 0.5 M EDTA, pH 8 and 1 μl 10 mg/ml proteinase K, mix gently using a wide-bore pipette tip and incubate for 1 h at 37°C.

9. Add 1 μl 0.1 M PMSF, mix with a wide-bore pipette tip and transfer each ligation mix onto a dialysis filter floating on ~30 ml H₂O in a petri dish. Dialyze for 2 h at 4°C. The volume increases during that time. Retrieve the liquid using a pipette tip cut at an angle. Pool the samples in one tube and estimate the total volume.

10. For each 300 μl pooled ligation mix float one membrane filter on 30 ml 30% (w/v) PEG in 0.5X TE, pH 8 in a petri dish. Add 100 μl per filter and dialyze at 4°C. After about 1 h add more. Continue the dialysis, adding more ligation mix until a total of 300 μl has been added to each filter. Continue the dialysis keeping an eye on the shrinking drop size. When the volume gets below ~100 μl, recover as much liquid as possible using a pipette tip cut at an angle. Pool the dialyzed and concentrated ligation mix in a 1.5-ml tube and store at 4°C.

11. Analyze 2 × 2 μl by electroporation (see step 3). If the titer is satisfactory carry out additional electroporations. Transformations carried out on the same day can be pooled after the 1-h incubation at 37°C. Add 1/5 volume 80% glycerol mix and freeze small aliquots in screw-cap cryovials at –80°C.

12. Thaw one aliquot and plate on 22×22 LB-Cam (15 mg/L)/sucrose (5% w/v) plate. Pick and analyze 96 transformants by *Not*I digestion and pulsed-field gel electrophoresis before plating and picking the entire BAC library.

5. Hints and Tips

5.1 Choice of Starting Material

Before embarking on a big genomic library project, particularly BAC-library construction, it is worth inquiring within the research community what strain and sex to use to prepare the source DNA. For example, some vertebrate research communities prefer the heterogametic sex (*i.e.,* the male for mammalians, but the female for birds) to study evolution of the Y chromosome. Others are more X-chromosome oriented and prefer a library where this sex chromosome is not under-represented.

Another important decision to make concerns the material from which to isolate the DNA for cloning. Sperm, for example, is considered germline DNA, whereas other tissues (*e.g.,* lymphatic tissues) may have undergone somatic rearrangements. For some species (for example endangered ones) there may be no choice but to use a transformed cell line. Small species may have to be used "as is". Finally, one should be aware of the possibility of contamination with other organisms or genomes (parasites, food, prey, chloroplasts, mitochondria, etc.), particularly for organisms caught or collected in the wild. Fortunately, many contaminant species are from a distant phylogenetic kingdom and therefore easily detectable by end-sequencing a small number (100-200) of clones derived from each batch of genomic DNA.

5.2 Alternative Partial-Digestion and Cloning Schemes

The fosmid library protocol as described includes a physical size-selection step and is rather liberal and wasteful in terms of starting material. An alternative cloning strategy is to dephosphorylate the partially digested DNA (to prevent co-ligation artifacts), to skip the preparative electrophoresis and instead rely solely on *in vitro* lambda packaging for size selection. This strategy allows fosmid or cosmid cloning from minute amounts of starting material, for example from flow-sorted human chromosomes (Gingrich *et al.*, 1996). For detailed protocols for cosmid cloning consult other laboratory manuals (Evans, 1997; Sambrook and Russell, 2001).

The protocols above employ two different methods to control the extent of digestion with restriction enzyme: limiting enzyme concentration (fosmid protocol) and limiting $MgCl_2$ concentration (BAC protocol). Both work well for *Mbo*I, and the choice between them is largely a matter of personal preference. For fine tuning the conditions, we have also used a hybrid approach, *i.e.*, varying the enzyme concentration at the near-optimal $MgCl_2$ concentration. The best method to control partial digestion with *Eco*RI is methylase competition, that is to set up reactions with different ratios of *Eco*RI restriction enzyme and *Eco*RI methylase. For further discussion on this topic and additional partial-digestion protocols see Riethman *et al.* (1997).

5.3 Library QC

To determine the number of genome equivalents cloned in an arrayed library one has to determine the total number of clones picked, the percentage of "empty" (*i.e.*, non-growing) wells, the percentage of recombinant clones and the average insert size. Some big library projects extend over several months, require more than one ligation mix and many days of electroporation (or transduction), plating and colony picking. One way to assess the quality of such libraries is to analyze one clone from each 384-well library archive plate by *Not*I (or PI-*Sce*I) digestion and pulsed-field gel electrophoresis and calculate the statistical genome coverage from this representative sample of clones.

The real test for actual genome coverage is to probe the library with 5-10 single-copy probes and determine the number of clones identified by each probe. As the probe may or may not be truly single-copy, the clones identified should be analyzed by a second, independent method, for example restriction-enzyme fingerprinting, to verify that they are from a single locus.

End-sequencing clones from the library is helpful to corroborate the percentage of recombinant clones in the library as determined on a pulsed-field gel. More importantly, it provides valuable information about the cloned inserts, for example whether the starting DNA contained other genomes (see above) or whether end-sequences would be informative for mapping purposes (which is not the case if the restriction enzyme used for partial digestion cuts frequently in repetitive elements).

6. Relevant Websites

The following websites provide a wealth of useful information including maps and sequences of cloning vectors, library QC data for existing libraries, protocols and ordering information regarding hybridization filters or individual clones:

http://www.chori.org/bacpac/

http://www.genome.clemson.edu/

http://www.genome.arizona.edu

http://www.tree.caltech.edu/

The following websites of US government agencies contain programmatic information on the public efforts in the US to fund the construction of BAC libraries for a variety of different organisms:

http://wwwgenome.gov/10001844

http://www.nsf.gov/pubs/2001/nsf01145/nsf01145.html

References

Birren, B. and Lai, E. 1993. Pulsed Field Gel Electrophoresis: A Practical Guide. Academic Press, San Diego.

Birren, B., Mancino, V. and Shizuya, H. 1997. Bacterial Artificial Chromosomes. In: Genome Analysis: A Laboratory Manual. B. Birren, E.D. Green, S. Klapholz, R.M. Myers, H. Riethman and J. Roskams, eds. Cold Spring Harbor Laboratory Press, Cold Spring Harbor, New York. Vol. 3. p. 241–295.

Burke, D.T., Carle, G.F. and Olson, M.V. 1987. Cloning of large segments of exogenous DNA into yeast by means of artificial chromosome vectors. Science. 236: 806–812.

Collins, J. and Hohn, B. 1978. Cosmids: A type of plasmid gene-cloning vector that is packagable *in vitro* in bacteriophage lambda heads. Proc. Natl. Acad. Sci. USA. 75: 4242–4246.

Coulson, A., Sulston, J., Brenner, S. and Karn, J. 1986. Toward a physical map of the genome of the nematode Caenorhabditis elegans. Proc. Natl. Acad. Sci. USA. 83: 7821–7825.

Evans, G.A. 1997. Cosmids. In: Genome Analysis: A Laboratory Manual. B. Birren, E.D. Green, S. Klapholz, R.M. Myers, H. Riethman and J. Roskams, eds. Cold Spring Harbor Laboratory Press, Cold Spring Harbor, New York. Vol. 3. p. 87–201.

Fleischmann, R.D., Adams, M.D., White, O., Clayton, R.A., Kirkness, E.F., Kerlavage, A.R., Bult, C.J., Tomb, J.-F., Dougherty, B.A. and Merrick, J.M. 1995. Whole-Genome Random Sequencing and Assembly of Haemophilus influenzae Rd. Science. 269: 496–512.

Foote, S., Vollrath, D., Hilton, A. and Page, D.C. 1992. The human Y chromosome: overlapping DNA clones spanning the euchromatic region. Science. 258: 60–66.

Frengen, E., Weichenhan, D., Zhao, B., Osoegawa, K., van Geel, M. and de Jong, P.J. 1999. A Modular, Positive Selection Bacterial Artificial Chromosome Vector with Multiple Cloning Sites. Genomics. 58: 250–253.

Gingrich, J.C., Boehrer, D.M., Garnes, J.A., Johnson, W., Wong, B.S., Bergmann, A., Eveleth, G.G., Langlois, R.G. and Carrano, A.V. 1996. Construction and characterization of human chromosome 2-specific cosmid, fosmid and PAC clone libraries. Genomics. 15: 65–74.

Green, E.D. and Green, P. 1991. Sequence-tagged site (STS) content mapping of human chromosomes: theoretical considerations and early experiences. PCR Methods Appl. 1: 77–90.

Green, E.D. and Olson, M.V. 1990. Chromosomal region of the cystic fibrosis gene in yeast artificial chromosomes: a model for human genome mapping. Science. 250: 94-98.

Hudson, T.J., Stein, L.D., Gerety, S.S., Ma, J., Castle, A.B., Silva, J., Slonim, D.K., Baptista, R., Kruglyak, L., Xu, S.-H., *et al.* 1995. An STS-based map of the human genome. Science. 270: 1945–1954.

Kim, U.-J., Birren, B.W., Slepak, T., Mancino, V., Boysen, C., H.-L., K., Simon, M.I. and Shizuya, H. 1996. Construction and Characterization of a Human Bacterial Artificial Chromosome Library. Genomics. 34: 213–218.

Kim, U.-J., Shizuya, H., de Jong, P.J., Birren, B. and Simon, M.I. 1992. Stable propagation of cosmid sized human DNA inserts in an F factor based vector. Nucleic Acids Res. 20: 1083–1085.

Luo, M., Wang, Y.-H., Frisch, D., Joobeur, T., Wing, R.A. and Dean, R.A. 2001. Melon BAC library construction using improved methods and identification of clones linked to the locus conferring resistance to melon Fusarium Wilt (Fom-2). Genome. 44: 154–162.

Marra, M., Kucaba, T., Sekhon, M., Hillier, L., Martienssen, R., Chinwalla, A., Crocket, J., Fedele, J., Grover, H., Gund, C., *et al.* 1999. A map for sequence analysis of the Arabidopsis thaliana genome. Nature Genet. 22: 265–270.

Marra, M.A., Kucaba, T.A., Dietrich, N.L., Green, E.D., Brownstein, B., Wilson, R.K., McDonald, K.M., Hillier, L.W., McPherson, J.D. and Waterston, R.H. 1997. High Throughput Fingerprint Analysis of Large-Insert Clones. Genome Res. 7: 1072–1084.

Myers, E.W., Sutton, G.G., Delcher, A.L., Dew, I.M., Fasulo, D.P., Flanigan, M.J., Kravitz, S.A., Mobarry, C.M., Reinert, K.H.J., Remington, K.A., *et al.* 2000. A Whole-Genome Assembly of Drosophila. Science. 287: 2196–2204.

Olson, M., Hood, L., Cantor, C. and Botstein, D. 1989. A common language for physical mapping of the human genome. Science. 245: 1434–1435.

Olson, M.V., Dutchik, J.E., Graham, M.Y., Brodeur, G.M., Helms, C., Frank, M., MacCollin, M., Scheinman, R. and Frank, T. 1986. Random-clone strategy for genomic restriction mapping in yeast. Proc. Natl. Acad. Sci. USA. 83: 7826–7830.

Osoegawa, K., Woon, P.Y., Zhao, B., Frengen, E., Tateno, M., Catanese, J.J. and de Jong, P.J. 1998. An Improved Approach for Construction of Bacterial Artificial Chromosome Libraries. Genomics. 52: 1-8.

Riethman, H., Birren, B. and Gnirke, A. 1997. Preparation, Manipulation, and Mapping of HMW DNA. In: Genome Analysis: A Laboratory Manual. B. Birren, E.D. Green, S. Klapholz, R.M. Myers and J. Roskams, eds. Cold Spring Harbor Laboratory Press, Cold Spring Harbor, New York. Vol. 1. p. 83–248.

Sambrook, J., Fritsch, E.F. and Maniatis, T. 1989. Molecular Cloning: A Laboratory Manual. Cold Spring Harbor Laboratory Press, Cold Spring Harbor, New York.

Sambrook, J. and Russell, D.W. 2001. Molecular Cloning: A Laboratory Manual. Cold Spring Harbor Laboratory Press, Cold Spring Harbor, New York.

Shizuya, H., Birren, B., Kim, U.-J., Mancino, V., Slepak, T., Tachiri, Y. and Simon, M. 1992. Cloning and stable maintenance of 300-kilobase-pair fragments of human DNA in Escherichia coli using an F-factor-based vector. Proc. Natl. Acad. Sci. USA. 89: 8794–8797.

Venter, J.C., Adams, M.D., Myers, E.W., Li, P.W., Mural, R.J., Sutton, H.O., Yandell, M., Evans, C.A., Holt, R.A., Gocayne, J.D., *et al.* 2001. The Sequence of the Human Genome. Science. 291: 1304–1351.

Zeng, C., Kouprina, N., Zhu, B., Cairo, A., Hoek, M., Cross, G., Osoegawa, K., Larionov, V. and de Jong, P.J. 2001. Large-Insert BAC/YAC Libraries for Selective Re-isolation of Genomic Regions by Homologous Recombination in Yeast. Genomics. 77: 27–34.

From: *Genome Mapping and Sequencing*
© 2003 Horizon Scientific Press, Wymondham, UK

4

The Use of YACs in Mapping and Sequencing

Michael A. Quail and Alan Coulson

Abstract

The ability to clone long stretches of DNA has been an essential feature of the development of genome analysis technologies over the twenty years that culminated in the completion of a draft sequence of the human genome in 2001. Yeast artificial chromosomes (YACs) have played a central role in this development mainly due to their capability for carrying inserts of exogenous DNA up to 2Mb in size. Although many of the apparent attributes of YACs have been usurped by P1 artificial chromosomes (PACs) (Ioannou *et al.*, 1994) and bacterial artificial chromosomes (BACs) (Shizuya *et al.*, 1992), YACs generally speaking offer the most comprehensive clone coverage of complex genomes, and hence are invaluable for generating highly complete clone-based physical maps. Despite being relatively difficult to purify (being linear and for practical purposes indistinguishable from host chromosomes) they are, being propagated in a eukaryotic host, able to clone particular eukaryotic sequences that will not be supported in a bacterial host. Consequently, although inserts can be unstable and chimaerism can be a problem, particularly in larger YACs, they are an important resource for the completion of genome sequencing projects.

1. Introduction

Up until 1987, DNA cloning was limited to vectors such as plasmids, phage lambda and cosmids, all propagated in bacteria, and the largest piece of foreign DNA that could be maintained was around 40kb. This all changed when David Burke, working in the laboratory of Maynard Olson, combined two existing technologies, namely the use of pulsed field gel electrophoresis (PFGE) to separate large DNA molecules (Schwartz and Cantor, 1984) and results from research into the elements that are required for chromosome maintenance in *Saccharomyces cerevisiae*, to develop the yeast artificial chromosome, or YAC. This work involved ligation of yeast and human DNA sequences of up to 130kb with vector arms (containing selectable markers, autonomously replicating sequence, centromere and telomere seeding sequences from Tetrahymena) and their subsequent stable maintenance as a chromosome within yeast (Burke *et al.*, 1987). After further research and optimisation YACs were constructed with inserts of >1Mb and sometimes approaching 2Mb in size. This offered great potential for the construction of physical maps of large genomes as it greatly reduced the number of clones (and so the amount of work) that were required for genome coverage. Subsequently YAC libraries were constructed for several of the larger eukaryotic genomes including human, mouse, *Arabidopsis*, rat, *C. elegans*, zebra fish, Drosophila and many others (for an exhaustive list see Green *et al.*, 1999). Four separate libraries were constructed that cover the human genome including the CEPH 'megaYAC' library (Albertsen *et al.*, 1990), ICI (Anand *et al.*, 1989), Washington University (Brownstein *et al.*, 1989) and ICRF (Larin *et al.*, 1991) YAC libraries, and maps that cover most of the human chromosomes have been constructed (e.g. Wang *et al.*, 1999) using combinations of hybridisation, fingerprinting and STS content mapping techniques. Similar maps are available for several other large genomes including mouse (Nusbaum *et al.*, 1999), rat (Cai *et al.*, 2000), rice (Saji *et al.*, 2001), *Arabidopsis* (Camilleri *et al.*, 1998), and *D. discoideum* (Kuspa and Loomis, 1996).

Nevertheless, the use of YACs is limited, as they have three major shortfalls. Firstly YAC clones are prone to deletion and rearrangement, though this problem has been partially eliminated by development of vectors with colorimetric selection for rearranged inserts (see *Construction of YAC libraries*). YAC libraries are also notorious for being chimaeric, indeed it is estimated that 30-40% of the CEPH megaYAC library, which has 7 genome equivalents and average inserts of 918kb, is chimaeric. Finally the last major obstacle to their widespread use is the physical similarity of YACs to the host chromosomes. *Saccharomyces cerevisiae* has 16 linear chromosomes ranging in size from 225kb upto 1.9Mb. Yeast artificial chromosomes, once introduced into a cell, are not easily separable from these endogenous chromosomes as they are also linear and of very similar size. The only practical way to purify them is to cut them from an agarose gel after separation of the chromosomes by CHEF or PFGE (see **Protocol 5**). Even this is not straightforward as the YAC often is so similar in size to one of the 16 host chromosomes that it cannot be separated and the researcher has to transfer that YAC into a 'window' strain (Hamer *et al.*, 1995) having altered host chromosome karyotype in the region where there YAC migrates (**Protocol 6**). This is in contrast to the ease with which the more recently introduced bacterial large-insert cloning systems (BACs, PACs and P1s (Quail, 2003a)) can be isolated. Up to 300kb of exogenous DNA can be cloned into BACs and PACs, and 100kb into P1s, and as these are all closed circular molecules that are much smaller than their host genomic DNA, they can easily be prepared using standard alkaline lysis miniprep protocols. This makes the bacterial clones more laboratory friendly and amenable to high throughput methodology. However, one major advantage that YACs have over bacterial clones is that the yeast host appears to be more tolerant of unstable, repetitive and extreme DNA sequences than *E. coli* (Coulson *et al.*, 1988). As a result they are extremely useful both for giving large insert clone coverage in organisms whose DNA cannot be maintained in *E. coli* and in filling the gaps between bacterial clone contigs in other genomes. Attempts to construct a BAC library for the malaria parasite, *Plasmodium falciparum*, proved unsuccessful and this was attributed to the high (>80%) A/T content of its genome. However YAC libraries of *P. falciparum* were successfully made, mapped (Thompson and Cowman. 1997) and utilised in the malaria genome sequencing project (Bowman *et al.*, 1999). This has also been the case for another organism with an A/T rich genome, *Dictyostelium discoideum* (Kuspa and Loomis, 1996). Most organisms do not have such extreme bias in base composition but do contain a large number of elements that obviate their cloning and propagation in *E. coli* and so YACs have a large role to play in gap-filling. Coulson *et al.*, (1988) estimated that only around 90% of the *Caenhorhabditis elegans* genome could be represented in bacterial clones and found YACs to fill the uncovered regions between. Since then several integrated maps have been published (*e.g.* Schalkwyk *et al.*, 2001) that incorporate both bacterial and YAC clones, and at the time of writing coverage of bacterial clones across the human genome has been exhausted and remaining gap closure efforts are focused on YACs.

One area where YACs have been useful is in functional gene studies and transgenics (Huxley, 1994), where the power of yeast genetics coupled with the large insert capabilities of these vectors have made them invaluable. Most genes in higher eukaryotes have a multiple-exonic structure, often with long introns. As a result the size of human genes in particular, (with their control regions), can exceed the insert capacity of bacterial cloning systems making the use of YACs preferable for gene function studies.

A gene knockout of almost every gene in *S. cerevisiae* is available allowing complementation studies. Sequences contained within YACs can easily be modified whilst in the yeast cell, by homologous recombination and a variety of methods including sphaeroplast-mediated transformation and microinjection have been used to transfer YACs into mammalian cells to create transgenic animals and cell lines, and for functional studies. More recent developments however, have seen the creation of mammalian and even human artificial chromosomes (Harrington *et al.*, 1997) which may in time replace the use of YACs in transgenics.

Currently YACs are proving to be a very useful tool for genomics. This chapter outlines some of these uses and provides protocols for the major methods involved.

2. Strategies and Real World Applications

2.1 The Physical Mapping of YACs

The techniques used for selection of and mapping of YACs can be classified into hybridisation, PCR-based STS content, and fingerprinting methods.

Because they carry such large inserts, it is possible to array in a manageable number of microtitre plates, individual YACs from libraries representing many fold coverage of large genomes. Thus clones are easily identified by library and plate address, and information can be collated from many labs using the same replicated resource. Similarly, membrane-based clonal DNA arrays can be made in large numbers from these archived libraries.

Screening of these filter grids with radiolabelled probes is straightforward. However, the complexity of the library and the number of probes to be analysed may be such as to make hybridisation screening impractical. PCR-based screening of DNA pools (**Protocol 1**) may be the preferred option.

Sequence tag sites (STSs) (Olson *et al.*, 1989) are designed such that a pair of oligonucleotide primers will generate a unique PCR product from the genomic DNA template. Hence a unique subset of the STS's that represent the entire genome will be generated from an individual YAC template. The STS's can be derived from a variety of sources, such as ESTs (expressed sequence tags) or from random sequencing of small-insert templates derived from FACS sorted chromosomes, for example. These STS's can also be used as hybridisation probes and for RH (radiation hybrid) mapping.

The YAC DNA pools can be designed in a variety of multidimensional schemes to reduce the necessary number of PCR reactions. Typically, these will involve multiple plates, all the wells of individual plates, and row and column pools of individual plates. Not uncommonly, a combination of PCR and hybridisation analyses will be used.

When searching for overlapping and extending YACs (walking), the most efficient probes are derived from the termini of the insert. These termini can be isolated and sequenced in several ways.

Vectorette PCR (Riley *et al.*, 1990. **Protocol 2**) has been widely used. This allows the amplification of fragments adjacent to a known sequence, and is thus ideal for the isolation of an insert terminus adjacent to the vector cloning site. This is a two-step procedure. The clone is digested, with an appropriate enzyme(s), and the vectorette adapter is ligated to all fragments. Then PCR amplification is performed with primers specific to the vector cloning site and the vectorette. Specificity is achieved by virtue of a 'bubble' of non-complementary sequence in the region of the vectorette priming site. Either or both ends of the insert can be isolated.

Insert ends can also be isolated by inverse PCR (Silverman, 1996). The principle here is that clone fragments are circularised (self-ligation of fragments is encouraged by appropriate dilution of the restriction digest). Those insert fragments that are circularised in conjunction with flanking vector can be amplified from a pair of vector-specific primers (arranged 'tail to tail' in the intact clone).

Thirdly, insert ends can be directly sequenced using universal primers specific for flanking vector sequences (Coulson *et al.*, 1991a). To date, this has only been achieved using radioactive sequencing, and consequently fluorescent sequencing is carried out on ends that have been rescued as described above.

The fingerprinting of YACs is complicated by the difficulty of separation of the YAC DNA from the physically indistinguishable chromosomes of the yeast host. This can only be achieved by pulsed field gel electrophoresis, and is therefore much more laborious than the relatively straightforward alkaline purification methods used for circular bacterial clones such as BACs. Hence fingerprinting methods have to be used that specifically generate characterising products from the YAC in a background of other DNA. Most commonly, this is achieved by taking advantage of frequently occurring species-specific repeat sequences. For example, the human genome contains just over 1 million Alu repeat sequences (International Human Genome Sequencing Consortium, 2001). Sequences flanked by Alu repeats can amplified using Alu-specific PCR primers (Nelson *et al.*, 1989). Such amplification of YAC-containing yeast DNA will give a characteristic pattern of products when the radiolabelled fragments are separated on a high resolution gel. Digitisation and computational analysis of the fragment patterns from related YACs will reveal partially matching patterns and hence clone overlaps (Coulson *et al.*, 1986).

2.2 YACs and Sequencing

YACs have been used to varying degrees in different sequencing projects. Bacterial clones have always been used in preference to YACs where possible. YACs have several drawbacks as sequencing templates. Firstly, although they may be the only source of DNA for a particular region, the sequencing can frequently be inefficient because the redundant region of the YAC, that overlaps previously sequenced bacterial clones, may be very large. Ideally, the size of the YAC will be reduced by *in vivo* homologous recombination methods (Markie, 1996) or subcloning of the YAC into an alternative bacterial vector. Alternatively, the entire YAC may be shotgun sequenced but only the relevant gap-spanning region finished. A second problem is that some YACs will inevitably co-migrate with host chromosomes on pulsed field gels. This can be overcome by transfer of the YAC to a host strain in which the chromosomes have been manipulated to provide a 'window' in the appropriate size range in the host electrophoretic karyotype. A set of these strains has been constructed (Hamer *et al.*, 1995) which allows the purification of YACs of all sizes (**Protocol 6**). Thirdly, YACs require careful purification (double CHEF/PFGE) in order to reduce the amount of contaminating host DNA to acceptable levels. This is particularly a problem when subcloning from A/T rich YAC inserts since, in practice, moderately G/C rich DNA tends to clone more efficiently. As a result vector arm and yeast chromosomal DNA are proportionally over-represented in subclone libraries from A/T rich YAC clones, and can sometimes amount to two-thirds of the library. In such cases sequencing of the YAC insert as a whole is economically prohibitive, but as an alternative a 'binning' strategy (Bowman *et al.*, 1999) can be employed whereby YAC libraries are sample sequenced allowing contigs from a chromosome shotgun that show matches to these reads to be grouped together.

Typically high-throughput DNA sequencing involves a random shotgun phase followed by directed finishing. Libraries for shotgun DNA sequencing must be of significant titre to provide enough coverage of the DNA being sequenced and must be as representative as possible. Basic statistics predict that a 1-fold sequence coverage of a DNA molecule, that is 200kb of sequence for a 200kb YAC, will provide 63% of the total sequence. 37% of the sequence will not be obtained since random sampling will mean that some regions are sequenced more than once.

Generally the aim is to sequence an 8-fold coverage during the shotgun phase of sequencing to get a theoretical 99.97% of the available sequence. For a 200kb YAC with average read lengths of 400 bases, libraries will be required of at least 2000 plasmid clones, which are subsequently sequenced in both directions.

That of course assumes that a library is representative. To maximise representation, the DNA to be sequenced is typically fragmented using a physical method (for a review see Quail, 2003b) such as sonication, syringe shearing, nebulisation, or by passage through a French Pressure cell. Such techniques are likely to be maximally random as their efficacy is not dependent on the occurrence of particular sequences. However it is important to bear in mind that there will still be some bias since the G-C bond is stronger than the equivalent A-T bond and some DNA sequences containing secondary structure may be more or less exposed to physical forces than one might expect. Following fragmentation the DNA is left with ragged termini which can be repaired and made blunt-ended by treatment with any of a group of enzymes including: T4 DNA polymerase, Klenow polymerase, Bal31 exonuclease, mung bean nuclease. Sambrook and Russell (2001), describe the action of each of these in detail.

More elaborate methods have been described for library construction that involve the isolation of vector with single insert molecules (*e.g.* the V+I method, Fleischmann *et al.*, 1995), which have the potential advantage of minimising chimaeric and non-recombinant clones. However direct ligation as described (**Protocol 10**) has the advantage that it is highly efficient and so only limited amounts of starting material (such as that available following YAC isolation from PFGE), are required and whilst non-recombinants will occur with low frequency, chimaeric clones are very rare indeed.

2.3 YACs and the C. elegans Project

The progress of the *C. elegans* sequencing project (The *C. elegans* Genome Sequencing Consortium 1998) provides a good, if somewhat extreme, example of the importance of YACs to large-scale projects. Although the use of YACs in this project pre-dated the development of BACs, it is highly probable that BACs would not have made the use of YACs unnecessary.

The 100Mb genome of *C. elegans* is unusual in that the chromosomes have no centromere. They are holocentric. That is to say that the chromosomal elements which form the kinetochore attachment sites are spread throughout the length of the chromosome. Thus, while in monocentric chromosomes such as those of *Drosophila* and humans the highly repetitive centromeric regions have been largely ignored at the sequence level, in *C. elegans* there was no choice but to map and sequence through these sequences in order to produce a complete analysis of the genome.

The initial phase of the mapping of the *C.elegans* genome was the fingerprint analysis of 17,000 cosmids, which assembled into some 700 contigs (Coulson *et al.*, 1986). Being bacterial multicopy plasmids, cosmids are prone to the deletion of repetitive sequences, sometimes to the point of fatality. Some sequences will be unclonable. Hence, cosmid coverage of the genome was found to be extremely non-random, with long stretches, particularly towards the extremities of the chromosomes, completely devoid of cosmid representation.

The *C. elegans* YAC libraries total approximately 11,500 clones. The library that is the largest part of this averages about 200kb insert size, with the remainder averaging about 600kb. These YACs were used to link the cosmid contigs in various ways (Coulson *et al.*, 1988, 1991b). The contig number was reduced to about 100 by hybridisation of radiolabelled YACs (PFGE purified on a relatively small scale) to gridded cosmid filter sets (a minimal subset of cosmids was selected to represent all contigs), and by the inverse hybridisation of cosmids to YAC filter grids. Links between YACs, in the absence of cosmid probes, were detected by hybridisation to

YAC grids of probes derived from the direct sequencing of YAC ends (such overlaps were confirmed by PCR on the relevant YAC templates) and the use of EST probes. Approximately 3,500 YACs are positioned on the physical map, and about 20% of the genome is represented only in YACs. Two gaps remain in the physical map that are not represented in any variety of library.

The majority of the sequence of the *C. elegans* genome was determined from 2,500 cosmid templates. Some cosmid bridges were spanned by fosmids (at 40kb more efficient than sequencing a YAC), but 22% of the genome was sequenced from 280 YAC clones.

2.4 Construction of YAC Libraries

Subsequent to the first cloning into yeast artificial chromosomes by Burke *et al.*, in 1987, the techniques for YAC library construction have become well established and libraries have been made for a variety of different organisms (for detailed methodology and reviews see Larin *et al.*, 1996 and Green *et al.*, 1999). Typically this is a multi-stage process as highlighted in Figure 1.

A typical, and probably the most widely used, YAC vector, pYAC4 (Burke *et al.*, 1987), is shown diagrammatically in Figure 2. In its parental form pYAC4 is circular and has the β-lactamase gene and bacterial origin of replication, allowing its propagation in, and subsequent easily preparation from *E. coli*. It also contains the essential elements for stable maintenance of linear chromosomes in yeast namely; telomere elements that seed telomere formation, a centromeric element providing mitotic and meiotic functionality, an autonomously replicating sequence (ARS element) and auxotrophic markers (*URA*3 and *TRP*1) allowing selection and maintenance of YAC carrying strains. Thus the pYAC4 vector contains two YAC arms either site of an *Eco*RI cloning site in *sup*4, with a *HIS*3 marker, flanked by *Bam*HI sites, between the telomere termini. The arms can be prepared by digesting the pYAC4 plasmid with *Eco*RI and *Bam*HI followed by dephosphorylation using calf intestinal phosphatase to prevent self-ligation of vector arms and stuffer. Having the cloning site within the *sup*4 suppressor tRNA can give a convenient colorimetric detection of YAC recombinants if the YAC is transformed into a yeast host (such as AB1380) that carries the ochre mutation in the *ADE*2 gene. This is because yeast cells that are unable to synthesise adenine gain a red colouration and cloning into *sup*4 disrupts suppression resulting in a block in adenine

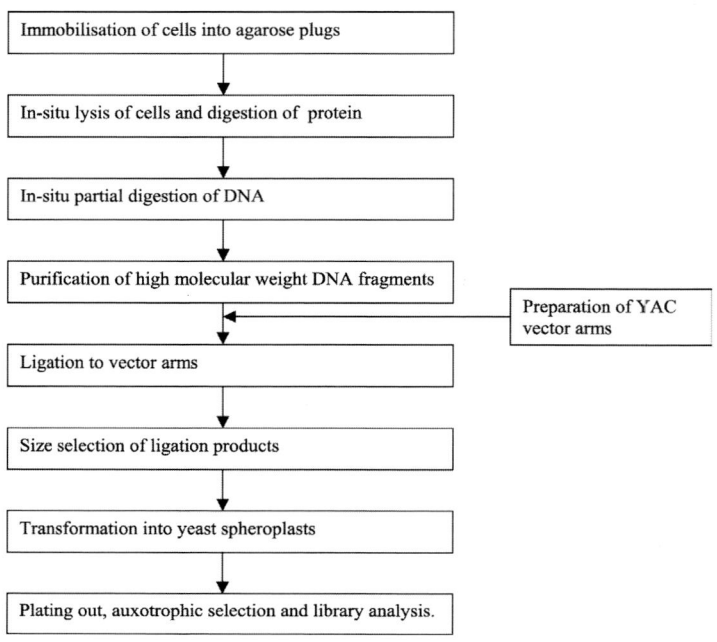

Figure 1. Flow diagram illustrating a typical YAC library construction experiment.

synthesis. This selection in reverse has been used in other YAC vectors to monitor YAC stability. For example YAC vectors pCL99 and pCL100 have a wild-type *ADE2* gene and so when maintained in an *ADE2* mutant strain (that normally gives red colonies) will give white colonies as long as the YAC is present. Loss of the YAC, deletions and rearrangements are therefore distinguished by red colonies or red sectors within colonies as the wild-type gene is disturbed. For details of these and other common YAC vectors see Loehrlein and Davis (1997).

YAC libraries are constructed *in vitro* by the ligation of exogenous DNA to suitably prepared YAC vector arms. Cloning of large inserts requires intact chromosomal DNA as a starting point. This is best achieved by lysing immobilised cells and is commonly done in small agarose blocks called 'plugs'. Cells (or protoplasts) are mixed with agarose and allowed to set in a mould. By incubating the plugs in successive solutions the cells are lysed and the proteins are degraded by proteinase K digestion. Subsequently molecular biology reactions such as restriction digests can be performed *in situ* by incubating the plugs in a solution containing the enzyme and its required buffer. When *Eco*RI is to be used as the cloning enzyme trial

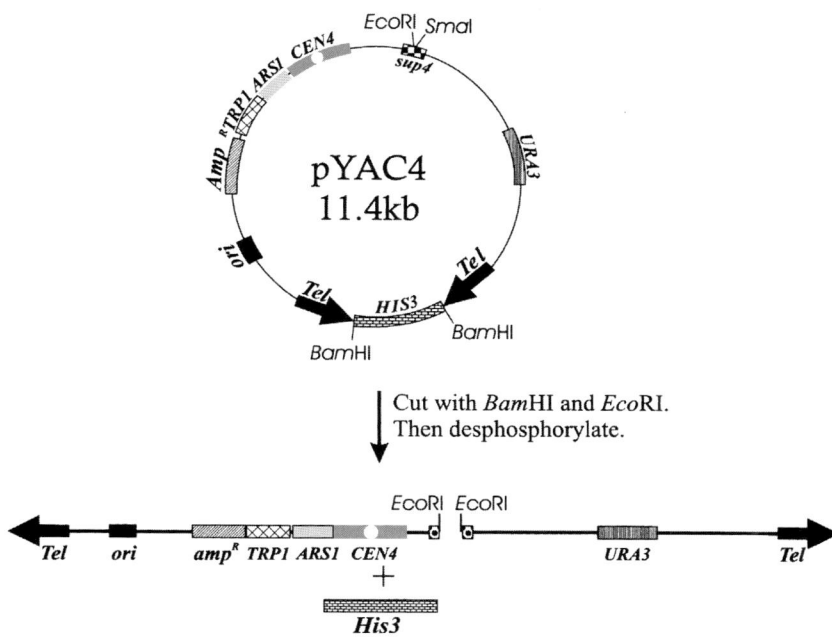

Figure 2. A diagrammatic illustration of the YAC vector pYAC4. The vector can be propagated as an 11.4kb circular plasmid in *E. coli* and is subsequently prepared for cloning by digestion with *Eco*RI to open up the cloning site within the sup4 suppressor tRNA and with *Bam*HI to remove the HIS3 stuffer fragment. This yields the stuffer fragment (which can be ignored) and the two vector arms.

competitive digests are performed by incubating single plugs with a set concentration of *Eco*RI methylase and increasing concentrations of *Eco*RI. If an alternate enzyme is to be used for which no cognate methylase is available digestions are performed for a set time period with a range of enzyme concentrations. The results of these digests are then analysed following PFGE/CHEF separation of the DNA in each plugs and the optimal digestion conditions noted (those resulting in the bulk of the DNA being within the desired library insert size range). Several plugs are then digested under these optimal conditions and DNA separated by PFGE/CHEF in low-melting point agarose. At this point a gel slice is cut from the gel corresponding to the desired insert size range. This agarose slice is then melted together with vector arms and incubated at a lower temperature, typically 37°C or room temperature, with ligase and ligase buffer. After stopping the reaction with EDTA the molten agarose-ligation mixture is allowed to set in the wells of a low-melting point CHEF or PFGE gel and subjected to a second round of size-selection. Fractions are cut from the gel, digested with β-agarase (NEB) after melting at 65°C, before being introduced into the required yeast host strain using spheroplast-mediated transformation. Recombinants are then selected by plating on selective media, *e.g.* synthetic -uracil plates for pYAC4 with AB1380 (or other -URA strain) as host.

2. Protocols and Techniques

Protocol 1. YAC DNA Mini-Preparation for Library Screening by PCR

Equipment and Reagents

- YAC library stored as glycerol stocks in 96 well plates at -80°C.

  Note : This method can easily be adapted for use with libraries which have been archived in 384 well plates. In which case a 384 pronged 'hedgehog is required.'

- SCE: 1M sorbitol, 0.1M sodium citrate pH 5.8, 10mM EDTA pH7.5
- YPD: 10g yeast extract, 20g peptone, 900 ml water. Adjust pH to 5.8 before autoclaving. After autoclaving, add 100ml of filter sterilised 20% (w/v) glucose solution. For solid media add 17-20g bacto-agar before autoclaving.
- AHC: 6.7g yeast nitrogen base without amino acids, 10g acid hydrolysed casein, 2ml of 0.5% (w/v) adenine hemisulphate, 900ml water. Adjust pH to 5.8 before autoclaving. After autoclaving, add 100ml of filter sterilised 20% (w/v) glucose solution. For solid media add 17-20g bacto-agar before autoclaving.
- Lyticase (Sigma, L2524)
- β-mercaptoethanol, RNase, SDS, isopropanol and potassium acetate (Sigma)
- 50mM Tris.HCl pH8.0, 20mM EDTA, 1% SDS
- 70% ethanol
- TE: 10mM Tris.HCl pH8.0, 1mM EDTA
- Specific oligonucleotide primers for screening
- Shaking incubator
- PCR machine

Method

1. Thaw glycerol plates on ice until just starting to melt. Using a 96-prong 'hedgehog' or similar device stamp out copy of plate onto AHC agar plate. Incubate at 30°C for 48 hours.

2. Inoculate 96 x 5ml YPD cultures, each from a single colony, (corresponding to a single well on a microtitre plate). Incubate at 30°C overnight.

3. Make row and column pools.

   Rows: Pool 1ml culture from each overnight in row A, repeat this for the other rows. Each 8ml total

   Column: Pool 1ml from each overnight in column 1; repeat this for the other columns. Each 12ml total.

   This gives 20 preps for each plate. 8 rows and 12 columns.

4. Centrifuge at 2000rpm, 4°C for 5mins. Discard supernatant.

5. Resuspend pellet in 0.75ml distilled water and repeat step 4.

6. Resuspend pellet in 0.9ml SCELM. (SCELM is prepared freshly by mixing, for each plate, 20ml SCE, 5mg lyticase and 80µl β-mercaptoethanol).

7. Incubate at 37°C for 90mins.

8. Centrifuge at 2000rpm for 5mins.

9. Resuspend pellet in 0.75ml of 50mM Tris pH 8.0, 20mM EDTA, 1% SDS.

10. Incubate at 65°C for 30mins.

11. Add 225µl of 5M potassium acetate, mix and leave on ice for 60mins.

12. Centrifuge for 10mins in microfuge at 13,000 rpm.

13. Ethanol precipitate the supernatant by adding 1ml ethanol at room temperature and centrifuging at 13,000 rpm for 20mins.

14. Resuspend pellet in 450µl TE + RNase (70µl of 10mg/ml Nase/ 10ml TE).

15. Incubate at 37°C for 30mins.

16. Add 500µl isopropanol and centrifuge for spin 10mins at 13,000 rpm.

17. Wash pellet with 70% ethanol and dry.

18. Dissolve pellet in 75µl H_2O.

19. You now have 20 DNA preps covering each plate. Determine the DNA concentration in each by diluting 1µl into 1ml of water and measuring absorbance at 260nm using a quartz cuvette. Assuming an A_{260} value of 1.0 corresponds to a DNA concentration of 50 µg/ml, work out the DNA concentration for each DNA prep. Pool equal concentrations of each row and column prep to give single plate preps, so that you now have 21 main DNA stocks per plate.

20. Make working dilutions of 2ng/µl from main DNA stocks.

21. Screen the library by PCR using primers designed against a marker of choice starting with amplifications using plate pools, or groups of plate pools, as template and then finally narrowing down

positive clones to individual well positions via amplification using row and column pools obtained from the same plate as those that gave a positive product in the first round. Primers are typically 25 bases in length, designed to have Tm values >60°C, picked from non-repetitive regions and are selected to give products between 500 and 1000bp. In a typical screen where one has a library covering 20 plates each with 96 wells, screening is performed over two stages. Thus stage 1 involves 20 PCRs on templates that each cover all the wells of each plate. A small sample of each PCR is then run on an agarose gel and those that contain a product of the correct size and identified. Stage 2 then involves 20 PCRs for each plate that gives the correct product, with row and column pools for that plate as template. If row A pool and column 7 pool give products then the desired clone is A7 from that plate of the library.

Hint: It is a good idea to include both positive and negative controls during the PCR. A negative control involving PCR using DNA from the yeast host strain as template and a positive control being DNA containing the marker of interest *e.g.* genomic DNA or a plasmid clone.

Protocol 2. Recovery of YAC Insert Termini by Vectorette PCR

This protocol is a modified version of that described by Riley *et al.*, 1990. It uses several YAC specific primers that recognise sequences in the pYAC4 vector but can be extended to use for other YAC vectors simply by designing appropriate vector specific primers. The technique uses 3 sets of vector primers, outer primers (vecC and vecD) at around position - 180 with respect to and pointing towards the cloning site, inner primers (vecF and vecG) for secondary PCR at around position - 30 with respect to and pointing towards the cloning site and sequencing primers pYACL and pYACR at around −17.

Equipment and Reagents

- Yeast DNA plug prepared from YAC strain (see **Protocol 4**)
- T0.1E: 10mM Tris.HCl pH8.0, 0.1mM EDTA
- PMSF: Phenylmethylsulphonylfluoride (Sigma)
- Vectorette oligos (for blunt ended ligations): Oligos should be HPLC or PAGE purified prior to use. Greater efficiencies can be obtained if the first oligo is 5' phosphorylated but this is not absolutely necessary.

^{5'}CAA GGA GAG GAC GCT GTC TGT CGA AGG TAA GGA ACG GAC GAG AGA AGG GAG AG^{3'} and ^{5'}CTC TCC CTTT CTC GAA TCG TAA CCG TTC GTA CGA GAA TCG CTG TCC TCT CCT TG^{3'}

- 25mM NaCl
- Ligase buffer: 50mM Tris.HCl pH7.6, 10mM $MgCl_2$, 1mM DTT
- T4 DNA ligase (NEB)
- ATP 100µM stock (Amersham)
- PCR primers:
  vecC ^{5'}CAC CCG TTC TCG GAG CAC TGT CCG ACC GC^{3'}
  vecD ^{5'}ATA TAG GCG CCA GCA ACC GCA CCT GTG GCG^{3'}
  vecE ^{5'}CGA ATC GTA ACC GTT CGT ACG AGA ATC GCT^{3'}
  vecF ^{5'}GTT GGT TTA GGC GCA AGA C^{3'}
  vecG ^{5'}GTC GAA CGC CCG ATC TCA AG^{3'}
- Sequencing primers
  pYACL ^{5'}AAT TTA TCA CTA CGG AAT TC^{3'}
  pYACR ^{5'}CCG ATC TCA AGA TTA CGG AAT TC^{3'}
- *Taq* DNA polymerase and buffer
- 5mM dNTP mix
- Perfect Match PCR enhancer (Stratagene)
- *Eco*RI and buffer

Method

1. Take half of a YAC plug. Incubate in 10ml T0.1E at 50°C for 20 mins. Then for plugs that have been prepared using proteinase K, incubate in T01.E + 100µM PMSF for 20mins at room temperature. Then incubate for 2 × 20mins in T0.1E at 50°C and finally for 2 × 20mins at room temperature.
2. Incubate plug at 37°C for 30 minutes in 100µl of 1x restriction enzyme buffer.
3. Replace buffer with fresh and add 20 units of appropriate enzyme (*Rsa*I for pYAC4). Incubate overnight at 37°C.
4. Recover plug and cut into three. Blot dry two of the thirds and store at 4°C and run one third through a 1% agarose minigel alongside a fraction of an undigested plug, to check for digestion.
5. Incubate one of the stored plug fractions in 1x ligation buffer for 1 hour at 4°C.

6. Anneal vectorette oligos by mixing equimolar amounts in 25mM NaCl, incubating at 65°C for 5 minutes and cooling slowly to room temperature. Dilute to 1pmol/µl in 25mM NaCl.
7. Aspirate off buffer and replace with 100µl of fresh ligation buffer.
8. Add 10µl of annealed vectorette oligo solution (1pmol/µl).
9. Heat at 65°C for 15 mins.
10. Equilibrate at 37°C for 5 mins. Add ATP to a final concentration of 1mM and 20 units of T4 DNA ligase. Incubate at 37°C for 1 hour.
11. Add 400µl of T0.1E and store in aliquots at 4°C.

Primary PCR

12. In separate PCRs amplify left and right YAC ends using primer combinations vecC and vecE for left hand end and vecD and vecE for right hand end. PCR contains 5µl of 100ng/µl oligo stock solution, 2µl of 5mM dNTP mix, 5µl of vectorette library, 1µl of perfect match PCR enhancer, 5µl of 10x *Taq* buffer and 2.5 units of *Taq* DNA polymerase in a 50µl reaction volume.
13. Temperature cycle as follows: 1 cycle of 94°C for 5 mins, then 38 cycles of 93°C for 1 min, 65°C for 1 min and 72°C for 3 mins, then a final cycle at 72°C for 5 mins.

Results of the PCR are now analysed by digestion with *Eco*RI

14. To 8µl of PCR add 1µl of 10x *Eco*RI buffer and 1µl of *Eco*RI. Incubate at 37°C for 1 hour.
15. Run digested PCR through a 2.5% agarose minigel alongside a sample of undigested PCR.
   Two or more bands should be visible following digestion. One band of constant size through all YACs will result from the vector segment upto the *Eco*RI site and the other of unique size will be from the end of the YAC insert. The left hand end vector fragment (from vecC to the *Eco*RI site) will be 287bp and the right hand end vector fragment (from vecD to the *Eco*RI site) will be 172bp.
16. If the primary PCR was specific the insert termini can be sequenced at this stage by using primers pYACL or pYACR which are at positions -17 and -20 with respect to the pYAC4 multi-cloning site, with an aliquot of undigested primary PCR as template.

Secondary PCR

17. If the primary PCR was non-specific a second round of PCR using inner vector primers is performed by taking 1µl of primary PCR as

template and primer combinations vecF and vecE for left hand end and vecG and vecE for right hand end. PCR contains 5μl of 100ng/μl oligo stock solution, 2μl of 5mM dNTP mix, 5μl of vectorette library, 5μl of 10x *Taq* buffer and 2.5 units of Taq DNA polymerase in a 50μl reaction volume.

18. Temperature cycle as follows: 1 cycle of 94°C for 5 mins, then 20 cycles of 93°C for 1 min, 59°C for 1 min and 72°C for 3 mins, then a final cycle at 72°C for 5 mins.

19. Analyse the results by digestion with *Eco*RI as in steps 14-15. Since primers vecF and vecG are located at –40 and –29 with respect to the EcoRI site, the vector band should give fragments of these sizes and the unique band will have come from the ends of the insert.

20. The insert termini can be sequenced at this stage by using primers pYACL or pYACR with an aliquot of undigested secondary PCR as template.

Protocols 3-5. Preparation and Handling of YAC Plugs

YAC DNA is typically purified by electrophoresis of yeast plugs prepared from the desired YAC carrying yeast strain. That is agarose slices are cut from a CHEF or pulsed-field gel whereupon the YAC DNA has been electrophoretically separated from the chromosomes of the host yeast strain. Briefly this involves culturing the yeast YAC strain (**Protocol 3**), embedding the yeast in agarose with lysis and proteolysis *in situ* to facilitate chromosomal integrity (**Protocol 4**), followed by two rounds of pulsed-field gel electrophoresis (**Protocol 5**). If YACs comigrate with host chromosomes they can be transferred to strains with altered karyotype (**Protocol 6**).

Protocol 3. Culturing of YAC Strains

Equipment and Reagents

- 20% (v/v) glycerol in deionised water
- -URA broth: (Per litre) 8g Difco yeast nitrogen base without amino acids, 55mg adenine sulphate, 55mg tyrosine, 11g Casamino acids vitamin assay. For solid -URA medium, also add 22g agar prior to autoclaving.
- 20% (w/v) glucose in deionised water
- 5mg/ml tryptophan in deionised water
- 5mg/ml leucine in deionised water

Method

1. Yeast strains containing YACs are best stored frozen in a 20% glycerol solution at -80°C. When a YAC is to be prepared keep the sample on dry ice and scrape a small amount of the relevant glycerol from the surface. Streak several times over a plate of the appropriate selective growth medium (for AB1380(pYAC4) this is normally a supplemented minimal medium without uracil *e.g.* -URA).

2. Grow for 48 hours at 30°C then pick a single standard sized colony and streak back and forth over a fresh plate of the same medium so as to provide a large block of cells that can be used as an inoculum. Likewise grow this second plate for 48 hours at 30°C.

3. Inoculate the cell mass from this second plate into 200ml of -URA broth, supplemented with 20ml of 20% glucose, 2ml of 5mg/ml tryptophan and 4ml of 5mg/ml leucine, in a 500ml sterile flask. Incubate at 200rpm, 30°C for 24 hours. This should provide enough biomass to prepare a sequencing library. Scale up if insufficient DNA is obtained.

Protocol 4. Preparation of Yeast Plugs

Equipment and Reagents

- Bench-top centrifuge
- 50mM EDTA pH8.0, 0.1M EDTA pH8.0 and 0.5M EDTA pH9.0
- Lyticase solution: 1mg Lyticase (Sigma) in 1ml SCE buffer with 50µl 2-mercaptoethanol
- SCE buffer: 1M sorbitol, 0.1M sodium citrate, 0.1M EDTA pH 7.0
- Proteinase K (Stratagene)
- Sarcosyl NL30 (Sigma)
- 1% (w/v) lithium dodecyl sulphate (Sigma)

Method

1. Harvest yeast cells by centrifugation at 6,000 rpm for 10 mins at 4°C (for this we use 250ml Sorvall model 03937 polypropylene bottles in a Sorvall RC-5B centrifuge fitted with a Sorvall GSA rotor). Remove supernatant and resuspend in 50ml of 50mM EDTA pH8.0. Respin to pellet cells then repeat this wash step.

2. Remove as much supernatant as possible then resuspend the cell pellet in 1ml of 50mM EDTA pH 8.0. Add 0.8ml of freshly made

lyticase solution, and leave at room temperature for 5 minutes.

3. Add 6ml of molten 1% low-melting point agarose in 0.1M EDTA. Mix gently taking care not to introduce air bubbles, and pour into a 9cm petri dish plug template (see Figure 3). Alternatively a BioRad plug mould can be used. Allow to set.

4. Remove parafilm and add 20ml of 0.5M EDTA pH 9.0 and 1.5ml of 2-mercaptoethanol. Cover with parafilm and incubate overnight at 37°C.

5. Next day aspirate this solution out of the dish and add 20ml of 0.5M EDTA pH9.0 containing 20mg proteinase K and 0.67ml sarcosyl NL30. Incubate overnight at 50°C.

6. Aspirate fluid out of dish and place dish over plug grid (see Figure 3). Cut agarose into plugs with a razor blade by cutting along the lines of the grid. Add 20ml of a 1% solution of lithium dodecyl sulphate and incubate overnight at 37°C.

7. Aspirate fluid out of dish. Add 20ml 0.5M EDTA pH9.0 and leave overnight at room temperature. Next morning aspirate off solution

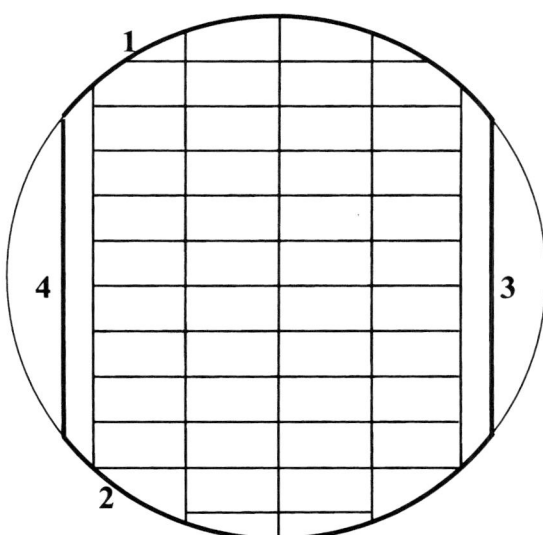

Figure 3. Diagram of YAC plug grid. Enlarge to size of petri dish to be used. Follow grid lines in central area for cutting out plugs. When pouring yeast-agarose mix (Protocol 4. step 3), affix 7.5mm wide strips of double thickness parafilm in positions demarcated by numbered lines 1-4, keep mixture in central part of plate so avoiding wastage.

and add 20ml of fresh 0.5M EDTA pH9.0 and store in fridge until required.

Protocol 5. CHEF Electrophoresis Isolation of YAC DNA

Equipment and Reagents

- CHEF apparatus
- TBE buffer: 100mM Tris, 100mM boric acid, 2mM EDTA pH8.3
- YPD: (Per litre) 10g yeast extract, 20g peptone. For solid medium add 20g bacto-agar. Adjust pH to 5.8 and autoclave. Add sterile glucose to a final concentration of 0.2% w/v, after cooling.
- 10mg/ml ethidium bromide solution (Sigma)
- TAE buffer: 40mM Tris, 20mM sodium acetate, 1mM EDTA pH8.2
- Long-wavelength UV transilluminator
- TE buffer: 10mM Tris.HCl pH8.0, 1mM EDTA

Method

This protocol is written for electrophoresis using a BioRad CHEF but should be transferable to other makes of CHEF/PFGE apparatus with only slight modifications.

1. Prepare gel by mixing 200ml of 0.5 X TBE buffer with 2g low-melting point agarose. Melt, and whilst cooling assemble gel tray and equilibrate at 4°C. Pour gel into tray, place comb in position (use a comb that gives wells that are just a fraction thinner than your plugs so as to give a tight fit) and allow to set for at least 1 hour at 4°C.

2. During this period soak yeast plugs to be loaded, in 0.05 X TBE changing solution once.

3. Remove comb, fill wells with 0.5 X TBE and carefully insert one yeast plug in each well. Reserve the first well for a reference plug of the *S. cerevisiae* host strain, we use the standard YAC host strain AB1380 for which a number of window strains are available (see **Protocol 6**). This strain will need to be grown on a complete yeast growth medium such as YPD and the plug should be prepared in a similar manner to that used above for YAC strains. The inclusion of a reference will allow easy discrimination of the YAC DNA band from that of the yeast chromosomes (the sizes of which are approx: 225kb, 295kb, 375kb, 450kb, 580kb, 680kb, 745kb, 785kb, 815kb, 930kb, 1.11Mb, 1.64Mb and 2.5Mb respectively).

4. Fill CHEF chamber with 4 litres of 0.5 X TBE. Mount gel tray into recess and run at 6V/cm, 14°C, with pulse and run times varying according to the size of the YAC to be separated as follows:

Size of YAC (kb)	Initial pulse	Final pulse	Run
time (seconds)	time (seconds)		time (hours)
<500kb	24	54	33
500-800	58	74	42
800-1100	79	115	42
1100-2000	112	205	44
All chromosomes	60	120	24

Note: These conditions are those that have been found to be optimal on our equipment. If performing this experiment for the first time it is advisable to try a number of different conditions on your apparatus to find those that give optimal separation. **HINT:** The autoalgorithm function of the BioRad CHEF mapper is a useful tool facilitating the optimisation of separation conditions.

5. After 5 hours pause the run and cut away that area of agarose that surrounds the wells, cutting as close to the wells as possible. This well help to minimise background smearing.

6. After the run remove the gel on its base place into a container containing approx. 500ml of distilled water. Add 50µl of 10mg/ml ethidium bromide stock solution and gently agitate for 2 hours.

7. Pour off the ethidium bromide solution and wash the gel with 1 litre of 0.5 X TAE for 30 mins.

8. Place gel onto a long wavelength transilluminator and excise the YAC DNA gel strip with a sterile scalpel.

Note: The UV emitted from a long wave transilluminator, to be used for DNA extraction from agarose, should be greater than 320nm. CAUTION: Many models of transilluminator have two settings, short and long wavelength, on these models the long wavelength setting is often approx. 280nm. If only a short wavelength unit is available cut out as much radiation as possible by placing a sheet of soda glass under the gel.

9. To achieve maximal YAC DNA purify cut YAC DNA gel strip into plug-like sizes and use to load a further CHEF gel and run as described above.

10. Soak the YAC DNA gel slices in 250ml TE and store at 4°C until ready to proceed with DNA extraction step.

Protocol 6. A Method for the Transfer of YACs into Window Strains (Hugerat *et al.*, 1994)

Isolation of YACs is often hindered by their close migration to one of the 16 endogenous yeast chromosomes during electrophoresis. To overcome this problem a number of mutants of the most common YAC host strain, AB1380, have been constructed that have had particular chromosomes fragmented, leading to a series of clearings or windows within their normal banding pattern on PFGE (Hamer *et al.*, 1995).

Strain	Area of window (kb)
YLBW1	150-280
YLBW2	250-450
YLBW3	310-590
YLBW4	450-680
YLBW5	590-755
YLBW6	680-950
YLBW7	810-1120
YLBW8	985-1640
YLBW9	1140-2000

When necessary YACs can be transferred into these strains using the kar-cross method, described below.

Equipment and Reagents

- AHC broth: (Per litre) 6.7g yeast nitrogen base without amino acids, 10g acid casein hydrolysate, 20mg adenine hemisulphate. For solid medium add 20g bacto-agar. Adjust pH to 5.8 and autoclave. Add sterile glucose to a final concentration of 0.2% w/v, after cooling.
- 3µg/ml cycloheximide solution
- PCR machine
- PCR reagents: *Taq* polymerase, buffer, dNTPs and mineral oil
- Test primers: ⁵'AGTCACATCAAGATCGTTTATCC³',
  ⁵'GCACGGAATATGGGACTACTTCG³' and
  ⁵'ACTCCACTTCAAGTAAGAGTTTG³'

Method

1. Inoculate 25 ml YPD broth (see **Protocol 5**) with the desired window strain. Inoculate 25ml AHC broth with the YAC strain. Grow for 24 hours at 30°C, 250rpm.

2. Estimate culture density of each using a haemocytometer. Mix such a volume that corresponds to 5×10^6 cells from each culture in a 50ml tube. Pellet cells by centrifugation (5 min at 4000 rpm), remove the supernatant and resuspend in 1ml YPD medium. Leave stationary at 30°C for 6 hours.

3. Spread 0.3ml and 0.5ml of this mixture over separate thickly poured, AHC agar plates containing 3g/ml cycloheximide. Incubate plates at 30°C for 4-5 days. After 3 days diploid cell colonies will form. Grow on and pick the slower growing diploids for further analysis.

4. To determine whether YAC transfer has been successful, the mating type of a dozen or so colonies are analysed by *MAT*-specific PCR (Huxley *et al.*, 1990).

i) Using sterile yellow tips, pick colonies into 50µl distilled water.

ii) Transfer 5µl of each to a sterile tube/microtitre plate well in which PCR is to be performed. As controls include a blank, a known *MAT*a strain *e.g.* YAC clone in AB1380 and a known *MAT*α strain *e.g.* the original window strain used.

iii) To each tube add 45µl PCR reaction mix (1x PCR buffer, 100M dNTPs, 0.5 units *Taq* polymerase and 1pmol of each of the three test primers.

iv) Overlay with mineral oil (if necessary) and incubate for 35 cycles of 92°C - 1min, 58°C - 2 min, 72°C - 2 min.

Note: These conditions are optimised for Perkin Elmer 480, 9600 models and may have to be modified slightly for other thermocyclers.

v) Analyse products of PCR by agarose gel electrophoresis. There are three possible outcomes.

PCR product(s)	MAT type	Conclusion
404bp	MATα	YAC has transferred into window strain
404bp and 544bp	MATα/a	Diploid
544bp	MATa	original AB1380 YAC strain

vi) Confirm usefulness of transfer by growing up at least two MATα colonies and analysing on a PFGE gel alongside AB1380 YAC strain and parent window strain. The results of a typical window transfer are shown in Figure 4.

Protocol 7. YAC DNA Isolation from Low-Melting Point Agarose Gel Slices

DNA of sufficient quality for library preparation can be easily isolated

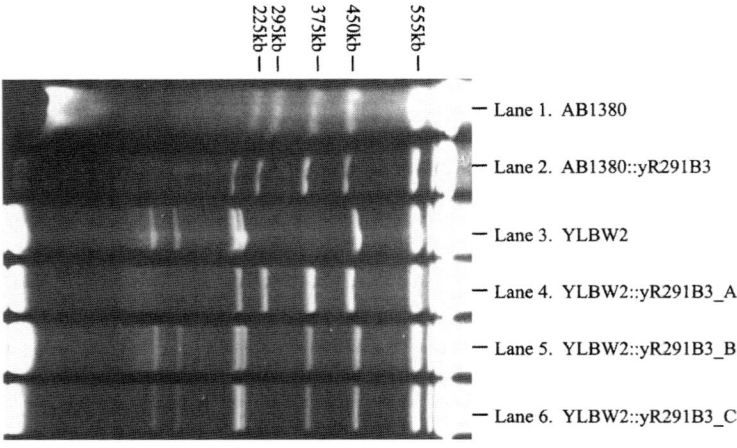

Figure 4. Ethidium-bromide stained CHEF gel showing the results of a successful window transfer. yR291B3 a 375kb YAC from the human X chromosome is seen to co-migrate with yeast chromosome III as this band is stained significantly brighter in the YAC carrying strain (Lane 2) compared with the wild-type strain with no YAC (Lane 1). When this YAC strain is crossed with window strain YLBW2 (Lane 3) which has an altered karyotype in this region, and three resulting colonies are analysed two outcomes are observed. Colony A (lane 4) has not transferred and the karyotype is identical to Lane 2, but colonies B and C (Lanes 5 and 6) clearly show window transfer with a YAC band at around 375kb in the region vacated by wild-type chromosome III.

from agarose gel slices, of isolated YACs, by digesting away the agarose matrix with (β-agarase (NEB)) and then extracting with phenol to remove contaminants.

Equipment and Reagents

- β-agarase and buffer (NEB)
- Ultrapure phenol (Invitrogen)
- Butanol (Sigma)
- 1M NaCl
- 1mg/ml glycogen (Sigma) in deionised water
- Ethanol (absolute and 70%)
- TE buffer: 10mM Tris.HCl pH8.0, 1mM EDTA

Method

1. Soak the gel slice in 1X β-agarase buffer for 30 minutes on ice. Drain and repeat this wash.

2. To allow the β-agarase enzyme access to the agarose matrix, the gel plug must first be dissipated by either incubating at 65°C for 10 minutes to melt the agarose, or by forcing the plug through a 23g needle 4-5 times. In our experience melting at 65°C is more efficient but can lead to denaturation of A/T rich DNA sequences and thus the under representation of such sequences in the resulting library. We therefore routinely use the following "cold extraction" protocol. Load the agarose plug into the back of a 5ml syringe barrel fitted with 23g needle. Force into the barrel of another 5ml syringe. Repeat 4 times, eventually forcing the agarose into a preweighed sterile tube. Reweigh and calculate volume of agarose by subtraction.

3. Add 1/9th volume of 10x β-agarase buffer and 8 units of β-agarase per ml. Mix and incubate at 37°C for 2 hours.

4. Add an equal volume of TE-equilibrated phenol and vortex for 1 min. Incubate on ice for 5 mins prior to centrifugation at 6500rpm to separate phases.

5. Take the upper aqueous phase and reduce down to approx. 300μl by successively extracting with an equal volume of butanol.

Note: Note that butanol forms the upper layer in this mixture and that if too much butanol is added all the aqueous layer will be absorbed into it. In such an event gradually add TE until an aqueous phase can be observed, the DNA should segregate back into the aqueous layer.

6. Remove the aqueous layer to a 1.5ml eppendorf tube and ethanol precipitate by adding 0.1 volumes of 1M NaCl, 10μl of 1mg/ml glycogen and 2.5 volumes of ice cold ethanol. Leave at -20°C overnight, or at -70°C for 30mins.

7. Pellet DNA by centrifugation, wash with 70% ethanol and dry. Resuspend pellet in 21μl TE, then ascertain yield by running 1μl of prep, alongside markers of known DNA concentration (we use 50ng of lambda *Hin*dIII digest) through a standard agarose minigel (*e.g.* 0.8% agarose in TBE).

Protocols 8-10. Construction of Shotgun Sequencing Libraries

The following protocols outline the subcloning of a YAC as a series of smaller fragments for the purposes of shotgun sequencing. Although this can be done with single-standed vectors such as M13, YAC shotgun libraries are typically prepared in plasmid vectors. This is for two main reasons. Firstly plasmid ligations are generally more efficient (circa 20x more efficient compared to ligation into M13), leading to larger libraries. This is a critical factor here as often only small amounts of DNA (100ng or so) may be recovered after CHEF separation and gel extraction. Secondly plasmids are especially useful during sequencing projects as they can be sequenced from both ends such that bridging clones that span gaps can clearly be identified.

Protocol 8 describes fragmentation of the YAC DNA and end-repair to make the fragments blunt-ended and therefore suitable for cloning. **Protocol 9** provides a method for preparing "ligation ready", cut and phosphatased, vector. It is written for preparation of *Sma*I cut pUC18 but can be readily transferred to any vector, that has a multiple cloning site containing a recognition sequence for a restriction enzyme that generates blunt ends, by just substituting the enzyme and changing the incubation conditions appropriately. **Protocol 10** describes ligation of the YAC fragments into this vector to generate the shotgun library.

Protocol 8. Insert Preparation Using Sonication and Mung Bean Nuclease

Equipment and Reagents

- 10x MBB buffer: 50% (v/v) glycerol, 10mM $ZnCl_2$, 0.5M NaCl, 0.3M sodium acetate pH 5.0
- Sonicator
- DNA molecular weight markers
- Agarose gel electrophoresis equipment
- Mung Bean Nuclease (Amersham, 156U/µl)
- 1M NaCl
- Pellet paint (Novagen)
- Absolute ethanol
- TE and TAE (see **Protocol 5**)
- Dye solution: 10% (w/v) ficoll 400 (Sigma), 0.1% w/v bromophenol blue in 1xTBE
- AgarAce (Promega)

Method

1. Pipette YAC DNA (around 10µg (only obtainable by pooling the material gained from several extractions) will give the best results though with care as little as 100ng for plasmid libraries will be

sufficient), into a sterile 1.5ml eppendorf tube. Add 6µl of 10x MBB buffer and water to give a final volume of 60µl. Place on ice.

2. Microfuge to settle contents and fragment DNA by sonication keeping the sample cold throughout to prevent denaturation of DNA (as an alternative to sonication an alternative fragmentation method such as nebulisation may be used in which case the fragmented DNA may need to be concentrated by ethanol precipitation before proceeding with the remainder of the protocol. Details of nebulisation can be found at the web site of Bruce Roe's group at the University of Oklahoma: http://www.genome. ou.edu/protocol_book/protocol_partII.html#II and descriptions of other alternate DNA fragmentation techniques in Quail, 2003b.). We use a sonicator (Model CL4, Misonix Inc, Farmingdale, NY) fitted with a cup-horn probe on top of which the sample is suspended in a 4°C waterbath. Cooling is maintained by siting of the whole apparatus in a cold room. Sonicate at 15% of full power for 2 x 10 seconds, at a distance of 1mm from the probe surface. Incubating on ice for 1 min and briefly microfuging to settle contents, between sonications.

3. Run a 1µl aliquot of each sample alongside a suitable marker (we use a mixture of lambda digested with *Hin*dIII and pBR322 digested with *Bst*NI (NEB)) on a 0.8% agarose minigel. There are three possible outcomes:

    (i) complete sonication (no sign of high mw DNA, smear between 4kb and 500bp).

    (ii) near complete sonication (smear and faint high mw DNA)

    (iii) unsonicated (faint smear and substantial high mw DNA)

    In the case of (ii) resonicate for 5 secs, further checking optional. In the case of (iii) resonicate again for 2 × 10 secs and recheck on a minigel before proceeding.

4. Add 0.3µl of mung bean nuclease to each tube, mix gently, briefly microfuge to settle contents and incubate at 30°C for 10 minutes.

5. Add 141µl of water, 20µl of 1M NaCl, 1µl of pellet paint and 550µl cold ethanol. Leave at -20°C overnight or at -70C for 30mins. Pellet DNA by centrifugation at 13,000g, 4°C for 30 minutes, wash with 1ml ice-cold ethanol and dry.

6. Resuspend DNA pellet in 6.25µl TE, 0.75µl 10x TAE and 2µl dye solution. Run out through a 0.8% low melting point agarose, 1x TAE gel (with 2µl ethidium bromide per 50ml gel volume). Leave one-track gaps between clones. We run a 15 x 10cm gel at 50V, 70mA for 2 hours.

7. On long wave UV transilluminator cut out 0.6-1kb, 1-1.4kb. 1.4-2kb and 2-4kb agarose blocks and place in fresh 1.5ml eppendorf tubes. Melt at 65°C for 5mins or pass through 26g needle to fragment. Equilibrate at 42°C, add 1 unit AgarAce (Promega) per 200µl volume and incubate at 42°C for 20mins.

8. Extract with TE-equilibrated phenol and ethanol precipitate, adding 1/10th volume 1M NaCl and 1µl of pellet paint as described previously.

9. Pellet DNA by centrifugation at 13,000g, 4°C for 30 minutes, wash with 1ml ice-cold ethanol and dry. Resuspend in 3µl of $T_{0.1}E$.

Protocol 9. Preparation of Dephosphorylated *Sma*I-Cut Vector

Inserts prepared in **Protocol 8**, are blunt-ended with terminal 5' phosphates. Therefore any vector that has suitable priming sites for sequencing and has been digested to leave a blunt end, then dephosphorylated to prevent vector self-ligation, can be used. As an alternative to using this protocol dephosphorylated pUC18-*Sma*I of excellent quality can be purchased from Oncor Inc. (Gaithersburg MD), but if a custom vector is to be used then the use of this protocol is advised

Equipment and Reagents

- *Sma*I, Calf intestinal phosphatase (10U/µl), buffer 3 and buffer 4 (NEB)
- 5M NaCl
- Phenol (Invitrogen)
- Ethanol
- TE and TAE (see **Protocol 5**)
- 0.5M EGTA
- Dye Solution (see **Protocol 8**)
- 10mg/ml ethidium bromide solution (Sigma)
- Agarose gel electrophoresis equipment
- 10mM Tris pH8.0, 10mM EDTA
- 1M NaCl
- Butanol

Method

1. Start with 20µg of plasmid DNA that has been purified by passage through two caesium chloride gradients then extensively dialysed against TE.

Note: Sambrook and Russell, 2001, contains an excellent description of plasmid purification using caesium chloride gradients. If using a vector DNA preparation obtained from a commercial supplier this is usually not necessary.

2. Add: 20μl NEB buffer 4

2μl *Sma*I (NEB)

water to 200μl

3. Incubate at room temperature for 30 minutes, no longer, as overdigestion results in damage to the DNA termini that manifests itself as a high background of false recombinants (false whites). Take a 0.5μl aliquot for analysis.

4. To the main digest add 4μl of 5M NaCl and 200μl phenol. Mix and then microfuge for 15 minutes at 4°C. Recover the aqueous layer, add 500μl ethanol and incubate at room temperature for 25 minutes.

5. Analyse the 0.5μl aliquot by running through a 0.8% agarose minigel alongside size markers and undigested vector. Ideally the digest should appear almost complete.

6. Pellet the DNA from the main digest in a microfuge at 4°C for 30 minutes. Wash with 1ml 70% ethanol then air-dry. Resuspend in 5μl TE.

7. Add: 20μl NEB buffer 3

173μl water

2μl Calf Intestinal phosphatase (10U/μl, NEB)

8. Incubate at 37°C for 1 hour then stop by adding 20μl of 0.5M EGTA and incubating at 65°C for 10 minutes. Cool to room temperature and extract with 220μl phenol, centrifuging at room temperature for 5 minutes to separate the phases.

9. Recover the aqueous layer and add 25μl of 10x TAE and 75μl dye solution. Load into the wells of a 250ml, 0.4% agarose-TAE gel. We use a model EH300 preparative electrophoresis apparatus (Cambridge Electrophoresis, Cambridge, U.K.) which has a gel bed area of 20 x 17cm. Run gel overnight at 28V, with 20μl of 10mg/ml ethidium bromide in the TAE running buffer.

10. Carefully place the gel onto a piece of glass on a long wavelength UV transilluminator (>320nm). With a sterile scalpel cut round the linearised plasmid band. Other bands corresponding to uncut plasmid or concatomers may be seen and should be avoided.

11. Place the gel slice corresponding to the linearised vector band, at the well end of the gel tray of a preparative electrophoresis tank (EH300 type), and cast a 250ml 0.4% Low-melting point agarose

TAE gel around the gel slice. Allow gel to set then add buffer to gel tank and run gel at 4°C in cold room, overnight at 28V. There should be no need to add further ethidium bromide as there was sufficient bound to the DNA from the first gel stage.

12. Place on long wavelength transilluminator and cut out linearised vector band. Place in a preweighed 50ml sterile tube (Falcon type) and reweigh to estimate gel volume. Add 0.5 volumes of 10mM Tris pH8.0, 10mM EDTA and 30µl 1M NaCl, then melt at 65°C for 5-10 mins.

13. Split between several 2ml Safelock eppendorf tubes and extract twice with an equal volume of phenol, once with an equal volume of phenol:chloroform and then repeatedly with butanol or isobutanol to reduce total volume to 300µl (pooling tubes when possible).

14. Add 700µl ethanol, mix and microfuge for 30mins at 4°C. Wash with 1ml of 70% ethanol then air-dry.

15. Resuspend in 200µl TE. At this stage the vector should be approx. 20ng/µl. Store at 4°C whilst testing in a ligation, preferably with a fragment that is known to ligate well. If sufficient recombinants are obtained and background is acceptable, then split the prep into 10µl aliquots in 0.5ml tubes and store at -20°C. If test ligations give unacceptable results then repeat prep; lengthening the period of *Sma*I digestion to reduce blue background or concentrating the prep to give more recombinants.

Protocol 10. Construction of pUC Libraries

Equipment and Reagents

- T4 DNA ligase (Roche, 5U/µl) and buffer
- pUC18 SmaI-CIP (40ng/µl)
- Electroporation apparatus
- Electrocompetent cells

Method

1. Set up ligations as follows, in 0.5ml tubes:

DNA solution	3µl
pUC18 SmaI-CIP	0.3µl
10X ligase buffer	0.4µl
Ligase	0.3µl

2. Mix with pipette by pipetting up and down slowly. Microfuge very briefly to settle contents then incubate overnight at 12-14°C.

3. Next morning incubate ligation at 65°C for 10 minutes to denature the ligase and dilute to 50µl with sterile distilled water. Store frozen.

Note: Keep treated ligations cold (preferably frozen), since there is no buffering in the solution.

4. Test libraries for quality and titre, by electroporation of a 0.2µl aliquot into high efficiency electrocompetent cells (such as DH10B (Invitrogen) or Electro Ten-Blue (Stratagene) that give >10^{10} transformants per µg of DNA.

4. Relevant Web Sites

1. http://www.qbiogene.com/products/media/index.shtml

   QBiogene. Supplier of an extensive range of yeast media.

2. http://www.protocol-online.org/prot/Molecular_Biology/Cloning/YAC_Cloning/index.html

   Extensive set of links to YAC and general yeast protocols.

3. http://www.sanraffaele.org

   Organisation providing screening services for CEPH and ICI human YAC libraries.

4. http://www.cephb.fr/services/
   http://www.hgmp.mrc.ac.uk/geneservice/index.shtml

   Order from the CEPH collection of human YAC clones.

5. http://www.hgmp.mrc.ac.uk/geneservice/index.shtml

   Order from the WI/MIT collection of Mouse YAC clones.

6. http://rgp.dna.affrc.go.jp/publicdata/physicalmap99/YACall.html

   Information on the rice YAC map.

   http://www.nal.usda.gov/pgdic/Probe/v1n1_2/theor.html

   An essay on the use of YACs in genome research.

7. http://www.resgen.com/include/menus/yacs_menu.php3

   Order from the Resgen collection of Human,

	Rat, mouse and zebra fish YAC clones and resources.
8. http://www.fruitfly.org/about/materials/ob.yac.clones.html	Order Drosophila YAC clones.
9. http://www.atcc.org	Repository of clones and strains.
10. http://www.fgsc.net/fgn41/yac1.html	Protocols for use with *N. crassa* YAC library but generally applicable.
11. http://zebrafish.mgh.harvard.edu/yac/	Details of the Harvard Zebra fish YAC library.
12. http://www.agron.missouri.edu/cgi-bin/sybgw_mdb/mdb3/Clone+Library/83191	Details of the Maize YAC library.
13. http://www.nhm.ac.uk/hosted_sites/schisto/resources/YAClibrary.html	Details of the *Schistosoma mansoni* YAC library.
14. http://www.wormbase.org/	Database of *C. elegans* genomics and post genomics

References

Albertsen, H.M., Abderrahim, H., Cann, H., Dausset, J., Le Paslier, D., and Cohen, D. 1990. Construction and characterization of a yeast artificial chromosome library containing seven haploid genome equivalents. Proc. Natl. Acad. Sci. (USA). 87: 4256–4260.

Anand, R., Villasante, A., and Tyler-Smith, C. 1989. Construction of yeast artificial chromosome libraries with large inserts using fractionation by pulsed-field gel electrophoresis. Nucleic Acids Res. 17: 3425–3433.

Bowman, S., Lawson, D., Basham, D., Brown, D., Chillingworth, T., Churcher, C.M., Craig, A., Davies, R.M., Devlin, K., Feltwell, T., Gentles, S., Gwilliam, R., Hamlin, N., Harris, D., Holroyd, S., Hornsby, T., Horrocks, P., Jagels, K., Jassal, B., Kyes, S., McLean, J., Moule, S., Mungall, K., Murphy, L., Barrell, B.G., *et al.* 1999. The complete nucleotide sequence of chromosome 3 of *Plasmodium falciparum*. Nature. 400: 532–538.

Brownstein, B.H., Silverman, G.A., Little, R.D., Burke, D.T., Korsmeyer, S.J., Schlessinger, D., and Olsen, M.V. 1989. Isolation of single-copy human genes from a library of yeast artificial chromosome clones. Science. 244: 1348–1351.

Burke, D. T., Carle, G. F., and Olson, M. V. 1987. Cloning of large DNA segments of exogenous DNA into yeast by means of artificial chromosome vectors. Science, 236: 806–812.

Cai, L., Lindpaintner, K., Browne, J., Gruetzner, F., Haaf, T., James, M.R., and Bihoreau, M. 2000. An anchored YAC-STS framework for the rat genome. Cytogenet. Cell Genet. 89: 168–70.

Camilleri, C., Lafleuriel, J., Macadre, C., Varoquaux, F., Parmentier, Y., Picard, G., Caboche, and M., Bouchez, D.1998. YAC contig map of *Arabidopsis thaliana* chromosome 3. Plant J. 5: 633–642.

Coulson, A., Kozono, Y., Lutterbach, B., Shownkeen, R., and Waterston, R. 1991b. YACs and the *C. elegans* genome. BioEssays. 13: 413-417.

Coulson, A., Kozono, Y., Shownkeen, R., and Waterston, R. 1991a. The isolation of insert-terminal YAC fragments by genomic sequencing. Technique. 3: 17–23.

Coulson, A., Sulston, J., Brenner, S., and Karn, J. 1986. Towards a physical map of the genome of the nematode *Caenorhabditis elegans*. Proc. Nat. Acad. Sci. Wash. 83: 7821–7825.

Coulson, A., Waterston, R., Kiff, J., Sulston, J., and Kohara, Y. 1988. Genome linking with yeast artificial chromosomes. Nature. 335: 184–186.

Fleischmann, R.D., Adams, M.D., White, O., Clayton, R.A., Kirkness, E.F., Kerlavage, A.R., Bult, C.J., Tomb, J.F., Dougherty, B.A., Merrick, J.M., *et al.* 1995. Whole-genome random sequencing and assembly of *Haemophilus influenzae* Rd. Science. 269: 496–512.

Green, E.D., Hieter, P., and Spencer, F.A. 1999. Yeast Artificial Chromosomes. In: Genome analysis: A laboratory manual. Volume 3, Cloning systems. Birren, B., Green, E.D., Klapholz, S., Myers, R.M., and Roskams, J. Eds. Cold Spring Harbor Laboratory Press, Cold Spring Harbor, New York. p297–565.

Hamer, L., Johnston, M., and Green, E.D. 1995. Isolation of yeast artificial chromosomes free of endogenous yeast chromosomes: construction of alternate hosts with defined karyotypic alterations. Proc. Natl. Acad. Sci. (USA). 92:11706–11710.

Harrington, J.J., Van Bokkelen, G., Mays, R.W., Gustashaw, K., and Willard, H.F. 1997. Formation of *de novo* centromeres and construction of first-generation human artificial minichromosomes. Nature Genetics.

15: 345–355.

Hugerat, Y., Spencer, F., Zenvirth, D., and Simchen, G. 1994. A versatile method for efficient YAC transfer between any two strains. Genomics. 22:108–117.

Huxley, C., Green, E.D., and Dunham, I. 1990 Rapid assessment of *S. cerevisiae* mating type by PCR. Trends Genet. 6: 236.

Huxley, C. 1994. Transfer of YACs to mammalian cells and transgenic mice. Genetic Eng. 16: 65–91.

International Human Genome Sequencing Consortium. 2001. Initial sequencing and analysis of the human genome. Nature. 409: 820–921.

Ioannou, P.A., Amemiya, C.T., Garnes, J., Kroisel, P.M., Shizuya, H., Chen, C., Batzer, M.A., and deJong, P.J. 1994. A new bacteriophage P1-derived vector for the propagation of large human DNA fragments. Nature Genetics. 1: 84–99.

Kuspa, A., and Loomis, W. F. 1996. Ordered yeast artificial chromosome clones representing the *Dictyostelium discoideum* genome. Proc. Natl. Acad. Sci. (USA). 93: 5562–5566.

Larin, Z., Monaco, P and Lehrach, H. 1991. Yeast artificial chromosome libraries containing large inserts from mouse and human DNA. Proc. Natl. Acad. Sci. (USA). 88: 4123–4127.

Larin, Z., Monaco, P and Lehrach, H. 1996. Generation of large insert YAC libraries. In: Methods in Molecular Biology, Vol. 54. YAC Protocols. D. Markie, ed. Humana press Inc, Totowa, NJ. p1–31.

Loehrlein, C., and Davis, R.W. 1997. Yeast Artificial Chromosome (YAC) Construction and Applications. In: Encyclopedia of molecular biology and molecular medicine. Volume 6. R.W. Meyers, ed. VCH, Weinheim. p291–302.

Markie, D., and Ragoussis, J. 1996. Genomic reconstruction by mitotic recombination of YACs. In: Methods in Molecular Biology, Vol. 54: YAC Protocols, Ed D.Markie Humana Press Inc., Totowa, NJ, USA.

Nelson, D.L., Ledbetter, S.A., Corbo, L., Victoria, M.F., Ramirez-Solis, R., Webster, T.D., *et al.* 1989 Alu polymerase chain reaction: a method for rapid isolation of human-specific sequences from complex DNA sources. Proc. Natl. Acad. Sci. (USA). 86: 6686–6690.

Nusbaum, C., Slonim, D.K., Harris, K.L., Birren, B.W., Steen, R.G., Stein, L.D., Miller, J., Dietrich, W.F., Nahf, R., Wang, V., Merport, O., Castle, A.B., Husain, Z., Farino, G., Gray, D., Anderson, M.O., Devine, R., Horton, L.T. Jr., Ye, W., Wu, X., Kouyoumjian, V., Zemsteva I.S., Wu, Y., Collymore, A.J., Courtney, D.F., *et al.* 1999. A YAC-based physical map of the mouse genome. Nature Genetics. 4: 388–393.

Olson, M., Hood, L., Cantor, C., and Botstein, D. 1989. A common language for physical mapping of the human genome. Science. 245: 1434–1435.

Quail, M.A. 2003a. DNA Cloning. In: Encyclopedia of the Human Genome. Ed Cooper, D.N. Nature Publishing, London.

Quail, M.A. 2003b. DNA: mechanical breakage of. In: Encyclopedia of the Human Genome. Ed Cooper, D.N. Nature Publishing, London.

Riley, J., Butler, R., Ogilvie, D., Finniear, R., Jenner, D., Powell, S., Anand, R., Smith, J.C., and Markham, A.F. 1990. A novel, rapid method for the isolation of terminal sequences from yeast artificial chromosome (YAC) clones. Nucleic Acids Res. 18: 2887–2890.

Saji, S., Umehara, Y., Antonio, B.A., Yamane, H., Tanoue, H., Baba, T., Aoki, H., Ishige, N., Wu, J., Koike, K., Matsumoto, T., and Sasaki, T. 2001.A physical map with yeast artificial chromosome (YAC) clones covering 63% of the 12 rice chromosomes. Genome. 1: 32–37.

Sambrook, J., and Russell, D.W. 2001. Molecular Cloning: A laboratory manual. 3rd edition. Cold Spring Harbor Laboratory Press, Cold Spring Harbor, NY.

Schalkwyk, L.C., Cusack, B., Dunkel, I., Hopp, M., Kramer, M., Palczewski, S., Piefke, J., Scheel, S., Weiher, M., Wenske, G., Lehrach, H., and Himmelbauer, H. 2001. Advanced Integrated Mouse YAC Map Including BAC Framework. Genome Research. 11: 2142–2150.

Schwartz, D.C. and Cantor, C.R. 1984. Separation of yeast chromosome-sized DNAs by pulsed field gel electrophoresis. Cell. 37: 67–75.

Shizuya, H., Birren, B., Kim, U.J., Mancino, V., Slepak, T., Tachiiri, Y., and Simon, M. 1992. Cloning and stable maintenance of 300-kilobase-pair fragments of human DNA in *Escherichia coli* using an F-factor-based vector. Proc. Natl. Acad. Sci. (USA). 18: 8794–8797.

Silverman, G.A. 1996. End-rescue of YAC clone inserts by inverse PCR. In: Methods in Molecular Biology, Vol. 54. YAC Protocols. D. Markie, ed. Humana press Inc, Totowa, NJ. p1–31.

The *C. elegans* Genome Sequencing Consortium. 1998. Genome Sequence of the nematode *C. elegans*: a platform for investigating biology. Science. 282: 2012–2018.

Thompson, J.K., and Cowman, A.F. 1997. A YAC contig and high resolution restriction map of chromosome 3 from *Plasmodium falciparum*. Mol. Biochem. Parasitol. 2: 537–542.

Wang, S.Y., Cruts, M., Del-Favero, J., Zhang, Y., Tissir, F., Potier, M.C., Patterson, D., Nizetic, D., Bosch, A., Chen, H., Bennett, L., Estivill, X., Kessling, A., Antonarakis, S.E., and van Broeckhoven, C. 1999. A high-resolution physical map of human chromosome 21p using yeast artificial chromosomes. Genome Res. 11: 1059–1073.

5

Fluorescence *In Situ* Hybridisation, Flow-Sorting and Related Technologies

Susan Gribble

Abstract

Molecular cytogenetics is an evolving scientific field that combines the techniques of both cytogenetics and molecular genetics. FISH-based molecular cytogenetic techniques allow chromosomal rearrangements to be investigated at a higher resolution than standard G-banding analysis. FISH and related technologies are used to identify cryptic chromosomal rearrangements, to clarify complex karyotypes and to define the position of translocation breakpoints. These investigations help to refine specific translocation breakpoint regions allowing them to be characterised at the molecular level by PCR and sequencing. In this chapter I describe the techniques that may be employed to identify both the genes involved and the potential causative mechanisms underlying chromosomal rearrangements.

1. Introduction

The first fluorescently detected *in situ* hybridisation (FISH) experiments were conducted in the early 1980s (Bauman *et al.*, 1980; Van Prooijen-Knegt *et al.*, 1982; Landegent *et al.*, 1984). FISH has many advantages over earlier *in situ* methods that were detected using radioisotopes. Aside from the obvious safety benefits, FISH is a flexible technique that uses fluorescence microscopes and dedicated image analysis software to rapidly provide images that are readily interpretable.

The basic stages of a FISH experiment include:
1. preparation and fixation of target DNA
2. preparation of probe DNA
3. labelling of probe DNA with hapten or fluorochrome conjugated nucleotides
4. denaturation of probe DNA and pre-annealing with Cot-1 DNA to block repetitive genomic sequences
5. denaturation of target DNA
6. incubation of denatured probe with denatured target, to allow specific sequences to anneal
7. removal of excess probe by washing and the detection of indirectly labelled probes using fluorescently conjugated antibodies
8. image acquisition using an epifluorescence microscope and light source, relevant filters, a CCD camera, controlled by image acquisition software

FISH is a powerful research tool with the flexibility to address a range of cytogenetic investigations. The resolution of FISH can be varied by the choice of DNA target. Hybridisation to metaphase chromosomes provides a resolution of approximately 2 Mb (Trask *et al.*, 1991) but by utilising the less condensed chromatin in the interphase nucleus and ultimately extended chromatin fibres, resolution of 50 kb and 5 kb respectively can be achieved (Heng *et al.*, 1992). The recent development of DNA microarrays may enable the whole genome to be interrogated at an even higher resolution (Solinas-Toldo *et al.*, 1997; Pinkel *et al.*, 1998; Albertson *et al.*, 2000).

The complexity of the DNA probe in a FISH experiment can also vary, from whole genomes as used in Comparative Genomic Hybridisation (CGH) (Kallioniemi *et al.*, 1992; du Manoir *et al.*, 1993), multicolour FISH (MFISH) (Speicher *et al.*, 1996), Spectral Karyotyping (SKY) (Schrock *et al.*, 1996), or specific genomic regions (YACS, BACS, PACS, cosmids and fosmids), as used in FISH mapping (Table 1). Another variable in FISH experiments is the choice of fluorochrome combination. Traditionally biotin (Langer *et al.*, 1981) and digoxigenin labelled nucleotides have been incorporated when labelling a DNA probe and

Table 1. The complexity of FISH DNA probes used in different FISH techniques, the derivation of the probe and a suggested method of labelling.

Technique	Probe DNA complexity	Source of probe	Labelling method
CGH	Genomic DNA 3289 Mb 3.2 Gb	Genomic DNA from sample extracted by conventional methods	nick translation or DOP-PCR (fragments 2000-500 bp)
M-FISH	All Chromosomes	Flow sorted chromosomes	DOP-PCR
Whole chromosome painting	Individual specific chromosomes	Flow sorted chromosomes, monochromosome cell hybrids	nick translation, chemical labelling, DOP-PCR
Array painting	Derivative chromosomes	Flow sorted and DOP-amplified	Random-Priming
Chromosome arm painting	Chromosome arms 45-150 Mb	Microdissected chromosome arms, DOP-amplified	DOP-PCR, nick translation
Chromosome band specific painting	Specific chromosome bands 5-10 Mb	Microdissected chromosome bands, DOP-amplified	DOP-PCR, nick translation
Alpha satellite DNA Centromeric probes	Tandem repeat sequence DNA specific for individual chromosomes Several Mbs High order repeats (171bp) tandemly repeated 100-5000 times	Plasmids 5 kb or 500 bp inserts	nick translation
Telomere specific FISH	Subtelomeric specific DNA Cosmid, P1 or PAC clones(approx. 100 kb)	100-300 kb from the tel ends	nick translation
Metaphase locus specific FISH	YACS 100 kb -1 Mb	ICRF HD1, ICRF 4X, ICRF 4Y, CEPH Mega , ICI , Washington University	Alu PCR, nick translation
	BACS 150 kb insert	RPCI-11, CIT978SK, CIT978SK, CEPH human BAC library, RCPI-13, Genome Systems	nick translation
	PACS 100 kb insert	RPCI-1,3,4,5, RPCI-6, RPCI-21	nick translation
	Cosmids 30-44 kb insert	LLOXNCO1, LL22NCO3, SC6cA, LA04NCO1, SC7cD, SCXcC, SC22cB, SCM4cB, LL22NC01	nick translation
	Fosmids 37.5 kb	SC6fA, SCXfC, CITF22,	nick translation
	Plasmids 5-10 kb	cDNA library of interest	nick translation
	λ Bacteriophage 25 kb		nick translation

Continued

Technique	Probe DNA complexity	Source of probe	Labelling method
	cDNA (as small as 500bp)	H9 human placental library, Pfizer human brain library, UP37 + monocyte library	nick translation, PCR
Fibre FISH	YACS, BACS, PACS, cosmids, fosmids As above	As above	nick translation
	Long range PCR products 4-25 kb	PCR products of interest	PCR

indirectly detected using avidin or antibodies conjugated to Texas Red or Fluorescein. More recently, both direct and additional indirect detection methods have been developed, using an increased range of nucleotide or antibody conjugated fluorochromes. This expansion of available fluorochromes with corresponding filter sets for the fluorescence microscope have enabled the development of modern sophisticated molecular cytogenetic techniques such as MFISH and SKY.

2. FISH Experimental Design

2.1 Selection of Target DNA

2.1.1 Human Metaphase Chromosomes

Normal human metaphase chromosomes are used as a target in FISH experiments to physically map the location of a clone containing human DNA (YACs, BACs) (Lawrence *et al.*, 1990; Collins *et al.*, 1995). They are also used as targets in CGH experiments to locate genomic imbalances that may be present in the sample DNA (Kallioniemi *et al.*, 1993; Forozan *et al.*, 1997; Hemminki *et al.*, 1997; Gribble *et al.*, 1999). FISH analysis of metaphase chromosomes harbouring abnormalities allows detailed analysis of the rearrangement (Schrock *et al.*, 1996; Speicher *et al.*, 1996).

Metaphase chromosomes are routinely prepared from peripheral blood samples or lymphoblastoid cell lines. Human blood is a potentially hazardous material, and its handling is subject to strict safety guidelines. Lymphoblastoid cell lines offer the advantage of being simpler to handle and provide an immortal supply of normal metaphase chromosomes. However, it should be borne in mind that the development of a human lymphoblastoid cell line and its subsequent passaging has the potential to introduce de novo genetic alterations unassociated with the original blood sample. Consequently some mapping experiments may require confirmation on a different cell line to validate the original result.

The initial step of a metaphase preparation (protocol 1) is to incubate cultured cells with Bromo deoxyuridine (BrdU). The BrdU is incorporated into the DNA in the S phase of the cell cycle and ensures the chromosomes maintain their length by preventing chromosomal condensation. Addition of Ethidium Bromide (EtBr) also prevents chromosomal condensation by intercalating into the chromosomal DNA. The mitotic index of a cell preparation is increased by the addition of colcemid which acts by inhibiting the formation of the mitotic spindle, arresting cells at metaphase. Cells are then incubated in a hypotonic buffer to help separate the chromosomes, and the swollen cells fixed in 3:1 methanol:acetic acid solution, which in conjunction with dropping onto a glass slide breaks the cell membrane and releases the chromosomes. It is essential at this stage that metaphase preparations are viewed under a phase contrast microscope to assess that, 1) the chromosomes are cytoplasm free, 2) the chromosomes are a good length and 3) there are minimal chromosome overlaps.

2.1.2 Human Interphase Nuclei

Human interphase nuclei are prepared from confluent fibroblast cell lines (*e.g.* MRC-5) (protocol 2). FISH signals present on replicating cells are observed as doublets due to the presence of two sister chromatids. In confluent cell lines, which are no longer undergoing replication, FISH probes are usually observed as a single signal. Interphase chromosomes are less condensed than metaphase chromosomes and provide a higher resolution target for FISH mapping (Trask *et al.*, 1989; Leversha, 1997). The chromosomes have been shown to remain in chromosome territories which means they are suitable targets for use in establishing the order of clones (Cremer and Cremer, 2001).

2.1.3 Extended Chromatin Fibres

Extended fibres increase the utility of FISH experiments to resolve FISH signals from clones located only 5 kb apart (Buckle and Kearney, 1993; Windle *et al.*, 1995). Clones separated by more than 500 kb do not usually hybridise to the same fibre and their relationship is beyond detection by fibre FISH. To prepare extended chromatin fibres, cells are dried onto a glass slide, lysed in alkaline buffer (some techniques use detergents) and physically stretched (protocol 3). Other protocols have been described which enable fibres to be prepared from high molecular weight DNA stored in agarose blocks (Heiskanen *et al.*, 1994) and also from fixed chromosomal suspensions (Fidlerova *et al.*, 1994). In fibre FISH experiments, indirectly labelled probes are favoured as they produce amplified fluorescent signals. The protocols used in our laboratory (protocol 3) produce fibres which are differentially stretched. An alternative technique, termed "molecular combing" produces equally stretched fibres and provides signals which may be quantified accurately (Michalet *et al.*, 1997).

2.1.4 DNA Microarrays

DNA microarrays have recently been developed to increase the sensitivity of Comparative Genomic Hybridisation (CGH). Conventional CGH can detect a chromosomal deletion of approximately 10 Mb (Parente *et al.*, 1997; Bentz *et al.*, 1998). Array CGH replaces chromosomal targets with DNA clones spotted onto glass slides (Solinas-Toldo *et al.*, 1997; Pinkel *et al.*, 1998; Albertson *et al.*, 2000). Array CGH has a greater sensitivity which is determined by the size and separation of the clones on the array. Currently typical arrays comprise clones which physically map at 1 Mb intervals along the human genome, providing a far greater resolution in CGH than that obtainable using metaphase chromosomes.

2.2 Selection and Preparation of FISH Probes

2.2.1 Flow-Sorting of Chromosomes

To prepare whole chromosome paints, multiple copies of individual chromosomes can be separated by flow sorting on a flow cytometer (Ross and Langford, 1997). To separate individual chromosomes, chromosome suspensions are stained with two fluorescent dyes, Hoechst 33258 and Chromomycin A3, which bind preferentially to AT-rich and GC-rich regions, respectively. The quantitative differences in the fluorescence characteristics of the labelled chromosomes are then analysed using the flow cytometer. By plotting the intensity of Hoechst 33258 against the intensity of Chromomycin A3, a flow karyotype is produced, which identifies individual chromosome types as specific data point clusters. The sorting process uses electrostatic deflection to direct specifically charged droplets of the sheath fluid, containing the chromosome of choice, into a collection tube.

Sorted chromosomes are then amplified by PCR and fluorescently labelled to produce specific chromosome paints. PCR products from different chromosome types can be combined prior to labelling with multiple fluorochromes to produce more complex multicolour FISH probes. Abnormal chromosomes which are unidentifiable by conventional cytogenetics can be flow sorted, labelled and hybridised onto normal chromosomes to identify the component chromosome regions, in a procedure termed Reverse Painting (Carter *et al.*, 1992; Nacheva *et al.*, 1995).

2.2.2 Microdissection of Chromosomal Regions

Chromosome arm painting probes (CAPs) are produced by microdissection (Guan *et al.*, 1996). Similarly, specific chromosome bands can be microdissected to produce band specific painting probes (BSPs) (Reid *et al.*, 2001). Abnormal chromosomes can be microdissected to enable further characterisation by reverse painting (Birnbacher *et al.*, 2001). Conventional microdissection methods, use glass needles to physically scrape the required chromosomal region from the sample preparation. Recent advances in technology have enabled chromosomal regions to be isolated using laser capture microdissection (LCM). LCM uses a focussed laser beam to ablate surrounding genomic material allowing the region of interest to be isolated and retained in a collection tube (http://www.palm-mikrolaser.com/index-nf.html, http://www.arctur.com).

2.2.3 Chromosome Mapping Using Regional Locus Specific Probes

FISH mapping has played a pivotal role in ordering the clones in the compiled tiling path used to generate the Human Genome sequence (International Human Genome Sequencing Consortium 2001). These sequenced clones offer a valuable resource for FISH mapping and their location and position can be readily identified using the Ensembl or other databases (Figure 1). A procedure for extracting BAC DNA is described in protocol 4.

Locus specific clones have found considerable use in diagnostic cytogenetics. FISH is often the method of choice for the detection of loss of chromosomal regions associated with specific microdeletion syndromes. For example, DiGeorge syndrome which manifests with congenital heart lesions, learning disorders and developmental delay is associated with a 2 Mb deletion on chromosome 22q. Specific DiGeorge probes are commercially available to identify the presence of the 22q deletion in affected patients (Larson and Butler, 1995; Novelli *et al.*, 1999; Berend *et al.*, 2000) (Figure 2).

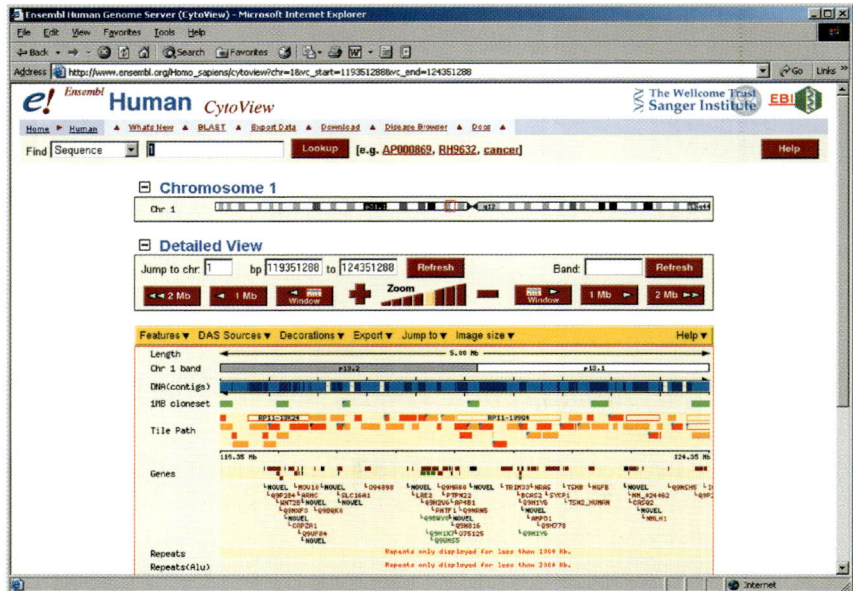

Figure 1. An example of Cytoview. The Ensembl web site is a useful resource to search the Golden Path clones used to assemble the human genome sequence. This view shows a region of the short arm of chromosome one, 1p13.2-1p13.1, the 1 MB clone set is presented in green, tiling path clones to cover the region are marked in red and orange and all genes in the region are shown below. An additional feature is the facility to dump the entire 1 MB clone set and all the genes in an area as a list, these buttons are not shown in this window. (From the Ensembl web site, a collaboration between the Sanger Institute and EBI.)

Chromosomal aneuploidy can be detected without the need to prepare metaphases on interphase nuclei using chromosome specific centromeric probes. This has the advantage of rapidly determining chromosome copy number in uncultured specimens, and is also useful for analysing samples which are mitotically inactive. Alpha satellite probes are often used for this purpose as its DNA is composed of high order repeats (171 bp) that are tandemly repeated 100-5000 times at chromosome centromeres. The total size of the DNA target is several megabases, and the subsequent FISH signals are large and easily visible on interphase nuclei preparations. Commercial centromeric probes are used to detect the presence of aneuplodies such as trisomy 21 associated with Downs syndrome (Spathas *et al.*, 1994; Weremowicz *et al.*, 2001).

Telomeric probes are used to detect chromosome translocations or telomeric microdeletions such as those associated with idiopathic mental retardation (Knight *et al.*, 1997; Gribble *et al.*, 2000). The ends of all human

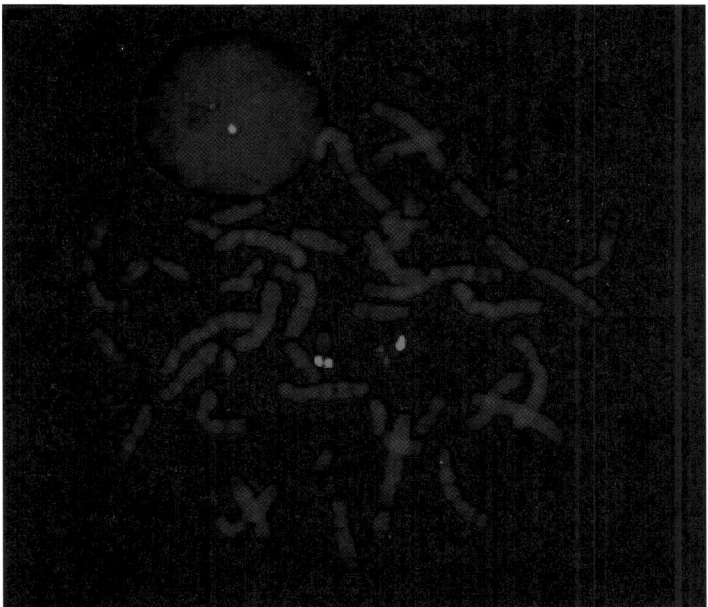

Figure 2. A metaphase chromosome spread from a DiGeorge patient peripheral blood sample. The metaphase is counterstained with DAPI after hybridisation with the Vysis LSI DiGeorge/VCFS Region Probe. The Vysis LSI DiGeorge/VCFS Region Probe is a two colour probe mixture that contains in SpectrumOrange TUPLE 1 (HIRA) probe (39 non-coding region of TUPLE 1, D22S553, D22S609 and D22S942) and in SpectrumGreen LSI ARSA (Arylsulfatase A) gene control probe that maps very close to the telomeric end of 22q at 22q13.3. Note the absence of a red signal on one chromosome 22 indicates deletion of the TUPLE 1 locus at 22q11.2. This region is deleted in DiGeorge patients. The image was captured using a Zeiss Axioscop microscope fitted with a photometrics cooled CCD camera and motorised filterwheel controlled by SmartCapture imaging software (Digital Scientific, Cambridge).

chromosomes are composed of a TG rich repeat, (TTAGGG)n, ranging from 2-15 kb in length. The chromosome unique telomeric sequences begin 100-300 kb from the telomere end and it is these regions that are used for clinical application.

2.3 Probe Labelling Methods

2.3.1 Degenerate Oligonucleotide Primed-Polymerase Chain Reaction, DOP-PCR

DOP-PCR is used to perform general amplification of small DNA samples such as flow sorted or microdissected chromosomes (Telenius *et al.*, 1992).

DOP-PCR (protocol 5) uses a primer, based on a partially degenerate sequence, which generally amplifies DNA from any source. The commonly used DOP primer is 6-MW, (5'-CCG ACT CGA GNN NNN NAT GTG G-3)'. DOP-PCR is carried out using two different annealing temperatures. In the preliminary cycles the annealing temperature is low enabling the 6-MW primer to bind at multiple sites and consequently amplify a good representation of the template DNA. In later cycles the annealing temperature is increased to specifically amplify the products of the earlier rounds. The DOP-PCR reaction is also used to incorporate fluorescently conjugated dUTPs, usually in a secondary PCR (protocol 6), which produces the chromosome paint.

2.3.2 Nick Translation

The most commonly used method to fluorescently label DNA probes is nick translation (protocols 7 and 8). Single strand breaks in the double stranded probe DNA extracted from BAC clones or genomic DNA, are introduced by DNAase I. It is this initial enzymatic digestion that determines the final probe fragment length. A second enzyme, DNA polymerase I then binds to the nicked DNA and sequentially replaces bases in a 5' to 3' direction. As the reaction mix contains hapten conjugated dUTPs, these are incorporated into the newly synthesised strand. For FISH, the products of a BAC clone nick translation reaction are required to be in the 50-500 bp range, however genomic DNA fragments to be used in a CGH hybridisation should be 500-2000 bp. The size of the probe is controlled by the concentration of the DNAase I or by the nick translation reaction time. Nick translation kits for the direct incorporation of fluorescently conjugated nucleotides, are available commercially. Direct labelling of DNA probes produces discrete hybridisation signals, accelerates the FISH process and lowers background fluorescence. The disadvantage of direct labelling is the cost of fluorescence conjugated nucleotides. Conversely indirect labelling produces larger hybridisation signals which may be slightly displaced due to amplification steps, requires lengthy detection steps, may have higher associated background but is less expensive.

2.3.3 Random Primer Labelling

Random primer labelling is suitable for small amounts of DNA as the labelling process can produce a 10-40 fold amplification of the probe. The reaction uses random primers (often octamers) which bind to denatured

DNA template. The primers are then extended by the Klenow large fragment of E. Coli DNA polymerase I in the presence of fluorescent nucleotides to produce a labelled probe. Random Primer labelling kits are available commercially (protocol 9).

2.3.4 Chemical Labelling of DNA

Besides traditional enzyme mediated labelling methods, chemical labelling systems have also been developed. For example, the ULYSIS labelling system (Molecular Probes/Kreatech Diagnostics) exploits the binding properties of platinum complexes with nucleic acids (http://www.probes.com/lit/bioprobes37/0.pdf). Labelling is typically performed in aqueous solution at neutral pH at 65ºC for 15 min. Under these conditions approximately one in every 10-20 bases is labelled. The platinum complexes are linked to marker compounds such as biotin, digoxigenin, Cy dyes or Alexa Fluor dyes. This DNA labelling system offers the advantages of being rapid and inexpensive, labelling the native DNA template and the same protocol can be used to label DNA of variable length.

2.4 Hybridisation and Detection of Annealed Probes

The basic process of hybridising DNA probe and DNA target are similar in all FISH experiments including CGH, MFISH or locus specific FISH. To denature the probe the DNA is usually heated at 65ºC in the presence of formamide (protocol 10). Cot-1 DNA is usually contained in the probe mix and during a preanneal step prior to hybridisation, binds to repetitive sequences in the probe DNA. Denaturation of the DNA target is also conducted in the presence of formamide at a high temperature. Target DNA can either be denatured using formamide denaturation buffer under a coverslip on a heated block, or by incubating target DNA immobilised on a glass slide in jars of formamide denaturation buffer in a water bath. The precise timing of denaturation will vary according to the target DNA as metaphase chromosome morphology is affected by denaturation time, whilst interphase nuclei or chromatin fibres can withstand longer denaturation times. After combining probe with target, hybridisation is typically performed overnight, however FISH with more complex DNAs for example CGH, may require longer time for complementary sequences to hybridise. Post-hybridisation stringency washes remove unbound probe which is controlled by changing the temperature or salt concentration of the wash buffer. Subsequent detection of hapten conjugated dUTPs is described in protocol 11. Finally, the DNA targets are stained with a counterstain, typically DAPI, to enable the sample to be visualised on the fluorescence microscope and to identify chromosomes from their banding pattern.

3. Real World Applications

3.1 Chromosome Rearrangements

Constitutional chromosome abnormalities are present in all body cell types, and may be inherited or de novo (Shaffer and Lupski, 2000). Acquired chromosome abnormalities develop in specific somatic tissues and may be involved in the progression of cellular malignancy. The karyotypes associated with malignant cells are frequently more complex than constitutional rearrangements. The genome can become structurally altered or imbalanced by a diverse range of chromosomal rearrangements:

- whole chromosome gain or loss
- interstitial deletions
- duplications
- chromosome amplifications
- intrachromosomal inversions
- reciprocal translocations

Interstitial deletions and duplications have been associated with a number of syndromes (reviewed in Shaffer and Lupski, 2000). For example the majority of hereditary neuropathy (HNPP) syndrome patients carry an interstitial deletion of chromosome 17(p12p12) and Charcot-Marie-Tooth disease type 1A has been associated with dup(17)(p12p12). Interstitial duplication and interstitial deletion may be the consequence of unbalanced reciprocal recombination events.

In malignant cells, chromosomal deletions frequently occur at sites of known tumour suppressor genes. Malignant cells may also present with chromosomal amplifications, and these regions may indicate the presence of oncogenes. These amplifications may occur in chromosomal structures called double minute chromosomes or homogenously staining regions (HSRs) (Wlodarska et al., 1994; Bruckert et al., 2000; Streubel et al., 2000).

Intrachromosomal inversions cause rearrangement of a specific chromosomal region in reverse orientation, resulting in the abnormal positioning of genomic sites. The most common inversion is the pericentric inversion which affects the heterochromatic region of chromosome 9 in more than 2% of the population.

Interchromosomal reciprocal translocation joins two regions of the genome not usually positioned next to each other (Rowley, 1999). Translocations can be either balanced or unbalanced and the reciprocal rearrangement products are frequently lost from the genome in highly rearranged cancer karyotypes. An apparently balanced reciprocal chromosome translocation, may actually be unbalanced due to the presence of cytogenetically undetectable DNA deletions or duplications (Sinclair *et al.*, 1997; Sinclair *et al.*, 2000). To identify these different chromosomal rearrangements, the appropriate molecular cytogenetic technique needs to be selected (Table 2).

3.2 Utilising Resources Generated from the Human Genome Sequencing Project: Strategy for Mapping and Sequencing Chromosome Breakpoints

The sequence of the human genome and the maps from which it was generated are easily accessible on the internet. Mapped clone resources which can be utilised by molecular cytogeneticists are also freely available from a number of research centres. Chromosome paints, chromosome arm paints and band specific paints are all useful for investigating large scale chromosome rearrangements and are available from the resource maintained by Professor Marriano Rocchi's laboratory in University of Bari, Italy (http://www.biologia.uniba.it/rmc). The same resource site offers alpha

Table 2. Molecular cytogenetic techniques to detect chromosome rearrangements.

Rearrangement	Technique
Balanced reciprocal translocation Cryptic balanced reciprocal translocation	G-Banding, Chromosome paints MFISH, Array painting
Unbalanced translocations	G-Banding, Chromosome paints, CGH
Pericentric inversions	G-Banding, Arm specific paints
Insertion	G-banding, Chromosome paints
Double minutes	Chromosome paints, CGH
Homogenously staining regions (HSR)	Chromosome paints, CGH
Telomeric deletion	Sub-telomeric probes
Telomeric translocation	Sub-telomeric probes
Duplications >5 Mb	G-banding, Chromosome paints and CGH
Duplications <5 Mb	G-banding, locus specific probes
Interstitial deletions >5 Mb	G-banding, Chromosome paints and CGH
Interstitial deletions <5 Mb	G-banding, Locus specific probes
Unidentified marker structures	CGH and MFISH

CGH=Comparative Genomic Hybridisation, MFISH=multicolour fluorescence *in situ* hybridisation satellite and telomeric clones for each human and murine chromosome. Other internet sites provide detailed listings of all the clones utilised in the process of compiling the human genome sequence. These sites include the National Center for Biotechnology Information (NCBI, http://www.ncbi. nlm.nih.gov/), and the University of California, Santa Cruz (http://genome. ucsc.edu/). The Ensembl database is an invaluable internet resource developed in collaboration by the Wellcome Trust Sanger Institute and the EBI. Ensembl incudes a genome view designed for molecular cytogeneticists entitled "Cytoview" (http://www.ensembl. org/Homo_sapiens/cytoview) which presents all Golden Path clones in an intuitive and user friendly format. These databases can be used to identify clones in chromosome regions of interest, the clones can then be obtained from a variety of resources at minimum cost, including the University of Bari (http://www.biologia.uniba.it/rmc/), BACPAC resources (http://www.chori. org/bacpac/) the Wellcome Trust Sanger Institute (http://www.sanger.ac. uk/cgi-bin/humace/CloneRequest), and the commercial company Research Genetics (http://www.resgen.com/ index.php3). These powerful resources are revolutionising the study of chromosome rearrangements and the mapping and characterisation of chromosome breakpoints.

3.2.1 Characterisation of Chromosomal Breakpoints

3.2.1.1 Mapping Chromosome Breakpoints by Conventional FISH Mapping

A routine diagnostic G-banding analysis of a chromosome translocation will identify the chromosome breakpoint band positions. However the precise determination of the breakpoints requires the use of FISH-based techniques. Human BAC clones, selected at 1 Mb intervals along the Golden Path, provide an ideal starting resource for FISH mapping experiments to refine the position of a chromosomal breakpoint (http://www.ensembl.org/Homo_sapiens/cytoview). Once localised between two clones in the 1 Mb set the breakpoint can be further refined easily using tiling path clones within the 1 Mb interval. This analysis will result in the identification of a breakpoint spanning clone, the initial step in the molecular characterisation of the breakpoint.

3.2.1.2 Mapping Breakpoints by Array Painting

A novel alternative to conventional FISH mapping is breakpoint identification using array painting, a technique combining FISH and array technologies (manuscript in preparation). Array painting uses human BAC clones arrayed onto a glass slide (Fiegler et al., 2003) which can be hybridised (painted) with fluorescently labelled derivative chromosomes. This approach offers the advantage of performing the equivalent of many FISH mapping experiments in a single hybridisation. The principle of array painting to determine chromosome breakpoints is explained in Figure 3. In brief, chromosome BAC clones selected at 1 Mb intervals, are DOP-PCR amplified and arrayed onto a glass slide. Derivative chromosomes are flow sorted to separate them from their normal homologues, and amplified in a DOP–PCR reaction. In the example presented in Figure 3, amplified derivative chromosome DOP-PCR products from the reciprocal translocation are labelled differentially with Cy3 and Cy5 in a random-primed labelling reaction. The derivative chromosomes are hybridised onto the array in the presence of COT-1 DNA. The DNA array is then analysed to determine the relative fluorescent ratios which are plotted against chromosome position. The transition from a low ratio value to a high ratio value indicates the location of the chromosomal breakpoint.

Array painting can be used at increased resolution using tiling path clones positioned within the region of interest. The fluorescence ratio values indicate BAC clones which contain the breakpoint spanning sequence.

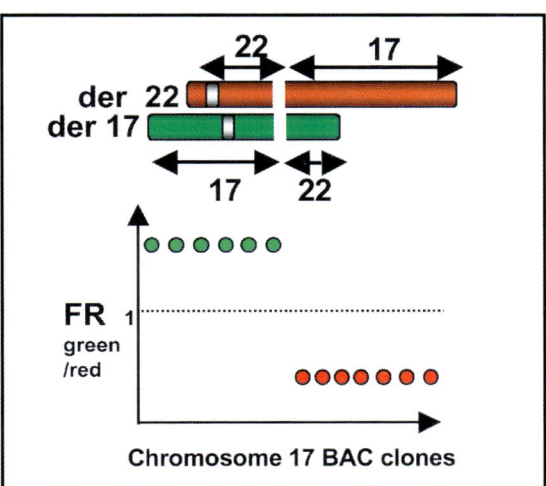

Figure 3. A schematic representation of Array Painting. The ideograms at the top represent two differentially labelled derivative chromosomes the products of a t(17;22). The labelled derivative chromosomes are cohybridised onto a genomic microarray and the fluorescence ratios for chromosome 17 clones are shown plotted against clone position. The transition from a high fluorescence to a low fluorescence marks a chromosome breakpoint. If a clone spans the breakpoint the fluorescence ratio will be close to one.

3.2.1.3 Refining Breakpoints using Fibre FISH

Hybridising two differentially labelled breakpoint spanning BAC clones onto patient derived extended chromatin fibres can refine the position of chromosome breakpoints. Using this procedure, hybridisation patterns showing joined signals from each BAC will be produced one from each derivative chromosome breakpoint. The fluorescent signal patterns obtained using fibre FISH, indicate the approximate position of the breakpoint within the spanning BAC clone DNA.

3.2.1.4 STS PCR Breakpoint Mapping on Flow Sorted Derivative Chromosomes and DNA Sequencing of the Junction Fragment

We have successfully combined flow sorting of derivative chromosomes and STS PCR mapping to further refine chromosome translocation breakpoints. Primers are designed to amplify STS specific PCR fragments of 80-100bp at approximately 1 kb intervals across the region of interest.

Each derivative chromosome is amplified using panels of STS primers located within each of the two chromosome breakpoint regions. The breakpoint is refined when two adjacent STS markers amplify only on different derivative chromosomes. Identification of the breakpoint flanking STS markers enables amplification across the junction fragment using the forward STS primer from one chromosome and the reverse STS primer from the other chromosome. The amplified product can be sequenced to locate the precise breakpoint position. This approach has been utilised successfully to identify the breakpoint sequences in a ptosis patient with a t(1;8) rearrangement (McMullan *et al.*, 2002).

4. Technology/Instrumentations/Automation

4.1 Image Aquisition

FISH experiments require sophisticated and expensive imaging equipment. Hardware requirements include an epifluorescence microscope (*e.g.* Axioplan 2, Zeiss) equipped with filters appropriate for the specific fluorophores used in the experiment. Advanced epifluorescence microscopes have the filters placed in a wheel that rotates into position under software control. Basic requirements should include filters for Texas Red, FITC and DAPI. A triple band pass filter set is particularly useful as this enables the simultaneous visualisation of Texas Red, FITC and DAPI. Narrow band pass filters are available which allow more defined wavelengths to be selected. This advanced filter set allows two fluorochromes that have close emission/excitation spectra to be used in the same experiment, (for example, Cy3 and Cy3.5 see useful web sites).

An advantage of the latest microscopes such as the Axioplan 2 is the presence of an 8 position filter wheel, enabling 8 different filters to be fitted and hence a higher number of different fluorophores to be visualised in a single experiment. This filter range is essential for MFISH experiments, which require the imaging of a minimum of 5 different fluorophores and a counterstain.

Less sophisticated microscopes utilise a single triple-band-pass mirror block with single excitation filters linked to a computer controlled filter wheel placed in front of the lamp house. With this arrangement grey level images for each of the selected fluorochromes can be aquired using sequential excitation through the same fixed filter block without the image shift associated with the manual changing of filter blocks.

Modern FISH microscope systems require charged-coupled device (CCD) cameras to capture the image. Typically a monochromic CCD is used under automatic software control. The image is compiled as the first filter block (or exciter) is moved into the excitation light path allowing the CCD to capture a monochrome image for the first fluorochrome. This procedure is repeated for each fluorochrome and the individual monochrome images are merged into a colour image on the computer screen. This process requires the use of commercially available dedicated FISH imaging software such as, SmartCapture from Digital Scientific (http://www.digitalscientific.co.uk), Genus/CytoVysion from Applied imaging (http://www.dssimage.com/genus.html, http://www.aicorp.com) and Isis from Metasystems (http://www.metasystems.de). Image software should include the following utilities:

- a fast and simple to use interface
- the ability to acquire multiple colour channels
- store raw images as well as processed images
- capability to support different cameras, filters and stages
- comprehensive image enhancement tools
- on screen video preview
- capability to batch capture (if large numbers of images need to be taken)
- time-lapse capture (if capturing movement)

- scripting for automating and customising
- the ability to export images to other programs

5. Protocols

Protocol 1. Preparation of Metaphase Chromosomes from Lymphoblastoid
Cell Lines

Lymphoblastoid cell lines should be handled in a class II biological safety
cabinet until step 13. Gloves should be worn.

Equipment and Reagents

- RPMI 1640 (Sigma R8758) supplemented with 16% FCS, 2 mM
  L-glutamine, 100U/ml penicillin and 100µg/ml streptomycin.
- 3mg/ml stock solution of BrdU
- 10mg/ml stock solution of ethidium bromide
- 10µg/ml stock solution of colcemid (GIBCO BRL)
- 75mM KCl hypotonic solution (5.6g/l)

Method

1. Grow the lymphoblastoid cell culture just to confluency and then
   sub-culture with 2 additional volumes of media.
2. Approximately 24 hours after subculture, add 100µl of BrdU (3mg/ml)
   to 20mls of culture and mix well. Final concentration : 15µg/ml.
3. Incubate 3 hours.
4. Add 20µl ethidium bromide (10mg/ml) to the culture. Final
   concentration: 10µg/ml. Colcemid is added at this stage to a final
   concentration of 0.02µg/ml (40µl in 20ml)
5. Incubate for 2 hours.
6. Transfer the contents of culture flasks to universals and centrifuge
   at 250 x g for 8-10 minutes.
7. Remove supernatant, trying to leave pellet as dry as possible.
   Loosen cell pellet thoroughly by flicking the base of the tube.
8. Add 10mls of 75mM KCl prewarmed to 37°C and carefully
   resuspend cells by gentle swirling. If clumps are visible, the pellet
   was not sufficiently resuspended before the KCl was added. Use a
   pasteur pipette to break up large clumps. Transfer the contents to a
   10ml screw-capped centrifuge tube.
9. Incubate at 37°C for 8-10 minutes.

10. Add 2-3ml freshly prepared ice-cold fixative, comprising 3 parts methanol to 1 part glacial acetic acid. Mix thoroughly by gentle swirling.
11. Centrifuge at 250 x g for 10 minutes.
12. Remove supernatant and loosen cell pellet thoroughly.
13. Add 10ml fixative and mix.
14. Centrifuge at 250 x g for 10 minutes.
15. Repeat steps 14-16 twice more.
16. Resuspend cells finally in a sufficient volume of fixative to give the desired cell density.
17. Store at -20°C until required.

Protocol 2. Preparation of Interphase Nuclei from Normal Fibroblasts

Equipment and Reagents

- DMEM (Sigma D5796) supplemented with 16% FCS, 2 mM L-glutamine, 100U/ml penicillin and 100μg/ml streptomycin.
- normal male fibroblasts (obtained from ECCAC or ATCC)
- trypsin-EDTA solution (Sigma, T3924), warmed to 37°C
- 75 mM KCL, prewarmed to 37°C

Method

1. Culture fibroblasts until confluent.
2. Leave undisturbed for a further 4–7 days to ensure that mitotic activity has reached a minimum.
3. Release cells using trypsin-EDTA and dilute with 5 volumes of fresh medium.
4. Transfer cells to 20 ml centrifuge tubes and centrifuge at 250g for 5 minutes.
5. Remove the supernatant and loosen the cell pellet by lightly flicking the base of the tube.
6. Add 10ml pre-warmed 75mM KCl. Resuspend cells.
7. Centrifuge at 250g for 5 minutes.
8. Repeat steps 5 and 6.
9. Incubate cells at 37°C for 10 minutes.
10. Add 2-3ml methanol/glacial acetic acid fixative and proceed as for protocol 1, step 10 onwards.

Protocol 3. Slide Preparation for DNA Fibres

Equipment and Reagents

- lymphoblastoid cell culture
- PBS
- Haemocytometer
- Lysis solution (5 parts 70mM NaOH, 2 parts absolute ethanol. To make 7ml, mix 4912.5µl sdH$_2$O, 87.5µl 4M NaOH and 2000µl 96% Ethanol. This needs to be made up fresh every time).
- Cadenza slide chamber (Shandon)
- Microscope slides stored overnight in 96% ethanol (not sonicated)

Method

1. Using a disposable pipette, take 2-3 ml of cell suspension from a healthy culture and centrifuge at 250 x g for 5 minutes.
2. Tip off supernatant and resuspend in 3mls of PBS and recentrifuge. (Repeat this step twice).
3. Re-suspend in 1ml PBS.
4. Count an aliquot of cells using the haemocytometer.
5. Dilute or concentrate cells (using PBS) to give a final concentration of approximately 2-3×10^6/ml.
6. Spread 10µl of cell suspension over a 1cm area on the upper part of a clean microscope slide (cleaned in ethanol).
7. Allow the slides to air dry at room temperature for approximately 30 minutes.
8. Fit a slide into a plastic Cadenza coverplate and clamp in a nearly vertical position using a bent metal rack.
9. Apply 150µl of lysis solution into the top of the cadenza/slide gap.
10. As the level drops below the frosted edge of the slide, add 150µl of 96% ethanol.
11. Allow to drain until meniscus stops falling (about 30 seconds).
12. Holding the edges, carefully lift the slide and cadenza unit out of the rack.
13. Pull the top of the slide back from the cadenza, allowing the meniscus to move slowly down the slide.
14. Air dry.
15. Fix in acetone for 10 minutes.
16. Mark the limits of the fibres with a pencil on the side of the slide. Leave for >24 hours before hybridising. Slides can be stored satisfactorily in a slide box at room temperature for several months.

Protocol 4. Bacterial Clone DNA Mini Prep

Equipment and Reagents

- Lysis buffer: 50mM glucose, 10mM EDTA, 25mM Tris pH 8.0.
- 4 M NaOH
- 10% SDS
- 3M sodium acetate, pH5.2
- 3M sodium acetate, pH7.0
- isopropanol
- RNase A, 10mg/ml
- Phenol/chloroform
- 70% ethanol

Method

1. Inoculate 10ml of TY media containing the appropriate antibiotic with 1µl of glycerol stock. Incubate o/n at 37°C in a shaker at 200 rpm. Antibiotics: 30µl of (10mg/ml) kanamycin (PACs and cosmids), 10µl of (12.5 mg/ml) chloramphenicol (BACs and fosmids) and 5µl of chloramphenicol (Caltech BACs).

2. Pellet 10ml overnight culture at 2000 x g for 10 mins. Take the tubes back to the hood and pour off the supernatant in a beaker containing virkon and invert tube on tissue to drain. (Pellet may be stored at 4°C if necessary). Leave virkon to stand for 30 mins and then dispose of as per local regulations.

3. Add 200µl lysis buffer and resuspend gently. Transfer into a 1.5ml microfuge tube. Stand at room temperature for 10 minutes.

4. Add 400µl fresh 0.2M NaOH/1% SDS, mix gently by inversion. Incubate on ice for 5 minutes (1ml = 50µl of 4M NaOH plus100µl 10% filtered SDS plus 850µl H_2O).

5. Add 300µl 3M sodium acetate pH5.2 and mix gently by inversion. Incubate on ice for 10 minutes. (This and all further steps can be carried out on the bench).

6. Microfuge at top speed for 5 minutes, transfer clear supernatant to a fresh microfuge tube. Repeat twice. (If the supernatant is still not clear, stand on ice for another 10-30 minutes, microfuge and transfer clear supernatant to a fresh tube).

7. Add 600µl isopropanol, mix gently, and incubate at -70°C for 10 minutes (or overnight at -20°C).

8. Microfuge at top speed for 5 minutes, remove supernatant by tipping it off, and invert the tubes on tissue to drain. (Do not air dry too long as the pellet will be difficult to resuspend.)

9. Resuspend pellet in 200µl 0.3M sodium acetate pH7 (1ml = 100µl 3M sodium acetate pH7 plus 900µl H_2O). Can also leave on ice for 20–30 minutes and then resuspend.

10. In the fume hood, add 200µl phenol/chloroform. Vortex and microfuge at top speed for 3 minutes. Transfer 150µl of the aqueous phase (top layer) to a fresh tube. Add 50µl 0.3M sodium acetate pH7 to the original tube containing the phenol/chloroform. Vortex and microfuge for 2 minutes at top speed. Transfer 50µl of the aqueous phase to pool with the other 150µl.

11. Add 200µl isopropanol to the aqueous pool and mix by inverting. Incubate at -70°C for 10 minutes (or overnight at – 20°C). Microfuge for 8 minutes at top speed. Remove supernatant by tipping off in to a petri dish and drain the tubes on a tissue.

12. Add 500µl ice-cold 70% ethanol, without disturbing the pellet. Microfuge at top speed for 5 minutes, remove supernatant by aspiration and dry pellet at 37°C for approx. 30 minutes.

13. Add 50µl TE containing 200µg/ml RNaseA (50µl TE plus 2µl RNaseA) and flick the tube to resuspend. Incubate at 55°C for 15 minutes in a waterbath.

14. Run 2µl per sample on 1% agarose gel.

Protocol 5. Primary DOP-PCR

Equipment and Reagents

- TAPS 2 buffer (250 mM TAPS (Sigma T5130) pH 9.3, 500 mM KCl, 20mM $MgCl_2$, 10mM dithiothreitol
- 2-mercaptoethanol (BDH)
- 5% Bovine serum albumin (Sigma A4628)
- 1% Polyoxyethelene ether W1 (Sigma P7516)
- Taq polymerase (Amplitaq, Perkin Elmer)
- dNTPs (Amersham Pharmacia Biotech 27-2035-01, equal mix of 10 mM dATP, dGTP, dCTP, dTTP)
- DOP primer (20µM)

Method

1. Prior to using TAPS2 buffer add BSA (33 µl/ml) and 2 ME (7 µl/ml)

2. UV all tubes, pipettes, tips, water, TAPS-2 buffer by placing in a UV hood for 20 mins before setting up reactions.
3. To a 0.5 ml microfuge tube containing 500 flow sorted chromosomes in 33 μl water add:

     5μl TAPS 2 buffer

     5μl DOP primer

     4μl dNTP mix

     2.5μl W1

     0.5μl Amplitaq

     mix and spin in microfuge

4. PCR overnight using following conditions

     Step 1   94°C for 10 minutes

     Begin cycle 9 times

     Step 2   94°C for 1 minute 30 seconds

     Step 3   30°C for 2 minutes 30 seconds

     Step 4   72°C for 3 minutes, slope 0.23°C/second

     End cycle 9 times

     Begin cycle 29 times

     Step 5   94°C for 1 minute

     Step 6   62°C for 1 minute 30 seconds

     Step 7   72°C for 3 minutes

     End cycle 29 times

     Step 8   72°C for 8 minutes

     Step 9   hold at 12°C

Protocol 6. Secondary DOP-PCR

Equipment and Reagents

- TAPS 2 buffer (250 mM TAPS (Sigma T5130) pH 9.3, 500 mM KCl, 20 mM $MgCl_2$, 10 mM dithiothreitol
- 2-mercaptoethanol (BDH)
- 5% Bovine serum albumin (Sigma A4628)
- 1% Polyoxyethelene ether W1 (Sigma P7516)
- Taq polymerase (Amplitaq, Perkin Elmer)
- dNTPs (Amersham Pharmacia Biotech 27-2035-01, equal mix of 10 mM each of dATP, dGTP, dCTP and 5 mM dTTP)
- DOP primer (20μM)

- 1mM Cy3-dUTP (Amersham Pharmacia Biotech PA 53022)

1. Prior to using TAPS2 buffer add BSA (33 µl/ml) and 2 ME (7 µl/ml)

2. UV all tubes, pipettes, tips, water, TAPS-2 buffer before setting up reactions

3. To a 0.5 ml microfuge tube add:
   2.5µl TAPS 2 buffer
   2.5µl DOP primer
   12µl sterile water
   2µl primary DOP reaction product
   2µl dNTP mix
   2.5µl Cy3-dUTP
   1.25µl W1
   0.25µl Amplitaq
   mix and spin

4. PCR overnight using following conditions
   Step 1   94°C for 10 minutes
   Begin cycle 34 times
   Step 2   94°C for 1 minute
   Step 3   62°C for 1 minutes
   Step 4   72°C for 1 minute, 30 seconds
   End cycle 34 times
   Step 5   72°C for 9 minutes
   Step 6   12°C for ever

Protocol 7. Titration of DNasel for Nick Translation

Equipment and Reagents

- Deoxyribonuclease I (D4527, Sigma)
- 10x nick translation buffer (0.5 M Tris-HCl pH 7.5, 0.1 M MgSO$_4$, 1 mM dithiothreitol, 500 µg/ml bovine serum albumin)
- enzyme diluent comprising 50% glycerol / 50% 2x nick translation buffer (v/v) (500µl glycerol, 100µl 10x NT plus 400µl H$_2$O, mix well)
- 14°C waterbath or thermocycler set at 14°C
- 1kb size marker (BRL)
- 0.5 M EDTA, pH 8.0

Method

1. Prepare a 1mg/ml stock solution of DNaseI in enzyme diluent. This can be stored at -20°C for several years.
2. Prepare a 1µg/ml working solution of DNaseI in enzyme diluent from the stock solution. Mix thoroughly by vortexing. This stock will remain stable for at least a year of regular use, if kept on ice when in use and stored at -20°C.
3. Prepare several DNaseI digestion reactions on ice, containing 2µg test DNA, 5µl 10x nick translation buffer, 1-2µl DNaseI working solution, made up to 50µl final volume with sterile distilled water. Compare different amounts of DNase l working solution and different DNA samples.
4. Incubate at 14°C.
5. After 30 minutes incubation, remove 10µl aliquots from each reaction and transfer to labelled 0.5ml microfuge tubes containing 1µl of 0.5M EDTA to inactivate the DNaseI. Store on ice.
6. Remove additional 10µl aliquots at 10 minute intervals, giving a total time range of 30 to 70 minutes.
7. Electrophorese all samples against a 1kb size marker on a 1% agarose gel.
8. Choose the DNaseI concentration and incubation time which give fragment smears with a size range of 200-700bp.

Protocol 8. Nick Translation

Equipment and Reagents

- 10x nick translation buffer (0.5M Tris.HCl pH7.5, 0.1M MgSO$_4$, 1mM dithiothreitol, 500µg/ml bovine serum albumin)
- 0.5mM dNTPs (2µl each of Pharmacia 100mM dATP, dCTP, and dGTP and 1194µl sterile distilled water. Store at -20°C)
- 1mM biotin-16-dUTP or digoxigenin-11-dUTP (Boehringer) or fluorochrome-dUTP
- DNaseI, titrated as protocol 7
- DNA polymerase I (10 U/µl, D9380, Sigma)
- DNA to be labelled (1 µg/µl)
- 0.5M EDTA, pH 8.0
- 3M sodium acetate pH 7.0

- absolute ethanol, stored at -20°C
- 70% ethanol, stored at -20°C

Method

1. For approximately 1μg of DNA set up a 25μl reaction. Determine the volume of DNA required to give 1μg and calculate the volume of sterile distilled water needed. Scale up as required.
2. Add in order to a 1.5ml microfuge tube on ice:
   xμl sterile distilled water (to make final volume to 25μl)
   2.5μl 10x nick translation buffer
   1.9μl dNTPs
   0.7μl dUTP (conjugated to a hapten)
   1μl DNaseI working solution (or other volume determined by titration and adjust the volume of sterile distilled water accordingly to keep the final volume at 25μl)
   0.5μl DNA polymerase (5 units)
   yμl DNA (1μg)
3. Mix well by lightly flicking the tube. Pulse microfuge briefly to bring down the solution.
4. Incubate at 14°C for the desired time (as determined by DNaseI titration).
5. Add 2.5μl of 0.5M EDTA (1/10th of the volume) to inactivate the enzymes. Keep on ice.
6. Add 2.5μl of 3M sodium acetate pH7 (1/10th of the volume) and 1ml of ice-cold absolute ethanol to precipitate the DNA. Mix thoroughly.
7. Incubate at -70°C for 30 minutes (or -20°C overnight).
8. Microfuge at top speed for 10 minutes. A white pellet should be clearly visible. Pour off supernatant and blot tubes.
9. Add 500 μl ice-cold 70% ethanol, without disturbing the pellet, and microfuge immediately for 2 minutes at top speed. Aspirate the supernatant immediately, leaving the pellet as dry as possible. The pellet will now be transparent and difficult to see. (The pellet tends to loosen and the labelled probe may be accidentally discarded.)
10. Air dry pellet (keep open tubes in the oven at 37°C), but do not over-dry (about 25 mins).

11. Add 10µl TE buffer and stand on ice for 10 minutes. Flick mix to resuspend probe. Store at -20°C until required.
12. Run 2µl per sample on a 1% agarose gel.

Protocol 9. Random Priming

Equipment and Reagents

- BioPrime Labelling Kit (Invitrogen 18094-011)
- 10X dNTP mix (0.5 mM dCTP, 2 mM dATP, 2 mM dGTP, 2 mM dTTP in TE buffer)
- 1mM Cy3-dCTP or 1mM Cy5-dCTP (NEN Life Science, NEL576, 577)
- Micro-spin G50 columns (Pharmacia Amersham, 275330-01)

1. Random label 0.6 µg DNA in a final reaction volume of 100 µl by mixing the DNA with 40 µl 2.5X Random Primers Solution and make up to 84 µl with water.
2. Denature DNA in a heat block for 10 min at 100°C, and immediately cool on ice.
3. The following reagents are added on ice:
   - 10 µl 10X dNTP mix
   - 4 µl Cy3 or Cy5 labelled dCTP (1 mM)
   - 2 µl Klenow Fragment
4. Mix gently but thoroughly.
5. Incubate the reaction at 37°C overnight and stop it by adding 10 µl of stop buffer.
6. Resuspend the resin in the G50 columns by gentle vortexing.
7. Loosen the cap one-fourth turn and snap off the bottom closure
8. Place the columns in a 1.5 ml screw-cap microcentrifuge tube for support. Alternatively, cut the cap from a flip-top tube and use this tube for support.
9. Pre-spin the columns for 1 minute at 735 x g. Use columns immediately after preparation to avoid the resin drying out.
10. Place the columns in a new 1.5 ml tube and slowly apply 50 µl to the centre of the angled surface of the compacted resin bed of each of the columns, being careful not to disturb the resin bed. Careful application of the sample to the centre of the bed is essential for good separation. Do not allow any of the samples to flow around the sides of the bed.

11. Spin the columns for 2 minutes at 735 x g. The purified samples are collected in the bottom of the support tube.
12. Discard the columns and retain the flow-through samples.
13. Combine the two samples and run 5 µl on a 1% agarose gel

Protocol 10. Hybridisation - Metaphase Spreads and Fibres

Equipment and Reagents

- hybridisation buffer (50% deionised formamide, 2xSSC, 10% dextran sulphate, 0.1% Tween 20, 10 mM Tris pH7.4; stored at -20°C)
- Cot1 DNA (GIBCO BRL)
- 70% formamide (70% formamide, 30% 2xSSC (v/v), stored at 4-8°C and re-used for a week)
- ice-cold 70% ethanol (in a hellendahl jar, stored at -20°C)
- 22×22mm coverslips, stored in 96% ethanol
- rubber cement (Cowgum)
- ethanol series : jars containing 70%, 70%, 90%, 90%,100% ethanol

Method

1. Pre-warm a coplin jar of 70% formamide to 65°C in a waterbath and set another waterbath at 37°C.
2. For a 22×22 coverslip, in a 0.5ml microfuge tube add:

   0.5µl labelled DNA (30-50ng)   or   2 × 0.5µl labelled DNA

   1µl Cot1 DNA (1µg)   2.0µl Cot 1 DNA

   11.5µl hybridisation buffer   10.0µl Hyb buffer
3. Vortex and pulse microfuge.
4. Denature probe mix at 65°C for 10 minutes.
5. Transfer to 37°C to pre-anneal for 15mins -3 hours (40 mins is an optimal time).
6. Denature slides in 70% formamide at 65°C for 1min 30secs for metaphases or 2mins for fibres.
7. Quench slides in 70% ice cold ethanol for 1minute.
8. Dehydrate through an ethanol series (70%, 70%, 90%, 90%, 100%) for 1minute in each. Air dry.
9. On a 37°C hot plate pipette probe mix onto slides and cover with a polished 22×22mm coverslip. Seal with rubber cement.

10. Incubate overnight at 37°–42°C. Ensure that there is a box of distilled water in the bottom of the incubator for humidity.

For repetitive probes: e.g. centromeric alpha satellite probes

Use a probe mix of 0.5ng labelled DNA and hybridisation buffer to final volume of 15μl.
Do not pre-anneal at 37°C, instead, quench on ice after denaturation at 65°C.

Protocol 11. Two Colour Detection of Biotin and Digoxigenin Labelled Probes

Equipment and Reagents

- 20 x SSC (3 M NaCl, 0.3 M Na-citrate)
- Wash solution (4×TNFM): 4×SSC, 0.05% Tween 20, 5% non-fat milk powder, filtered through several layers of Whatman No.4 filter paper (To make 500ml : 25 gms non-fat milk powder,100ml 20×SSC, 250μl Tween 20 and add sterile distilled H_2O to make the final volume to 500ml. Dissolve and filter)
- 3 coplin jars containing 2×SSC, warmed to 42°C
- 2 coplin jars of 50% formamide (50% formamide, 50% 2×SSC v/v) warmed to 42°C
- 1 coplin jar of 4×TNFM, warmed to 37°C
- immunochemical detection solutions, diluted in 4×TNFM, according to chosen detection strategy. Make 100μl per slide plus a minimum of 50μl excess.
- Layer 1. 4μg/ml avidin Texas Red DCS (Sigma A-2665) and 1:500-1000 dilution of mouse anti-digoxigenin-FITC (Sigma F3523)
- Layer 2. 4μg/ml biotinylated anti-avidin D (BA-0300) and 10μg/ml goat anti-mouse FITC conjugate (Sigma, F0257)
- Layer 3. 4μg/ml avidin Texas Red DCS (Vector)
  Incubate detection solutions for 10 minutes at room temperature, then microfuge for 10 minutes and use supernatant.
- 0.08μg/ml DAPI in 2×SSC in foil-covered coplin jar (8μl DAPI in 100ml 2×SSC)
- antifade mountant (Citifluor AF1).
- 4×SSC/0.05% Tween 20 (80ml of 20×SSC, 200μl Tween 20 and add sdH$_2$O to make the final volume to 400 ml. This can be kept at room temperature)

Method

1. Remove dried rubber cement from slides. Soak off coverslips in first jar of warmed 2xSSC (5-10 minutes).
2. Incubate slides in first 50% formamide at 42°C for 5 minutes.
3. Incubate slides in second 50% formamide at 42°C for 5 minutes.
4. Wash for 5 minutes each in the two jars of 2xSSC at 42°C.
5. Incubate slides in 4xTNFM at 37°C for 5 minutes or longer.
6. Drain each slide and apply 100μl of Layer 1 antibody mix.
7. Cover with a 25×50mm strip of Nescofilm. Incubate the slides in a humidified box at 37°C for 20-60 minutes. Do not allow the slides to dry out at any stage. (Meanwhile warm 3 jars of 4×TNFM to 42°C in the waterbath)
8. Wash slides in each of 3 jars of 4×TNFM at 42°C for 5 minutes each.
9. Drain each slide and apply 100μl of Layer 2 antibody mix. Cover with a strip of Nescofilm and incubate as before.
10. Wash slides in each of 3 jars of 4×TNFM at 42°C for 5 minutes.
11. Drain each slide and apply 100μl of Layer 3 antibody mix. Cover with a strip of Nescofilm and incubate as before.
12. Wash twice in 4×SSC/0.05% Tween 20 at room temperature.
13. Stain in 0.08μg/ml DAPI for 2-3 minutes.
14. Rinse in 2×SSC and dehydrate through an ethanol series. Air dry. Apply 20μl aliquots of antifade solution to clean 22×50mm coverslips (soaked in ethanol). Overlay with slides, blot and seal with nail varnish.

6. Troubleshooting

6.1 Chromosome Morphology

6.1.1 Poor Chromosome Morphology after Denaturation

If morphology has deteriorated during an experiment, *i.e.* the chromosomes had good morphology when viewed under phase before the *in situ* process but after denaturation the chromosomes were "fuzzy", the denaturation process was not optimal. Denaturation temperature should not be above 75°C. A temperature of 65°C and a denaturation time of 1 min 30 sec is adequate to denature metaphase spreads from lymphoblastoid cell lines. The denaturation time should be varied to find the optimum for good morphology while maintaining good FISH signals.

6.1.2 Poor Chromosome Morphology before Denaturation

Some samples have poor morphology even before they are denatured. To help retain chromosome morphology a formaldehyde pre-treatment is useful.

6.1.3 Poor Chromosome Banding

If chromosomes are too lightly stained in counterstain *e.g.* DAPI the chromosome banding will not be good and chromosome identification will be hampered. The concentration of staining solutions should be checked and possibly increased, alternatively the time of staining can be lengthened. If the counterstain is too weak the slide can easily be restained by soaking off a coverslip in 2 × SSC and restaining. Alternatively if counterstain is too bright then the concentration of staining solutions should be reduced or incubation time decreased.

6.1.4 Chromosomes are Inadequately Spread

Chromosomes can be encouraged to spread by dropping preparations onto glass slides in moist conditions. This can be achieved by putting slides in a chamber containing wet tissues, by breathing on a slide just before use or by dipping the slide in a beaker of water just before use. After adding sample to slide a drop of fix can be added almost immediately. By adding fix from a height *i.e.* an arms length above the bench spreading may also be achieved. If a sample has chromosomes embedded in cytoplasm they may never spread adequately.

6.2 Fluorescence Background Problems

6.2.1 Background is Uniform and in a Circle Behind the Metaphase

This is most likely due to the probe binding non-specifically to remains of cytoplasm (biotin probes are susceptible to this kind of background). Pre-treatments using Pesin, or Proteinase K and/or RNase are commonly used to clean chromosome preparations on slides prior to *in situ* experiments.

6.2.2 High Background Located on the Chromosomes

The signal is non-specific and not restricted to the site it should bind to. This could be due to the probe sequence being highly repetitive and non-specifically binding to multiple locations on the chromosomes.

Increasing the concentration of Cot-1 DNA will help to compete out repetitive sequences.

6.2.3 Background is not Localised to Chromosomes but Appears as Bright Signals Randomly Placed Across a Field of View

Nick translation probe fragment size may be too large. The DNA probe should be analysed on an agarose gel to assess the fragment size is correct *i.e.* 50-500 bp for locus specific probes.

6.3 No Detectable Fluorescent Signal

If there is no detectable fluorescent signal chromosomes may have been insufficiently denatured. Check chromosomes were denatured at the correct temperature for the correct amount of time. If the denaturation conditions are appropriate check that the detection system reagents are correct. Alternatively the DNA probe may not have labelled properly. The DNA polymerase I may be inactive preventing the incorporation of fluorescent nucleotides.

6.4 Low Signal Intensity

If the fluorescence signal is weak it is probable that the washes were too stringent. The experiment should be repeated with less stringent washes. Washing conditions can be conducted at a lower stringency by lowering the temperature or increasing salt conditions. However, this approach may well increase non-specific background.

7. Useful Web Sites

http://www.sanger.ac.uk/

The Wellcome Trust Sanger Institute homepage

http://www.zeiss.de/de/micro/home_e.nsf

Carl Zeiss, microscope manufacturer

http://www.chroma.com/

Filter manufacturers

http://www.digitalscientific.co.uk/

Digital Scientific, Imaging software company

http://www.dssimage.com/genus.html http://www.aicorp.com/

Genus/CytoVysion software from Applied Imaging, Imaging software company

http://www.metasystems.de

Isis imaging software from Metasystems Imaging software company

http://www.probes.com/lit/bioprobes37/0.pdf

ULYSIS chemical nucleic acid labelling kit

http://www.med.yale.edu/genetics/ward/tavi/FISHdyes.html

Pharmacia-Amersham and Molecular Probes, useful site with fluorophore excitation and emission wavelengths

http://www.vysis.com/0.asp?ProductID=153

Vysis Inc. Probe manufacturers

http://www.probes.com/servlets/spectra/

Molecular Probes, fluorophore spectra information

http://www.chori.org/bacpac

BACPAC resources

http://www.sanger.ac.uk/cgi-bin/humace/CloneRequest

Sanger Centre clone resources

http://www.resgen.com/index.php3

Research Genetics

http://www.ncbi.nlm.nih.gov/

NCBI, Views of chromosomes, maps and loci

http://genome.ucsc.edu/

Washington University, contains links to clone and accession maps of the human genome

http://www.biologia.uniba.it/rmc/

Resources for molecular cytogenetics

http://www.ensembl.org/Homo_sapiens/

EBI/Sanger centre, Ensembl database

http://www.ensembl.org/Homo_sapiens/cytoview

Ensembl, Cytoview

Acknowledgements

I particularly wish to thank Dr Nigel Carter for his help in writing this manuscript and Dr Patrick Tarpey for critical reading of this manuscript. I would also like to thank Kim Smith, Clinical cytogenetics Addenbrooke's Hospital, for the human DiGeorge chromosome sample and Vysis UK for the DiGeorge LSP probe.

References

Albertson, D.G., Ylstra, B., Segraves, R., Collins, C., Dairkee, S.H., Kowbel, D., Kuo, W.L., Gray, J.W., and Pinkel, D. 2000. Quantitative mapping of amplicon structure by array CGH identifies CYP24 as a candidate oncogene. Nat Genet. 25: 144–6.

Bauman, J.G., Wiegant, J., Borst, P., and van Duijn, P. 1980. A new method for fluorescence microscopical localization of specific DNA sequences by *in situ* hybridization of fluorochromelabelled RNA. Exp Cell Res. 128: 485–90.

Bentz, M., Plesch, A., Stilgenbauer, S., Dohner, H., and Lichter, P. 1998. Minimal sizes of deletions detected by comparative genomic hybridization. Genes Chromosomes Cancer. 21: 172–5.

Berend, S.A., Spikes, A.S., Kashork, C.D., Wu, J.M., Daw, S.C., Scambler, P.J., and Shaffer, L.G. 2000. Dual-probe fluorescence *in situ* hybridization assay for detecting deletions associated with VCFS/DiGeorge syndrome I and DiGeorge syndrome II loci. Am J Med Genet. 91: 313–7.

Birnbacher, R., Chudoba, I., Pirc-Danoewinata, H., Konig, M., Kohlhauser, C., Schnedl, W., and Haas, O.A. 2001. Microdissection and reverse painting reveals a microdeletion 6(q26qter) in a de novo r(6) chromosome. Ann Genet. 44: 13–8.

Bruckert, P., Kappler, R., Scherthan, H., Link, H., Hagmann, F., and Zankl, H. 2000. Double minutes and c-MYC amplification in acute myelogenous leukemia: Are they prognostic factors? Cancer Genet Cytogenet. 120: 73–9.

Buckle, V.J., and Kearney, L. 1993. Untwirling dirvish. Nat Genet. 5: 4–5.

Carter, N.P., Ferguson-Smith, M.A., Perryman, M.T., Telenius, H., Pelmear, A.H., Leversha, M.A., Glancy, M.T., Wood, S.L., Cook, K., Dyson, H.M., and et al. 1992. Reverse chromosome painting: a method for the rapid analysis of aberrant chromosomes in clinical cytogenetics. J Med Genet. 29: 299–307.

Collins, J.E., Cole, C.G., Smink, L.J., Garrett, C.L., Leversha, M.A., Soderlund, C.A., Maslen, G.L., Everett, L.A., Rice, K.M., Coffey, A.J., and *et al.* 1995. A high-density YAC contig map of human chromosome 22. Nature. 377: 367–79.

Cremer, T., and Cremer, C. 2001. Chromosome territories, nuclear architecture and gene regulation in mammalian cells. Nat Rev Genet. 2: 292–301.

du Manoir, S., Speicher, M.R., Joos, S., Schrock, E., Popp, S., Dohner, H., Kovacs, G., Robert-Nicoud, M., Lichter, P., and Cremer, T. 1993. Detection of complete and partial chromosome gains and losses by comparative genomic *in situ* hybridization. Hum Genet. 90: 590–610.

Fidlerova, H., Senger, G., Kost, M., Sanseau, P., and Sheer, D. 1994. Two simple procedures for releasing chromatin from routinely fixed cells for fluorescence *in situ* hybridization. Cytogenet Cell Genet. 65: 203–5.

Fiegler, H., Carr, P., Douglas, E.J., Burford, D.C., Hunt, S., Smith, J., Vetrie, D., Gorman, P., Tomlinson, I.P.M., and Carter, N. 2003. DNA microarrays for comparative genomic hybridisation based on DOP-PCR amplification of BAC and PAC clones. Genes Chromosomes Cancer. 36:361–374.

Forozan, F., Karhu, R., Kononen, J., Kallioniemi, A., and Kallioniemi, O.P. 1997. Genome screening by comparative genomic hybridization. Trends Genet. 13: 405–9.

Gribble, S., Andrews, K., Williams, D., Tillett, A., Bloxham, D., Proffit, J., Hackbarth, M., Grace, C., Green, A., and Nacheva, E. 2000. Fluorescence *in situ* hybridization detection of two telomeres on the short arm of a derived chromosome 16 in an infant with thrombocytopenia. Cancer Genet Cytogenet. 120: 99–104.

Gribble, S.M., Sinclair, P.B., Grace, C., Green, A.R., and Nacheva, E.P. 1999. Comparative analysis of G-banding, chromosome painting, locus-specific fluorescence *in situ* hybridization, and comparative genomic hybridization in chronic myeloid leukemia blast crisis. Cancer Genet Cytogenet. 111: 7–17.

Guan, X.Y., Zhang, H., Bittner, M., Jiang, Y., Meltzer, P., and Trent, J. 1996. Chromosome arm painting probes. Nat Genet. 12: 10–1.

Heiskanen, M., Karhu, R., Hellsten, E., Peltonen, L., Kallioniemi, O.P., and Palotie, A. 1994. High resolution mapping using fluorescence *in situ* hybridization to extended DNA fibers prepared from agarose-embedded cells. Biotechniques. 17: 928-9, 932–3.

Hemminki, A., Tomlinson, I., Markie, D., Jarvinen, H., Sistonen, P., Bjorkqvist, A.M., Knuutila, S., Salovaara, R., Bodmer, W., Shibata, D., de la Chapelle, A., and Aaltonen, L.A. 1997. Localization of a susceptibility locus for Peutz-Jeghers syndrome to 19p using comparative genomic hybridization and targeted linkage analysis. Nat Genet. 15: 87–90.

Heng, H.H., Squire, J., and Tsui, L.C. 1992. High-resolution mapping of mammalian genes by *in situ* hybridization to free chromatin. Proc Natl Acad Sci U S A. 89: 9509–13.

International Human Genome Sequencing Consortium 2001. Initial sequencing and analysis of the human genome. Nature. 409: 860–921.

Kallioniemi, A., Kallioniemi, O.P., Sudar, D., Rutovitz, D., Gray, J.W., Waldman, F., and Pinkel, D. 1992. Comparative genomic hybridization for molecular cytogenetic analysis of solid tumors. Science. 258: 818–21.

Kallioniemi, O.P., Kallioniemi, A., Sudar, D., Rutovitz, D., Gray, J.W., Waldman, F., and Pinkel, D. 1993. Comparative genomic hybridization: a rapid new method for detecting and mapping DNA amplification in tumors. Semin Cancer Biol. 4: 41–6.

Knight, S.J., Horsley, S.W., Regan, R., Lawrie, N.M., Maher, E.J., Cardy, D.L., Flint, J., and Kearney, L. 1997. Development and clinical application of an innovative fluorescence *in situ* hybridization technique which detects submicroscopic rearrangements involving telomeres. Eur J Hum Genet. 5: 1–8.

Landegent, J.E., Jasen in de Wal, N., Baan, R.A., Hoeijmakers, J.H., and Van der Ploeg, M. 1984. 2-Acetylaminofluorene-modified probes for the indirect hybridocytochemical detection of specific nucleic acid sequences. Exp Cell Res. 153: 61–72.

Langer, P.R., Waldrop, A.A., and Ward, D.C. 1981. Enzymatic synthesis of biotin-labeled polynucleotides: novel nucleic acid affinity probes. Proc Natl Acad Sci U S A. 78: 6633–7.

Larson, R.S., and Butler, M.G. 1995. Use of fluorescence *in situ* hybridization (FISH) in the diagnosis of DiGeorge sequence and related diseases. Diagn Mol Pathol. 4: 274–8.

Lawrence, J.B., Singer, R.H., and McNeil, J.A. 1990. Interphase and metaphase resolution of different distances within the human dystrophin gene. Science. 249: 928–32.

Leversha, M. 1997. Fluorescence *in situ* hybridisation. In: Genome mapping. P.H. Dear ed. Oxford University Press, Oxford p. 199–225.

McMullan, T.F.W., Crolla, J.A., Gregory, S.G., Carter, N.P., Cooper, R.A., Howell, G.R., and Robinson, D.O. 2002. A candidate gene for congenital bilateral isolated ptosis identified by molecular analysis of a de novo balanced translocation. Hum Genet. 110:244–250.

Michalet, X., Ekong, R., Fougerousse, F., Rousseaux, S., Schurra, C., Hornigold, N., van Slegtenhorst, M., Wolfe, J., Povey, S., Beckmann, J.S., and Bensimon, A. 1997. Dynamic molecular combing: stretching the whole human genome for high- resolution studies. Science. 277: 1518–23.

Nacheva, E., Holloway, T., Carter, N., Grace, C., White, N., and Green, A.R. 1995. Characterization of 20q deletions in patients with myeloproliferative disorders or myelodysplastic syndromes. Cancer Genet Cytogenet. 80: 87–94.

Novelli, A., Sabani, M., Caiola, A., Digilio, M.C., Giannotti, A., Mingarelli, R., Novelli, G., and Dallapiccola, B. 1999. Diagnosis of DiGeorge and Williams syndromes using FISH analysis of peripheral blood smears. Mol Cell Probes. 13: 303–7.

Parente, F., Gaudray, P., Carle, G.F., and Turc-Carel, C. 1997. Experimental assessment of the detection limit of genomic amplification by comparative genomic hybridization CGH. Cytogenet Cell Genet. 78: 65–8.

Pinkel, D., Segraves, R., Sudar, D., Clark, S., Poole, I., Kowbel, D., Collins, C., Kuo, W.L., Chen, C., Zhai, Y., Dairkee, S.H., Ljung, B.M., Gray, J.W., and Albertson, D.G. 1998. High resolution analysis of DNA copy number variation using comparative genomic hybridization to microarrays. Nat Genet. 20: 207–11.

Reid, A., Gribble, S.M., Andrews, K.M., Green, A.R., and Nacheva, E.P. 2001. Chromosome band specific FISH probes allow improved detection of terminal translocations in leukaemic metaphases. Leukemia. 15: 860–1.

Ross, M.T., and Langford, C.F. 1997. The use of flow-sorted chromosomes in genome mapping. In: Genome mapping. P.H. Dear ed. Oxford University Press, Oxford p. 165–184.

Rowley, J.D. 1999. The role of chromosome translocations in leukemogenesis. Semin Hematol. 36: 59–72.

Schrock, E., du Manoir, S., Veldman, T., Schoell, B., Wienberg, J., Ferguson-Smith, M.A., Ning, Y., Ledbetter, D.H., Bar-Am, I., Soenksen, D., Garini, Y., and Ried, T. 1996. Multicolor spectral karyotyping of human chromosomes. Science. 273: 494–7.

Shaffer, L.G., and Lupski, J.R. 2000. Molecular mechanisms for constitutional chromosomal rearrangements in humans. Annu Rev Genet. 34: 297–329.

Sinclair, P.B., Green, A.R., Grace, C., and Nacheva, E.P. 1997. Improved sensitivity of BCR-ABL detection: a triple-probe three-color fluorescence *in situ* hybridization system. Blood. 90: 1395–402.

Sinclair, P.B., Nacheva, E.P., Leversha, M., Telford, N., Chang, J., Reid, A., Bench, A., Champion, K., Huntly, B., and Green, A.R. 2000. Large deletions at the t(9;22) breakpoint are common and may identify a poor-prognosis subgroup of patients with chronic myeloid leukemia. Blood. 95: 738–43.

Solinas-Toldo, S., Lampel, S., Stilgenbauer, S., Nickolenko, J., Benner, A., Dohner, H., Cremer, T., and Lichter, P. 1997. Matrix-based comparative genomic hybridization: biochips to screen for genomic imbalances. Genes Chromosomes Cancer. 20: 399–407.

Spathas, D.H., Divane, A., Maniatis, G.M., Ferguson-Smith, M.E., and Ferguson-Smith, M.A. 1994. Prenatal detection of trisomy 21 in uncultured amniocytes by fluorescence *in situ* hybridization: a prospective study. Prenat Diagn. 14: 1049–54.

Speicher, M.R., Gwyn Ballard, S., and Ward, D.C. 1996. Karyotyping human chromosomes by combinatorial multi-fluor FISH. Nat Genet. 12: 368–75.

Streubel, B., Valent, P., Jager, U., Edelhauser, M., Wandt, H., Wagner, T., Buchner, T., Lechner, K., and Fonatsch, C. 2000. Amplification of the MLL gene on double minutes, a homogeneously staining region, and ring chromosomes in five patients with acute myeloid leukemia or myelodysplastic syndrome. Genes Chromosomes Cancer. 27: 380–6.

Telenius, H., Pelmear, A.H., Tunnacliffe, A., Carter, N.P., Behmel, A., Ferguson-Smith, M.A., Nordenskjold, M., Pfragner, R., and Ponder, B.A. 1992. Cytogenetic analysis by chromosome painting using DOP-PCR amplified flow-sorted chromosomes. Genes Chromosomes Cancer. 4: 257–63.

Trask, B., Pinkel, D., and van den Engh, G. 1989. The proximity of DNA sequences in interphase cell nuclei is correlated to genomic distance and permits ordering of cosmids spanning 250 kilobase pairs. Genomics. 5: 710–7.

Trask, B.J., Massa, H., Kenwrick, S., and Gitschier, J. 1991. Mapping of human chromosome Xq28 by two-color fluorescence *in situ* hybridization of DNA sequences to interphase cell nuclei. Am J Hum Genet. 48: 1–15.

Van Prooijen-Knegt, A.C., Van Hoek, J.F., Bauman, J.G., Van Duijn, P., Wool, I.G., and Van der Ploeg, M. 1982. *in situ* hybridization of DNA sequences in human metaphase chromosomes visualized by an indirect fluorescent immunocytochemical procedure. Exp Cell Res. 141: 397–407.

Weremowicz, S., Sandstrom, D.J., Morton, C.C., Niedzwiecki, C.A., Sandstrom, M.M., and Bieber, F.R. 2001. Fluorescence *in situ* hybridization (FISH) for rapid detection of aneuploidy: experience in 911 prenatal cases. Prenat Diagn. 21: 262–9.

Windle, B., Silvas, E., and Parra, I. 1995. High resolution microscopic mapping of DNA using multi-color fluorescent hybridization. Electrophoresis. 16: 273–8.

Wlodarska, I., De Wolf-Peeters, C., Dierick, H., Hilliker, C., Thomas, J., Mecucci, C., Cassiman, J.J., and Van den Berghe, H. 1994. Detection of amplified sequences at 5q11-->q13 in a homogenously staining region found by fluorescent *in situ* hybridization in a case of B-cell non-Hodgkin's lymphoma. Cytogenet Cell Genet. 65: 179–83.

From: *Genome Mapping and Sequencing*
© 2003 Horizon Scientific Press, Wymondham, UK

6

Assembling Physical Maps and Sequence Clone Selection

Andrew J. Mungall and Sean J. Humphray

Abstract

The majority of large genomes sequenced to date have made heavy use of a hierarchical mapping approach and this is the strategy in use by the International Human Genome Sequencing Consortium. Ideally, to minimise redundancy and cost, physical clone maps of the genome are developed ahead of large-scale genomic sequencing. However, the necessity to produce the sequence and the technology now available to achieve high-throughput sequencing dictates the pace and mapping strategy employed. In this chapter we discuss the resources and methods needed to generate physical clone maps of human chromosomes 6 and 9 at the Sanger Institute and the criteria for selecting clones to be sequenced. Large-insert bacterial clone maps were constructed by a combined restriction enzyme fingerprinting and landmark content analysis and were the substrate for genomic sequencing of the chromosomes. In addition the clones in the map are a lasting resource for future genomic analyses which include chromosome structure, comparative genomic hybridisation, gene inactivation and other functional genetics.

1. Introduction

One of the major goals of the Human Genome Project (HGP) in 1990 was the construction of a high-resolution physical map which could ultimately serve as the resource for generating the sequence itself (Watson and Jordan, 1989). Many bacterial genomes have been sequenced without a prior map by directly sequencing random shotgun reads of the whole genome, an approach not readily applicable to larger and more complex genomes. Anyone reading this chapter will no doubt be familiar with the analogy that mapping and sequencing a genome is like doing a huge jigsaw puzzle. The analogy is a good one. However, for the human genome this jigsaw puzzle has more than 50% almost identical pieces (repeats) which one could imagine as the pieces of blue sky in a landscape painting. As if the problem of repeats weren't great enough the assembly of the puzzle is confounded by the fact that thousands of individuals from around the world are involved in its reconstruction and therefore impressive co-ordination is required to complete the task. The clone-by-clone approach to mapping the human genome alleviates many of these problems since it greatly reduces the complexity of an assembly to a region of 100-200Kb, the average size of a P1-derived artificial chromosome (PAC) (Ioannou et al., 1994) or bacterial artificial chromosome (BAC) (Shizuya et al., 1992), as opposed to 3000Mb for the whole genome. It also allows for the effective distribution of effort in many laboratories, and funding agencies, across the globe.

The human genome is conveniently packaged into 24 units, 22 autosomes and the two sex chromosomes (XX or XY). In many cases single genome centres have committed to sequencing entire chromosomes. The Sanger Institute has mapped and sequenced all or significant regions of chromosomes 1, 6, 9, 10, 13, 20, 22 and X representing approximately one third of the DNA content of the human genome (Dunham et al., 1999; Bentley et al., 2001; Deloukas et al., 2001). In this chapter, chromosomes 6 and 9 will be used to illustrate the different strategies and methods used to map human chromosomes with the aim of sequencing. We will also discuss the assembly of physical maps and clone selection strategies for other vertebrate and invertebrate models where these differ from the human case. Of course the February 2001 publication of a working draft of the human genome received much publicity not least because it reported on two human genomes, one sequenced by the publicly funded International Human Genome Sequencing Consortium (International Human Genome Sequencing Consortium., 2001) making the sequence freely available to all and the second a private effort by Celera Genomics (Venter et al., 2001). The private assembly included the public data in the form of mapped BAC sequences deposited in GenBank, and hence also depends on the prior construction of physical maps. The ongoing sequencing of the mouse and zebrafish genomes has employed a hybrid strategy of whole genome shotgun sequence data anchored to a mapped scaffold of BACs, followed by finishing using shotgun sequencing of individual BACs. Thus physical maps are still a valuable resource in genome sequencing.

The availability of large numbers of markers, together with extensive bacterial clone resources provides a means to accelerate the process of mapping a chromosome and preparing bacterial clone contigs ready to sequence. The sections in this chapter follow the processes involved in physical clone mapping, from marker import and generation through to sequence clone selection (see Figure 1).

2. Genomic Resources

2.1 Sequence Tagged Site (STS) Import

Sequence tagged sites (STSs) are primer pairs derived from DNA sequence (Olson et al., 1989) and are therefore widely used in the physical mapping of all genomes. STSs are generally localized to a single position in the genome and assayed using PCR (Olson et al., 1989). The move away from using YACs as the basis for long range physical mapping due to their problems with chimaerism and deletions coincided with an increase in the use of radiation hybrid (RH) mapping (section 4 and chapter 2) to generate long range maps which could subsequently be used to order and orientate bacterial clones for sequencing (Mungall et al., 1996; Mungall et al., 1997). Since bacterial clones (typically BACs and PACs) are an order of magnitude smaller than YACs, a higher density of markers is required to achieve the same degree of contiguity. We estimated that a marker density of 15 per megabase (Mb) should be sufficient to achieve good coverage of clones on both chromosomes 6 and 9. Therefore chromosomes 6 and 9 with estimated sizes of 183Mb and 145 Mb respectively (Morton, 1991) would require 2745 and 2175 markers respectively to achieve the target of 15/Mb. In order to achieve maps which could integrate existing genetic and physical maps, markers were imported via anonymous ftp (URLs are given in section 10 of this chapter) from the Genome Database (GDB), Génèthon, Whitehead Institute (WI), Stanford Human Genome Center (SHGC), Cooperative Human Linkage Center (CHLC) and the RH database (RHdb). In the mid 1990s public databases did not contain the required densities of markers and available markers were generally biased towards gene sequences ie. Expressed Sequence Tags (ESTs) (Adams et al., 1991)

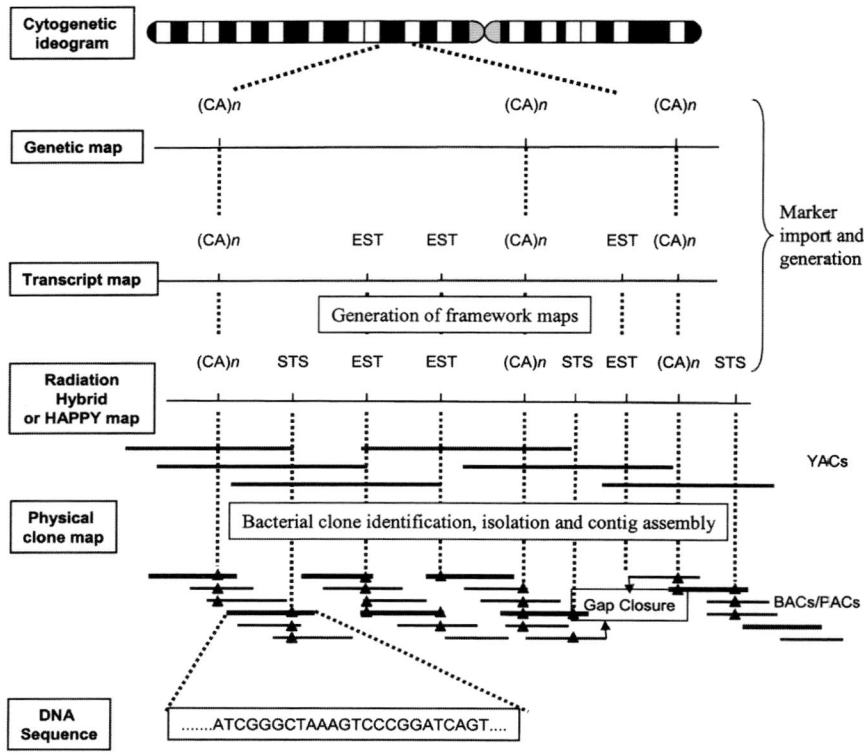

Figure 1. Hierarchical mapping approach to sequence an entire chromosome. The availability of large numbers of markers, together with extensive bacterial clone resources provides a means to accelerate the process of mapping a chromosome and preparing bacterial clone contigs ready to sequence. Microsatellites and other polymorphic markers used to generate genetic linkage maps and ESTs from regional and genome wide transcript maps were imported from public databases and anonymous ftp sites. To achieve high marker densities, required for bacterial clone screening and assembly, novel STS markers were generated in-house following single-pass sequencing of clones from libraries of digested flow-sorted chromosome DNA ligated into Bluescript plasmid. Successfully developed markers are ordered on a framework map such as a radiation hybrid, HAPPY or YAC map which serve to integrate all previously published maps. Markers in the framework map are then used to identify large-insert bacterial clones which are analysed by restriction enzyme fingerprinting and STS-content analysis to identify sequence-ready contigs. Contig gap closure is achieved by clone walking.

or polymorphic simple sequence repeat markers used in the construction of genetic linkage maps. In many cases markers derived from the same parent sequence had multiple different names and therefore DNA sequence homology searches using BLAST (Altschul *et al.*, 1990) were performed to ensure a non-redundant set of markers was used for mapping. Where no parent sequence was available for analysis, the primer pair was searched against a library of all other primer pairs using the program electronic PCR (ePCR) (Schuler, 1997). Non-redundant sets of markers for chromosomes 6 and 9 contained 3415 and 2565 markers respectively.

2.2 STS Generation

To achieve the 15 marker/Mb target we generated STSs in-house from single pass Sanger dideoxy sequencing of plasmid libraries containing flow sorted chromosome DNA. Chromosome 6 is easily flow-sorted and two libraries were generated. A HindIII digest and TaqI digest were used to counteract potential cloning biases of each enzyme. HindIII cuts on average every 4096 bp and cuts DNA in AT-rich regions whereas TaqI cuts on average every 256 bp in GC-rich regions. Due to the increased DNA insert lengths of the HindIII library, this gave sequences of better quality and suitability for primer design. In total 826 successful STSs were generated from the HindIII library and 514 STSs from the TaqI library. In contrast, chromosome 9 cannot readily be separated from chromosomes 10, 11 and 12 using flow-sorting and therefore HindIII digested DNA for all four chromosomes in the cluster was used for library generation. Since the Sanger Institute also planned to sequence chromosome 10, markers from such a flow-sorted library were used for this mapping project also. Markers mapping to chromosomes 11 and 12 were stored for distribution to other sequencing centres. Sequences were analysed for repeat content using RepeatMasker (see URL section) and searched against the non-redundant set of publicly available markers. Unique sequences free of repeats were subjected to primer design (PRIMER 3.0 Steve Rozen, Helen J. Skaletsky (1997), see useful URLs section) and the resulting STSs were primer tested (protocol 1) to confirm chromosome assignment and determine the optimal annealing temperature for the PCR assay.

2.3 Large-Insert Bacterial Clone Libraries

Having established a suitable set of non-redundant markers, these markers are used to identify large-insert bacterial clones spanning large regions of the chromosome which will serve as the substrates for shotgun sequencing (see chapter 10). At the onset of the chromosome 6 project PACs (Ioannou et al., 1994) were the substrate of choice for sequencing since they provided more insert stability than the larger inserts in YACs and yet could contain approximately 3-fold larger inserts than cosmids. More recently bacterial artificial chromosomes (BACs) (Shizuya et al., 1992) have been the substrate of choice. Like PACs, BACs are also propagated in an *E.coli* host and are therefore easy to culture and to extract DNA from. BACs can retain inserts of up to 300 Kb, though the average insert size in a typical library will usually be 150-200Kb. BAC vectors are smaller than their PAC counterparts and therefore offer a better insert to vector ratio, which is of critical importance when considering the shotgun sequencing of a large genome. The main libraries used in the mapping and sequencing of the human genome are detailed in Table 1.

3. Primer Testing

The aim of primer testing is to determine the optimal annealing temperature of STSs under standard PCR reaction conditions. PCR assays need to be robust to enable the large number of STSs in a large-scale mapping project to be tested efficiently. Each primer pair is tested at three temperatures for the amplification of total human genomic DNA, a mono-chromosomal hybrid DNA (*e.g.* For chromosome 6 GM11580, Coriell Cell Repository) for confirmation of the chromosome assignment, and hamster DNA, the hybrid background. Since STSs were to be screened against the Genebridge (GB4) panel of human-hamster radiation hybrids (Gyapay

Table 1. Commonly used large-insert bacterial clone libraries in the mapping and sequencing of the human genome. Data taken from (International Human Genome Sequencing Consortium., 2001) and http://www.sanger.ac.uk/HGP/methods/mapping/ info/lib-details.shtml.

Library name	Clone prefix	Vector type	Vector	DNA source	Enzyme digest	Average insert size (Kb)
Caltech B	CTB	BAC	pBAC108L	987SK cells	HindIII	120
Caltech C	CTC	BAC	pBeloBAC11	Human sperm	HindIII	125
Caltech D1 (CITB-H1)	CTD	BAC	pBeloBAC11	Human sperm	HindIII	129
Caltech D2 (CITB-E1	CTD	BAC	pBeloBAC11	Human sperm	EcoRI	
RPCI-1	RP1	PAC	pCYPAC2	Male, blood	MboI	110
RPCI-3	RP3	PAC	pCYPAC2	Male, blood	MboI	115
RPCI-4	RP4	PAC	pCYPAC2	Male, blood	MboI	116
RPCI-5	RP5	PAC	pCYPAC2	Male, blood	MboI	115
RPCI-6	RP6	PAC	pPAC4	Female, blood	MboI	135
RPCI-11	RP11	BAC	pBACe3.6	Male, blood	EcoRI	178
RPCI-13	RP13	BAC	pBACe3.6	Female, blood	EcoRI	166
CEPH	CEPH	BAC	pBeloBAC II	Lymphoblastoid cell-line	BamH1	130
Keio	KB	BAC	pBAC-Lac	Male, FLEB14-14	HindIII	110

et al., 1996) supplied by Research Genetics, it was important to obtain a specific human PCR product without co-amplification of a hamster derived sequence of the same length. Although STSs amplifying both human and hamster DNA of the same size are unsuitable for use in RH mapping (see below and chapter 2), these conserved STSs are of interest in subsequent bacterial clone library screening. In addition to determining the optimal annealing temperature of the primer pairs the amplification products are excised from the agarose gel and stored for subsequent PCR labelling to screen arrayed genomic libraries by hybridisation (Section 5.2).

4. Generating Framework Maps

Framework maps serve as long range scaffolds onto which bacterial clone contigs can be ordered and orientated. The construction of framework maps is largely PCR based and therefore amenable to automation and high throughput. PCR amplifiable markers on existing maps, such as the genetic or transcript maps of the 1990s can be integrated into a single framework map (Figure 1) and is therefore of added benefit to the scientific community. The rapid construction of a marker dense framework map is of critical importance to the subsequent assembly of physical clone contigs, the substrate for genomic sequencing.

4.1 Radiation Hybrid (RH) Mapping

RH mapping (see chapter 2) provides a means to localise any STS to a defined map position in the genome, using PCR. The method is amenable to automation and therefore high throughput and has been used to rapidly generate long-range maps of complex genomes (Hudson *et al.*, 1995; Schuler *et al.*, 1996; Deloukas *et al.*, 1998). Successfully developed markers from primer testing (section 3) are typed in duplicate against a panel of radiation hybrids. In the chromosome 6 and 9 projects at the Sanger Institute, 2802 and 1507 markers respectively were typed against the Genebridge 4 (GB4) panel of radiation hybrids. A target marker density of 1 per 70Kb was established to ensure that subsequent large-insert bacterial clone contigs gave good coverage with minimum requirement for chromosome walking to close gaps.

4.2 HAPPY Mapping

HAPPY mapping is an alternative method to generate framework maps and as with RH mapping allows the placement of markers across large regions or whole genomes (Konfortov *et al.*, 2000, Piper *et al.*, 1998). Here DNA is randomly sheared and the resulting fragments are diluted to yield aliquots, of approximately 1 haploid genome equivalent (Dear and Cook, 1993). These aliquots are generally produced in a 96-well format to aid throughput. Linkage between markers can be determined by the presence of shared PCR positives in the same aliquot. Unlinked markers are less likely to share positive pools.

The protocol can be adapted to resolve markers in the region of a few kilobases or megabases depending on the size of the fragments used to generate the mapping panel. In co-operation with Paul Dear's group at the MRC in Cambridge we have adapted parts of the technique (Dear, 1997) to aid in the elucidation of the malaria whole genome shotgun assembly.

We have successfully used this approach to help elucidate the whole genome shotgun assembly of chromosomes 6, 7 and 8 in *Plasmodium falciparum* (Hall, *et al.*, 2002). These chromosomes represent 5Mb of the genome and are very similar in size making isolation by gel electrophoresis difficult and thereby precluding a whole chromosome shotgun strategy. The HAPPY map enabled shotgun contigs to be oriented and binned, as well as confirming previously made joins.

5. Bacterial Clone Identification and Isolation

5.1 PCR Screening of Libraries

In the early stages of the chromosome 6 mapping project STSs from the RH map were used to screen 66 primary DNA pools from the RPCI-1, 3, 4 and 5 whole genome PAC libraries (http://www.chori.org/bacpac/) (Figure 2). Each pool contained the DNA from 3072 individual PAC clones, corresponding to a single high-density gridded library filter. The amount of purified DNA used should be empirically determined and is best performed by testing serial dilutions of pool DNA with a PCR assay known to amplify a sequence within one of the clones contained in that pool. Generally DNA isolated from a filter containing 3072 PAC colonies of approximately 1mm diameter by standard alkaline lysis protocol typically needs to be recovered in 1ml, with 5μl used in a 15μl PCR assay. This PCR based library screening method was effective in determining which library filters to include in subsequent hybridisation screening experiments and also gave independent experimental confirmation of landmark content analysis (section 7). The necessity for increased throughput ultimately gave rise to increased numbers of STS probes being pooled and hybridised to high-density library filters and therefore most if not all filters contained positive clones rendering the PCR selection of filters redundant.

5.2 Hybridisation Screening of Libraries

PCR products obtained from primer testing (Section 3) can be used as templates in radioactive PCR labelling (protocol 2). Overgo oligonucleotide

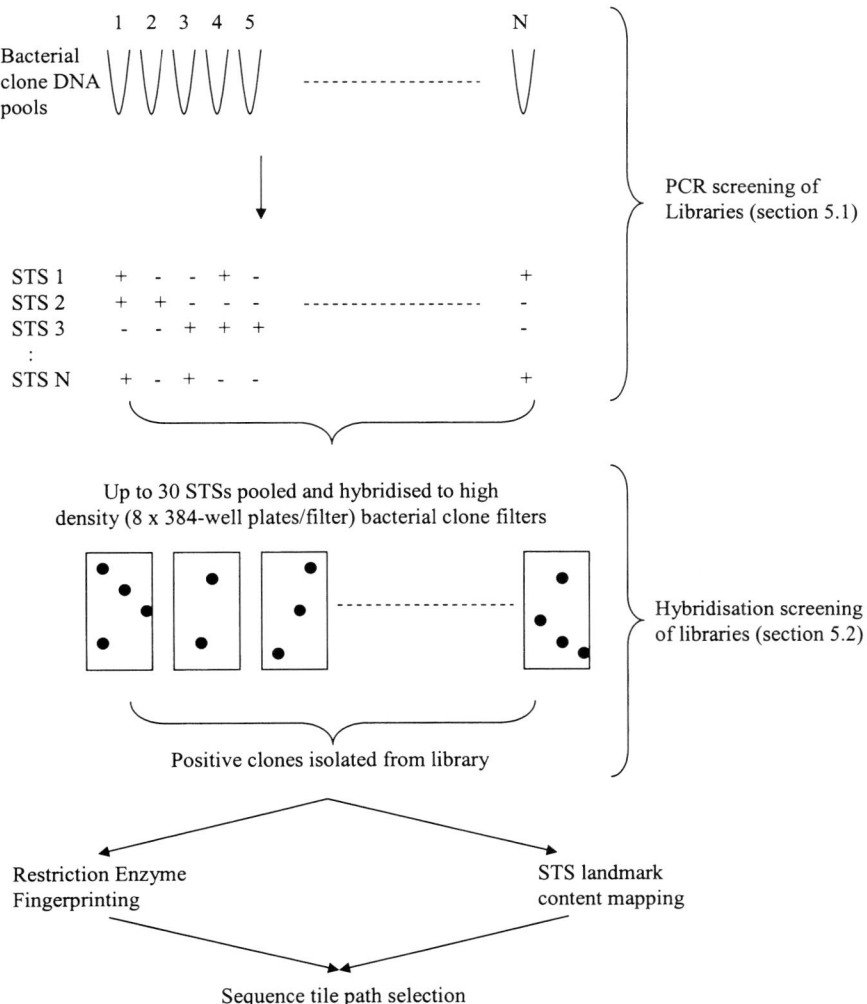

Figure 2. Bacterial clone library screening strategy.

probes (overgos) offer an alternative method for generating hybridisation probes with high specific activity and specificity (Ross *et al.*, 1999). Overgos are created by filling in the single-stranded overhangs of two overlapping oligonucleotides (typically 24-mers with an 8bp overlap) using radiolabelled nucleotides and Klenow DNA polymerase. The resulting double-stranded 40-mer overgo probe can be pooled with other probes and hybridised to immobilised DNA on nylon membranes (protocol 3). An advantage in using overgo probes over radiolabelled STSs is that they can be designed to short sequences such as those conserved between the genomes of different species (Thomas *et al.*, 2002). Pools of up to 30 radiolabelled probes are hybridised to an entire set of genomic library filters (protocol 3) (Figure 2). Following hybridisation and washing, up to 13 high density library filters, of dimensions 8cm x 12cm, can be accommodated in a 35.6 by 43.2 cm autoradiograph cassette.

6. Fingerprinting

The restriction enzyme digest fingerprinting method used as a basis for the maps and a template for clone-by-clone sequencing has evolved through the course of time. The earliest chromosome to be mapped at the Sanger Institute (chromosome 22, (Dunham *et al.*, 1999)) used a double digest radioactive end labelling method developed by Coulson and colleagues (Coulson *et al.*, 1986). This was later updated (Gregory *et al.*, 1997) to incorporate a fluorescent label which enabled the multiplexing of samples, and was used in the mapping of chromosomes 1, 6, 20 and X. Adoption of a single enzyme digest method (Marra *et al.*, 1997) and the emergence of the whole genome fingerprinting database (International Human Genome Mapping Consortium., 2001), meant the later chromosomes 9, 10 and 13 were mapped using the single enzyme method.

Most of the fingerprints used to assemble maps for chromosomes 9, 10 and 13, were generated at the Genome Sequencing Centre (International Human Genome Mapping Consortium., 2001). The clones, first identified through STS screening of bacterial clone libraries, were extracted from the whole genome fingerprint database using the programme 'get clones' (see Sanger ftp site, section 10). Contigs were reassembled within chromosome specific databases, where the inclusion of marker content data, and FISH localisation, facilitated assembly. Where necessary, additional fingerprinting was performed at the Sanger Institute (Humphray *et al.*, 2001) to augment the St. Louis clones. For chromosome 9, 88% (21417) of the fingerprints were incorporated from the Genome Sequencing Centre, with the remaining 12% (3054) being generated at the Sanger Institute.

See protocols 4, 5 and 6 for methods to generate restriction enzyme fingerprinting data.

6.1 Entering Data

Gel images from restriction digest fingerprinting experiments are processed and entered using a software package called Image.

Details on how to download and operate the Image software can be found at; http://www.sanger.ac.uk/Software/Image/

Image converts raw data into a set of normalised integers attached to each entered band and a series of gel traces for each clone. It is ideally suited for high throughput fingerprinting projects.

Although the Image software has some capacity to identify genuine bands, a good deal of manual manipulation is still required to produce the final set of data. This makes bandcalling one of the most labour intensive aspects of fingerprinting data entry. An automatic bandcalling package has recently been developed by D. Fuhrmann in St.Louis called Bandleader. Using the fragments in the standard lanes for orientation, Bandleader works through the sample lanes and selects suitable bands. The process is completely automated and greatly decreases the processing time for gels.

6.2 Analysis of Fingerprint Data

For both fingerprinting techniques FPC (FingerPrinted Contigs) (Soderlund *et al.*, 1997; Soderlund *et al.*, 2000) is used to analyse data output from IMAGE. There is a detailed overview of FPC in Chapter 7 of this book. Here we aim to give a brief overview of the procedures we follow to manually edit contigs prior to sequence clone selection.

For cost effective clone-by-clone sequencing it is necessary to identify a minimally overlapping set of clones from a fingerprinting contig. Too much overlap between clones leads to high redundancy. Non-overlapping sequence clones necessitate the identification of additional gap closure clones, sometimes to cover small regions. To ensure appropriate tile path selection it is often necessary to refine contigs generated from a global fingerprint build. Contigs should be checked for correct local ordering. Each clone should be next to the clones it shares the greatest degree of overlap with. The overlap with each subsequent clone should decrease in a step wise manor. Overlaps should as much as possible reflect the number of matched bands shared between two clones. To simplify sequence clone selection, clones which appear to be deleted, mixed or poorly entered should be buried within a clone with which they share a high overlap.

6.3 Picking Sequence Tile Paths

Once contigs have been fine tuned, potential merges can be interrogated. Lowering the overlap stringency in FPC by altering the cutoff and using Ends->Ends, is an automated way of comparing the ends of all contigs to one another. The output needs to be filtered, such that only where a number of clones from the ends of two contigs show overlap is a merge considered. Other evidence indicating a clone overlap can be flagged within FPC. For example the presence of shared markers, identified through hybridisation and PCR, or (if sequence is available) by ePCR (Schuler, 1997), can signify a merge. These data can be incorporated into FPC and can be used to supplement the fingerprint data. A slightly lower stringency of fingerprint overlap is accepted where clones share marker content data (see Figure 3). Once a merge has been verified and made, there may be the need for local ordering and some burying where the two contigs overlap. Merged contigs offer greater scope for tile path selection. Also by joining contigs the proportion of end clones being sequenced is reduced. This is important as end clones can lead to a high redundancy in sequence once contigs are merged.

Tile path clones should, as much as possible, fully represent the genomic DNA being sequenced. The absence of contig bands, or presence of extra bands in a clone may be indicative of a deletion or rearrangement and they should therefore be avoided. Minimal overlaps can be determined using the fingerprint data especially where size information is known, such as with a single enzyme digest. Also marker content data can be used to confirm overlaps in clones where fingerprint data alone indicates little overlap.

7. Landmark Content Analysis

Landmark content analysis of the clones in the physical map is of key importance in our strategy (Figure 1). Landmarks serve to anchor contigs to a framework map such as the RHmap or HAPPY map and therefore assist in the ordering and orientation of contigs across the chromosome, providing valuable information for subsequent gap closure (see below). Point landmarks offer much more specific information concerning a clone overlap than fingerprinting and since most landmarks consist of less than 300bp of DNA sequence, clones with small sequence overlaps can be

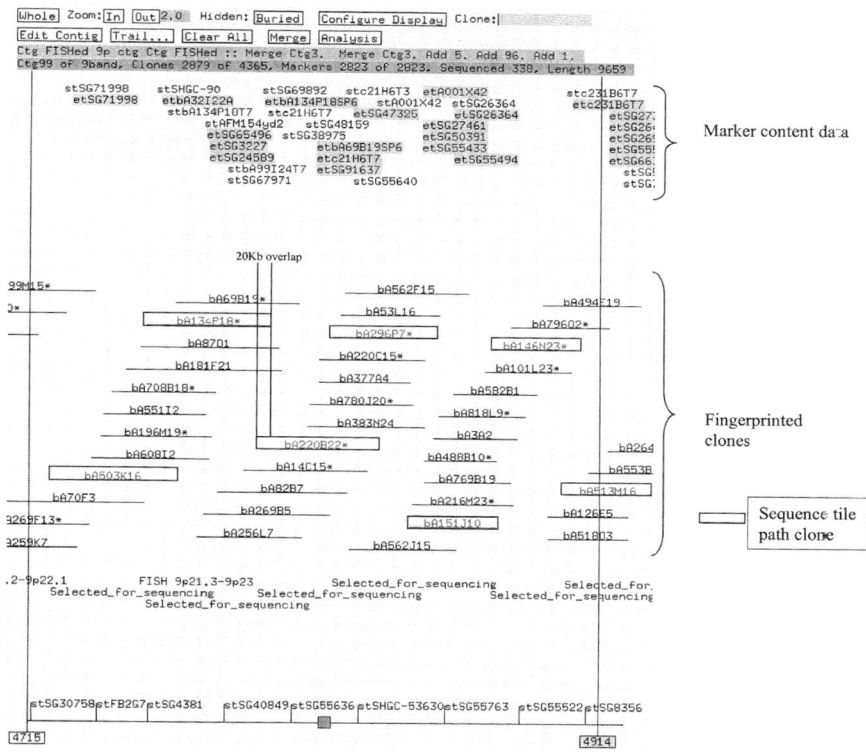

Figure 3. Sequence tile path selection. A selected sequence tile path through a bacterial clone contig on human chromosome 9p is shown. Minimally overlapping clones were selected with a combination of restriction enzyme fingerprint and shared STS content data.

detected. The same overlaps could not be detected using fingerprinting analysis alone.

Landmark (or STS) content mapping is achieved by either PCR or hybridisation. Single bacterial colonies can be tested directly in PCR (colony PCR, protocol 7) without the need to purify clone DNA. This is a rapid method to establish STS content of clones but it does not provide a comprehensive landmark content of the number of clones that can be tested in a single hybridisation experiment. Chromosome specific clones identified from library screening can be gridded at high density onto nylon membranes and these chromosome specific grids ("polygrids") then screened by individual STSs using hybridisation (protocol 3). All clones from a physical map of an average sized human chromosome can be accommodated on as few as 5 (8x12cm) filters (6144 [4x4x384] clones per filter). In a single experiment all clones can be tested for a particular STS which gives important clone overlapping information.

8. Gap Closure

Bacterial clone contigs assembled through the combined use of fingerprinting and landmark content analysis will inevitably contain gaps, either as a result of inadequate marker density used to screen the libraries or an under representation of certain genomic regions within the libraries screened. Construction of the RH map with an average marker density of 1 per 70Kb reduces the former problem. However, there will be certain biases within the marker sets used, for example EST bias to GC-rich regions of the mammalian chromosome. The screening of multiple genomic libraries and therefore genome equivalents may help to reduce the problem of under representation, but some sequences may be unstable within the vector host (*e.g.* tandem repeats) or harmful to the host when transcribed/translated. Chromosome walking is the most commonly used method for achieving gap closure. New probes are designed from sequences at the end of BACs/PACs flanking the gap and used to screen one or more libraries by hybridisation. The new clones identified are fingerprinted for integration with the existing contig assembly and analysis against other contig ends. In some cases a single round of walking may be sufficient to bridge the gap, for larger gaps subsequent rounds of probe design and library screening may be necessary. Vectorette probe generation is a cost effective and rapid means of generating clone end probes for walking (Riley *et al.*, 1990). Efforts to systematically generate BAC end sequences (see URLs, section 10) have greatly expedited the process of chromosome walking. The large scale fingerprinting of entire genomic libraries (International Human Genome Mapping Consortium., 2001) has also reduced the amount of experimental walking required to close gaps, and has generated a new technique, coined by some as, *in-silico* (computational) walking.

In cases where large-insert bacterial clones have not been found to span gaps in the physical map, after multiple rounds of screening, YAC libraries can be screened with flanking STSs. We have screened the ICI, ICRF and CEPH YAC libraries with chromosome 6 STSs and have successfully closed 10 gaps in the BAC/PAC physical map as a result. Many of the YACs identified from the ICI or ICRF libraries are not significantly larger than BACs and have therefore been shotgun sequenced directly (see chapter 4). It is important to verify that the YACs are not deleted or rearranged.

Establishing comprehensive landmark content data of the isolated YACs highlights those YACs containing deletions.

An important tool in determining the size of a gap in the physical map is the cytogenetic technique of fluorescent *in situ* hybridisation on DNA fibres (fibre-FISH). Two or more fluorescently labelled probes, typically BACs or PACs are hybridised to stretched DNA fibres on a glass slide. A number of individual fibres are observed and an average gap size given. This method has utility in sizing gaps of upto 300Kb and of course can also show clone overlaps (yellow signals where green and red signals overlap). Gaps of greater than 300Kb present a greater sizing problem although hybridisation of BACs/PACs to interphase nuclei or chromosomes in metaphase can give an upper limit for the size of the gap between two clones (see chapter 5).

9. Protocols

Protocol 1. Primer Testing

Equipment and Reagents

- MJ Research PTC-225 PCR machine or alternative thermocycler
- Costar Thermowell^{TM} 96 well PCR plates (M-type)
- Hybaid Omnimat for oil free PCR
- Eppendorf multistepper and Combitips (0.5ml [1 = 10 µl])
- Sigma BSA (A-2153) (1:10 dilution of 50mg/ml stock stored at 4°C)
- βME - Beta Mercapto-ethanol (M-7522) (1:20 dilution of stock)
- Sigma Human Placental DNA (D-3160), diluted to 12.5ng DNA/µl
- 10mM Tris/0.1mM EDTA ($T_{0.1}E$)/28% sucrose/cresol red (8mg/100ml)
- 5mM each dATP, dCTP, dGTP, dTTP (Amersham Pharmacia Biotech Inc., 27-2035-01)
- AmpliTaq DNA polymerase (Applied Biosystems manufactured by Roche, N808-0145) (5 units/µl)
- NEB buffer:　1.6g Tris
　　　　　　　0.44g Ammonium Sulphate (Enzyme grade)
　　　　　　　1.34ml 1M $MgCl_2$
　　　　　　　pH to 8.8 with HCl
　　　　　　　make up to 20ml with $T_{0.1}E$, then filter sterilise
- 96 well masterplate:
　　　　Ham = hamster DNA, 12.5ng/µl (CHOK1 or w3gh)
　　　　$T_{0.1}E$ = 10mM Tris/0.1mMEDTA

THD = Sigma human placental DNA (D-3160) 12.5ng/µl (total human DNA)

6 = chromosome 6 hybrid DNA, 12.5ng/µl (GM11580, Coriell Cell Repositories)

Ham	$T_{0.1}E$	Ham	$T_{0.1}E$	Ham	$T_{0.1}E$	Ham	$T_{0.1}E$	Ham	$T_{0.1}E$	Ham	$T_{0.1}E$
THD	6	THD	6	THD	6	THD	6	THD	6	THD	6
Ham	$T_{0.1}E$	Ham	$T_{0.1}E$	Ham	$T_{0.1}E$	Ham	$T_{0.1}E$	Ham	$T_{0.1}E$	Ham	$T_{0.1}E$
THD	6	THD	6	THD	6	THD	6	THD	6	THD	6
Ham	$T_{0.1}E$	Ham	$T_{0.1}E$	Ham	$T_{0.1}E$	Ham	$T_{0.1}E$	Ham	$T_{0.1}E$	Ham	$T_{0.1}E$
THD	6	THD	6	THD	6	THD	6	THD	6	THD	6
Ham	$T_{0.1}E$	Ham	$T_{0.1}E$	Ham	$T_{0.1}E$	Ham	$T_{0.1}E$	Ham	$T_{0.1}E$	Ham	$T_{0.1}E$
THD	6	THD	6	THD	6	THD	6	THD	6	THD	6

Method

1. 3 PCR plates as set out above are needed to test 24 primer pairs at 3 temperatures (typically 55°C, 60°C and 65°C). 5µl of DNA/$T_{0.1}E$ from the master plate are transferred into each well of the PCR plates using a pre-PCR (ie. not previously exposed to PCR product) multichannel pipette.

2. Make sufficient pre-mix for 288 reactions:

NEB buffer	432µl
dNTPs	432µl
BSA	144µl
βME	57.6µl
Taq polymerase	34.6µl
Cresol red/sucrose	1563.8µl

3. Aliquot 111µl into 24 tubes, then add 9µl of primer pair mix (100ng/µl) into each tube.

4. Dispense 10µl into each block of 4 wells on the 3 PCR plates using Eppendorf combitips.

5. PCR as follows: 94°C for 5 minutes, followed by 35 cycles at 94°C for 30 seconds, annealing temperature (55/60/65°C) for 30 seconds and 72°C for 30 seconds and finally followed by 1 cycle at 72°C for 5 minutes.

6. Using an eight channel pipette load 7µl of the PCR product from the first column of the microtitre plate into every second well of a 2.5% agarose gel, prepared in 1X TBE and ethidium bromide (250ng/ml). The products in the next column of the plate are then loaded between those of the first. Load 1Kb ladder marker

(Gibco-BRL), electrophorese at 200 Volts for approximately 45 minutes then photograph the gel on a UV transilluminator.

Load

- 1-12 55°C
- 1-12 60°C
- 1-12 65°C
- 13-24 55°C
- 13-24 60°C
- 13-24 65°C

7. Cut out gel slice if PCR product present at anticipated size in human and 6. Store in 100μl water at 4°C for subsequent use in PCR labelling (protocol 2).

Note: There is no need to use loading buffer when loading PCR samples because the cresol red/sucrose solution is in the reaction.

Protocol 2. PCR Labelling

Equipment and Reagents

- PCR thermocycler
- 0.5 ml microcentrifuge tubes (Eppendorf Safe-Lock, 0030 121.023)
- 5mM each dATP, dGTP, dTTP (Amersham Pharmacia Biotech Inc., 27-2035-01)
- α-^{32}PdCTP (3000 Ci/mmol, Amersham Pharmacia Bictech AA0005)
- AmpliTaq DNA polymerase (5 units/μl, Roche N808-0145)
- Mineral oil (Sigma, M-3516)

10X PCR buffer 1:

Component and final concentration	Amount to add per 10 ml
500 mM KCl	5 ml of 1 M
100 mM Tris-Cl, pH 8.3	1 ml of 1 M
15 mM MgCl$_2$	150 μl of 1 M
H$_2$O	to make 10 ml

Combine the components from sterile stock solutions and mix thoroughly. Aliquot and store at -20°C for 6-12 months.

Method

1. Amplify a sequence of interest by PCR (protocol 1).
2. Following separation by agarose gel electrophoresis, excise the PCR product within a small agarose block and place in 100µl of water in a microcentrifuge tube. Place at 4°C for at least an hour before use.

Note: The purified PCR product can be stored at 4°C for at least a year and still be useful for probe generation.

3. For each PCR product to be radiolabelled, combine the following:

Component and final concentration	Amount to add per 16.5µl
10X PCR buffer 1	2.5 µl
5 mM (each) mixture of dATP/dGTP/dTTP	1 µl
Taq DNA polymerase (5 units/µl)	0.25 µl
H$_2$O	12.75 µl

4. Add 1 µl of a mixture of the two PCR primers (each present at 100 ng/µl) to the 16.5µl pre-mix above.
5. Transfer 7 µl of the mixture from step 4 to a tube suitable for thermal cycling and 2.5 µl of the purified PCR product (from step 2, above) and 0.5 µl of [α]-^{32}P dCTP (3000 Ci/mmol).

Note: For some thermal cyclers, it may be necessary to overlay the samples with a drop of mineral oil prior to thermal cycling. If a mineral oil overlay is required, avoid transferring the oil with the sample when the radiolabelled probe is used.

6. Perform thermal cycling using 94°C for 5 mins followed by 20–25 cycles of 94°C, 30 secs; 55°C, 30 secs; 72°C, 30 secs and a final 5 mins at 72°C and then incubate the sample at 4°C until ready for use.

Note: When more than one PCR product is being radiolabelled, it is typically safe to thermal cycle all samples at the lowest annealing temperature required for any one PCR assay using the same thermal cycler.

Protocol 3. Hybridisation

Equipment and Reagents

- 15 or 50ml Falcon plastic tubes with screw caps
- Plastic sandwich box
- Orbital shaker/incubator

- Water bath
- 20X SSC (3M Sodium chloride, 300mM Tri-sodium citrate)
- 100X Denhardt's (2% Ficoll 400, 2% polyvinylpyrrolidone 40, 2% BSA)
- Dextran sulphate
- *N*-lauroyl sarcosine (sarkosyl)

Hybridization solution:

Note: When making hybridisation solution combine components, stir until dissolved (this may require slight warming of the solution), and filter through Whatmann 3MM to remove any insoluble material. Store at room temperature for up to 6 months.

Component and final concentration	Amount to add per 1 litre
6X SSC	300 ml of 20X SSC
10X Denhardt's	100 ml of 100X
50 mM Tris-HCl	50 ml of 1 M (pH 7.4)
1% (w/v) *N*-laurosyl sarcosine (sarkosyl)	100 ml of 10%
10% Dextran sulphate	200 ml of 50%
H_2O	to make 1 litre

0.5X SSC/1% sarkosyl wash solution:

Component and final concentration	Amount to add per 1 litre
0.5X SSC	25 ml of 20X SSC
1% (w/v) *N*-laurosyl sarcosine (sarkosyl)	100 ml of 10%
H_2O	to make 1 litre

Method

1. Place membrane filters in hybridization solution within a suitable container and incubate at 65°C for at least 2 hours with gentle agitation.

Note: Numerous approaches can be used for hybridizing small or large filters containing arrayed bacterial clone colonies. The specific logistical details heavily depend on the available hybridization equipment (*e.g.*, wet versus dry incubator, shaking incubator versus rotisserie system with bottles). However, numerous configurations can be used satisfactorily, especially since the hybridization analysis of bacterial clone colonies generally yields robust signals. For example, in the case of 8 × 12 cm filters, two or three filters can be rolled up with the colony side facing inwards and slid into a 15ml round-bottomed plastic test tube containing 10 ml of hybridization solution. Once wetted, the filters will unroll slightly. The tubes can be placed vertically in a shaking hybridization incubator. This approach is useful for hybridizing multiple independent probes to separate filters. For hybridization of a large number of 8 × 12 cm filters with a single probe, it is advisable to use a sealable plastic box or bag (which is slightly larger than the filters) containing 20-50 ml of hybridization solution. For larger filters (*e.g.*, 22 × 22 cm), correspondingly larger bags, boxes, or bottles are required, with a proportional increase in the amount of hybridization solution.

For hybridizations performed with round-bottomed plastic tubes, the probe should be added to the hybridization solution and the tube should then be capped and inverted several times to mix thoroughly. For hybridizations performed in plastic boxes, the filters should be first removed and the probe added to the hybridization solution. After gentle mixing, the filters should be returned to the solution one at a time. Similar routines should be adapted for other hybridization containers (*e.g.*, bags, bottles).

2. Prepare a radiolabelled probe(s) by standard methods. See PCR labelling (protocol 2).
3. Boil radiolabelled probe(s) for 5 minutes, snap-chill by placing on ice for 2-4 minutes, and add to the hybridization solution. Roughly 10^6-10^7 cpm of radiolabelled probe should be added per millilitre of hybridization solution.

Note: In the case of probes suspected to contain repetitive sequences, the probe should be boiled with excess sonicated source genomic DNA for 5 minutes after radiolabeling. The specific preparation (*e.g.*, sonication, amounts required) of this DNA will vary for different genomic sources. In the case of human DNA probes, add the following to the radiolabelled probe; 125 µl of 10 mg/ml sonicated, sheared human placental DNA (Sigma), 125 µl of 20X SSC, and H_2O to bring the total volume to 500 µl. The entire mixture should then be boiled for 5 minutes, snap-chilled on ice, and added to the hybridization solution. Of note, after boiling, some investigators incubate the radiolabelled probe/excess genomic DNA mixture at 65°C for 12 hours before adding it to the hybridization solution.

4. Incubate at 65°C overnight with gentle agitation.
5. Discard the hybridization solution and wash the filters twice in 1 litre of 2X SSC at room temperature for 5 minutes with gentle agitation.
6. Wash the filters twice in 1 litre of 0.5X SSC/1% sarkosyl wash solution at 65°C for 30 minutes with gentle agitation.
7. Wash the filters twice in 500 ml of 0.2X SSC at room temperature

for 5 minutes with gentle agitation.

Note: The last wash provides the appropriate level of stringency for most probes. If undesired amounts of non-specific background hybridization are encountered, more dilute SSC washes can be tried. Similarly, if the resulting hybridization signals are too weak, final washes with 1X SSC should be tried in subsequent experiments.

8. Place filters colony-side-up on Whatmann 3MM paper for a few seconds to remove excess solution and then immediately wrap in plastic wrap (Saran Wrap, Dow Chemical Company).

9. Perform autoradiography at room temperature overnight (or at -70°C if necessary).

Protocol 4. Bacterial Clone Preparation for Fingerprinting

Equipment and Reagents

- 1ml sterile deep well microtitre plates (Beckman Instruments Inc., Fullerton, CA.).
- Deep well microtitre plate caps (Beckman Instruments Inc., Fullerton, CA.).
- Sterile cocktail sticks or a 96-pin replicating tool (Denley, Labsystems Instruments, US).
- Orbital shaker (New Brunswick Scientific G24 environmental incubator, Edison, NJ).
- Combitip repeat dispenser (Eppendorf, Netherlands).
- 2x TY growth medium (15 g/l bacto-tryptone, 10 g/l yeast extract, 5 g/l NaCl, pH7.4)
- Appropriate selective agents depending on the cloning vector, *e.g.* chloramphenicol for BACs at a final concentration of 25ug/ml or 12.5ug/ml, for PACs/cosmids use kanamycin at a final concentration of 30ug/ml (Sigma Chemical company, MO., US).
- Centrifuge with microtitre plate holding rotor (Sorvall RT7, Du Pont Company Sorvall, Delaware US).
- Class II microbiological safety cabinet (Walker Safety Cabinets Limited, UK).
- Multichannel pipettes 5-50 µl and 50-300 µl and tips (Lab Systems, Finnpipette, Life Sciences International).
- U-bottom sterile microtitre plates and lids (Greiner Labortechnik).
- Vortex genie fitted with a microtitre plate hold (Scientific Industries Inc. Bohmenia, NY, US).

- 0.2μm Filter bottom plates (Millipore cat. no. MAGVN2250).
- Plate Sealers (Costar, Corning Incorporated, Corning, NY, US.)
- Solution I: GTE, 4.504 g glucose (BDH Laboratory Supplies, Poole England),

      10 ml 0.5M EDTA, 12.5 ml 1M Tris pH 8.0, make up to 500 ml with double distilled H_2O and filter sterilise and store at 4°C.
- Solution II: (make fresh each time)

      8.6 ml double distilled H_2O, 1 ml 10% SDS, 400 μl 5M sodium hydroxide (BDH Laboratory Supplies, Poole England), and
- Solution III: 3M potassium acetate at pH 5.5, store at room temperature.
- RNase A (Ribonuclease A Sigma Chemical company, MO., US).
- Ethanol (70% with water) and isopropanol (BDH Laboratory Supplies, Poole, UK).

Method

1. Using a Combitip repeat dispenser add 500 μl of 2x TY plus appropriate antibiotic to a deep-well microtitre plate.
2. Inoculate the media from one of the clone glycerol stocks using a 96-well inoculating tool. Incubate for 18 h in a 37°C shaking incubator at 300 rpm.
3. In a class II microbiological safety cabinet transfer 250 μl from the overnight growth to a clean microtitre plate and centrifuge at 1550 g for 4 min.
4. Discard most of the supernatant by inverting sharply, tap gently on tissue to absorb all of the supernatant. At this stage clone growth can be checked by scoring the presence of bacterial pellets.
5. To the bacterial pellets add; 25 μl of solution I, re-suspend pellets by vortexing gently (vortex genie set to 2); 25 μl of solution II mix by tapping the plate gently, leave at room temperature for 5 min. The solution should clear as the bacterial clones lyse.
6. Next add 25 μl of solution III and mix by gentle tapping and leave at room temperature for 5 min. The solution should become thick and cloudy.
7. Add the total well volume to a filter bottom plate. Place the filter plate on top of a microtitre plate containing 100 μl of isopropanol. Tape the two plates together and centrifuge at 1550g for 2 min at 20°C.
8. Remove the filter bottom plate checking there is no liquid left.

Leave the isopropanol microtitre plate at room temperature for 30 min then centrifuge at 1550g for 20 min at 20°C.

9. Discard the supernatant by inverting the plate. Remove surplus supernatant by gently tapping the inverted plate on to clean tissue paper, careful to avoid any disruption to the pellets. Dry DNA briefly.

10. Add 100 µl of 70% ethanol, mix by gentle tapping. Centrifuge at 1550 g for 10 min at 20°C. Remove the supernatant and dry as before. Repeat this wash. Dry again this time ensuring the pellet is transparent.

11. Prepare RNase A; 10 µl of 1 mg/ml RNase A per 1 ml of $T_{0.1}E$.

12. Add 5 µl of $T_{0.1}E$ with RNase A. Store samples at -20°C. DNA yield should be between 20-30 ng/µl, this can be checked by separating the DNA on a small 1% agarose gel.

Protocol 5. Fluorescent Fingerprinting of Bacterial Clones

Equipment and Reagents

- *Hin*dIII, (20 U/µl New England Biolabs Inc. US), NEB2 buffer (10x concentration New England Biolabs Inc. US), *Sau* 3AI (50 U/µl Amersham Life Sciences) BsaJ I (2.5 U/µl New England Biolabs Inc. US),

- Matrix standards for ABIs; NED matrix standard (cat. no. 402996 Perkin Elmer Applied Biosystems, Foster City, CA, US) and Fluorescence Amidite matrix standard kit, with TET, HEX and ROX (cat. no.401546 Perkin Elmer Applied Biosystems, Foster City, CA, US). (*see* Note 3)

- Fluorescently tagged dideoxy adenosine triphosphates, *e.g.* ddA-TET, -HEX and -NED Taq FS (cat. no. 4306379C Custom Fingerprinting Kit, 1 ml each of at 10 µM NED, TET and HEX, 3 ml of Taq FS, Perkin Elmer Applied Biosystems Foster City, CA, US) (*see* Note 2 and 3)

- Fluorescently tagged dideoxy cytosine triphosphate, *e.g.* ddC-ROX 5.08 µM (cat. no. 402118 Dye Terminator Cycle Sequencing Core Kit, Perkin Elmer Applied Biosystems Foster City, CA, US)

- ABI377 Sequencer (Perkin Elmer Applied Biosystems, Foster City, CA, US).

- Lambda DNA (500 ng/µl, New England Biolabs, Inc., US).

- Ethanol (96% and 70% with water, BDH Laboratory Supplies, Poole, UK).

- 0.3M sodium acetate.

- Centrifuge with microtitre plate holding rotor (Sorvall RT7, Du Pont

Company Sorvall, Delaware US).
- Hamilton repeat dispenser (Hamilton Company Reno, Nevada).
- Bench top centrifuge (Eppendorf 5415C, Netherlands).
- Blue dextran formamide dye, 9.8 ml deionised formamide, 200 μl 0.5M EDTA, 0.01 g of blue dextran dye.
- 66 well square tooth combs 0.2 mm (cat. no. 402183, Perkin Elmer Applied Biosystems, Foster City CA, US)
- TBE buffer (cat. no. EC860, 10x concentration National Diagnostics, Atlanta, GA, US)
- 37°C incubator (Economy Incubator Size I, Gallenkamp).
- 80°C oven (MINO/18/CLAD, Philip Morris Ltd. UK).

Method

1. Prepare a 5% denaturing acrylamide gel for an ABI377; 20 ml acrylamide, 140 μl 10 % ammonium persulphate, 28 μl TEMED.
2. In three 1.5 ml microfuge tubes (labelled TET, HEX, NED) set up three digest premixes, one for each fluorescent label. For one 96-well plate add to each tube, 25.5 μl $T_{0.1}E$, 24.5 μl NEB2 buffer, 5.0 μl *Hin*dIII (20 U/μL), 8.0 μl Taq FS (32 U/μl) and 3.0 μl *Sau*3AI (30 U/μl) then to the appropriate tube add 4.0 μl of one of the three ddA-dyes. Mix using a vortex, and spin in a bench top centrifuge.
3. Add 2 μl of the TET premix to the first 1/3 of the prepped DNA plate (A1-H4) using a Hamilton repeat dispenser. To the second 1/3 (A5-H8) add the HEX premix and to the last third (A9-H12) the NED premix, cover with a plate sealer. Mix the reaction by gentle agitation on a vortex, and spin the plate at 150 g for 10 s.
4. Incubate the reaction for 1 h at 37°C.
5. Using a multichannel pipette add; 7 μl 0.3M sodium acetate, 40 μl 96% ethanol, to each well.
6. Pool samples labelled with different dyes;
   a. to row 1 add rows 5 and 9 and mix,
   b. to row 2 and rows 6 and 10 and mix,
   c. to row 3 add rows 7 and 11 and mix
   d. and to row 4 add rows 8 and 12 and mix.
7. Precipitate the combined DNA by incubating at room temperature for 30 min in the dark.
8. Centrifuge the plate at 1550 g for 20 min at 20°C to pellet the DNA.
9. Discard the supernatant and dry the pellet by tapping the plate face down onto tissue paper.

10. Add 100 μl of 70% ethanol to each well, mix by gently tapping the plate and spin at 1550 g for 10 min at 20°C.

11. Discard the supernatant and dry the pellet as above ensuring that the DNA is clear.

12. Resuspend the DNA in 5 μl $T_{0.1}E$.

13. Prepare the marker, to a 1.5 ml microfuge tube add 70 μl $T_{0.1}E$, 10 μL NEB2, 6 μl lambda DNA (500 ng/μl), 6 μl BsaJ1 (2.5 U/μl), 4 μL TaqFS (32 U/μl), 4 μL ddC-ROX, incubate for 1 h at 60°C. To the reaction mix add 100 μl 0.3M sodium acetate and 400 μl 96% ethanol, leave at room temperature in the dark for 15 min, then at -20°C for 20 min. Spin in a bench top centrifuge at maximum for 20 min, discard the supernatant and dry the DNA pellet by tapping gently onto tissue paper. Wash the pellet by adding 200 μl 70% ethanol and spin in a bench top centrifuge at maximum for 5 min, discard the supernatant and dry pellet as before, this time ensuring the DNA is clear. Resuspend the pellet in 120 μl $T_{0.1}E$ and 120 μl blue dextran formamide dye.

14. Using a Hamilton repeat dispenser add 2μl of the marker mix to each sample, spin the plate to 150 g in a centrifuge. Denature the plate for 10 min at 80°C then load 1.25 μl of each sample on an ABI377, use ABI Prism Collection Software v1.1.

15. After data collection transfer the gel image to a UNIX workstation for entry into Image.

Protocol 6. Restriction Digest Agarose Fingerprinting

Equipment and Reagents
- Scanner;
  - Typhoon 8600 and ImageQuant solutions software for windows NT (Amersham Pharmacia Biotech).
  - SI Vistra Fluorescence (Molecular Dynamics, Sunnyvale, CA, US), PC, TAXAN 580plus LR XPS P90 (Dell Computers, Buffalo, NY, US).
- *Hin*dIII, (100 U/ul New England Biolabs).
- Vistra Green intercalating stain (Amersham Life Sciences UK).
- Plate Sealers (Costar, Corning Incorporated, Corning, NY, US.).
- Combitip repeat dispenser (Eppendorf, Netherlands).
- 12-channel Anachem 10-12A 1-10ul 12-channel multi-channel
- 8-channel Hamilton Company 8-channel syringe 84502/ 12-channel 84512
- Buffer re-circulation system;

- Low temperature bath/circulator 20 litres, LTD20G -30 to 100, Grant Instruments Camdridge UK.
- Modular drive pump and control pump, masterflex pump head, stainless steel mounting hardware, (Cole-Parmer Instrument Company, Illinois, USA)
- Tubing (Tygon).
- Or a Cold room regulated to 4°C can be used to run gels.
- Gel tanks, Gator Wide Format System model A3-1 (Owl Scientific US), additional gel casting trays (A3-UVT-1).
- 121-well comb/gel divider (per. comms. from Mandeep Sekhon, Washington University School of Medicine, Genome Sequencing Centre, St. Louis MO 63108, US).
- 1% agarose (SeaKem LE. FMC Bioproducts Rockland Maine US) gel, made with 1xTAE (0.04MTris-acetate, 0.001m EDTA).
- Loading dye (6x Buffer II for 10 ml, 1.5 g ficoll, 0.025 g bromophenol blue, 0.025 g xylene cyanol, 10 ml sterile water).
- Marker mix; Analytical Marker DNA, wide band (Promega, Madison, WI, US),
- DNA Molecular Weight Marker V (Roche Molecular Biochemicals, Switzerland).
- Bench top centrifuge (Eppendorf 5415C, Netherlands).
- Vortex genie fitted with a microtitre plate holder (Scientific Industries Inc. Bohmenia, NY, US).
- Orbital platform shaker, Belly Dancer (Storvall Life Sciences, NC, US).

Method

1. To digest a 96-well plate of prepared DNA (see protocol 4) set up one reaction mix in a 1.5 ml microfuge tube. Add 307 μl H_2O, 95 μl NEB2, 21 μl *Hind*III. Mix using a vortex, and spin in a bench top centrifuge. To each well add 4 μl of the reaction mix, using a combitip dispenser set to 1, with 0.2 ml tips, cover with a plate sealer.
2. Mix the reaction by gentle agitation on a vortex, and spin the plate to 150g for 10s.
3. Incubate at 37°C for 2 h.
4. Terminate the reaction by adding 2 μl of loading dye and seal the plate. Either load straight away or store at 4°C.
5. Make the marker in a 1.5 ml microfuge tube. For one gel add, 19.2

μl $T_{0.1}E$, 1.5 μl Analytical Marker DNA wide range, 0.2 μl Molecular Weight marker V and 4.2 μL 6x loading dye.

6. Prepare a 1% agarose gel in 1xTAE. (For a Gator Wide format gel tank, to 4.5 g agarose add 450 ml 1xTAE) (*Note:* for high throughput use two combs in each gel - to load two full 96well plates per gel).

7. Load 0.8 μl of the marker in the first well and then every fifth well. Then load 1μl sample between the marker lanes.

8. To run gels using a re-circulating system; set the low temperature bath to -5°C (for best results fill with half water and half antifreeze). Connect the tubing to the Gator gel tank and feed through the pump head fitted to the modular pump drive, so the buffer is re-circulating around the gel tank. Set the modular control to between 0 and 1 to ensure very slow buffer movement. Spool the tubing into the waterbath to chill the buffer and cool the gel during electrophoresis.

9. Run gels for 16 h at 85 volts and a buffer temperature of 7-8°C.

10. Alternatively run gels at 4°C in a cold room for 16 h at 85 volts.

11. Prepare a fresh stain mix for the gel; 5 ml $1M$ Tris HCL, 0.5 ml $0.1M$ EDTA, 50 μl Vistra Green, make up to 500 ml with H_2O. Store at 4°C, use within 48 h.

12. Stain for 30-45 min on a belly-dancer shaker. Wash briefly with H_2O to remove excess stain.

13. Scan the gel on a Typhoon 8600 or a FluorImager SI;

- Settings for Typhoon - acquisition mode - fluorescence, Emission filter -526nm, sensitivity - normal, laser - green, pixel size - 200microns, focal plane - Platen, PMT voltage - 800.

- Settings for FluorImager SI; Emission filter - 530nm, sensitivity - normal, excitation filter - 488 nm, dye - single label, pixel size - 100 microns, PMT voltage - 800, digital resolution - 16 bits.

Protocol 7. Colony PCR

Equipment and Reagents

- Sterile toothpicks or plastic disposable loops
- Bacterial clones streaked to single colony on LB agar plates containing appropriate antibiotic
- 96-well microtitre plate (as per protocol 1)
- $T_{0.1}E$ (as per protocol 1)

Method

1. Using a disposable, sterile toothpick or plastic loop, pick a small single bacterial colony into 100μl $T_{0.1}E$ in a microtitre plate.
2. Mix colony dilutions thoroughly using a multi-channel pipette and transfer 5μl to a fresh PCR microtitre plate.
3. Prepare PCR pre-mix as for primer testing (protocol 1) and aliquot 10μl to each of the colony dilutions to be tested.

10. Useful URLs

The Human Genome Project at the Sanger Institute:

http://www.sanger.ac.uk/HGP/

FTP sites

Markers at the Cooperative Human Linkage Center:

http://gai.nci.nih.gov/html-chlc/ChlcMarkers.html

Radiation Hybrid Database ftp directory:

ftp://ftp.ebi.ac.uk/pub/databases/RHdb/

The Genome Database:

http://gdbwww.gdb.org/

The Genethon public www site:

http://www.genethon.fr

The Wellcome Trust Sanger Institute anonymous ftp site:

ftp://ftp.sanger.ac.uk

The Stanford Human Genome Center public ftp site:

ftp://shgc.stanford.edu/pub/hgmc/

Data files release 12 (July 1997) from the Whitehead Institute Center for Genome Research:

http://www-genome.wi.mit.edu/ftp/distribution/human_STS_releases/july97/

Genomic resources

Caltech clone libraries:

http://www.tree.caltech.edu/lib_status.html

Coriell Cell Repositories:

http://locus.umdnj.edu/ccr/

Research Genetics:

http://www.resgen.com/index.php3

BAC and PAC clone libraries held at the Children's Hospital Oakland Research Institute:

http://www.chori.org/bacpac/

On-line databases and search engines

BAC end sequence database search at The Institute for Genomic Research:

http://www.tigr.org/tdb/humgen/bac_end_search/bac_end_search.html

Baylor College of Medicine Human Genome Sequencing Center search launcher:

http://searchlauncher.bcm.tmc.edu

Ensembl Genome Browser:

http://www.ensembl.org

Ensembl Human BLAST server:

http://www.ensembl.org/Homo_sapiens/blastview

National Center for Biotechnology Information (NCBI) BLAST server:

http://www.ncbi.nlm.nih.gov/BLAST

NCBI Human Genome Resources:

http://www.ncbi.nlm.nih.gov/genome/guide/human

Oligonucleotide design software at the Whitehead Institute (PRIMER 3.0 Steve Rozen, Helen J. Skaletsky (1997)):

http://www-genome.wi.mit.edu/genome_software/other/primer3.html

RepeatMasker:

http://ftp.genome.washington.edu/cgi-bin/RepeatMasker

University of California Santa Cruz (UCSC) Genome Bioinformatics:

http://genome.ucsc.edu

Acknowledgements

We'd like to thank Charlotte Cole, John Collins, Dave Beare, Ian Dunham, Simon Gregory and Gareth Howell for protocols.

References

Adams, M.D., Kelley, J.M., Gocayne, J.D., Dubnick, M., Polymeropoulos, M.H., Xiao, H., Merril, C.R., Wu, A., Olde, B., and Moreno, R.F. 1991. Complementary DNA sequencing: expressed sequence tags and human genome project. Science. 252: 1651–6.

Altschul, S.F., Gish, W., Miller, W., Myers, E.W., and Lipman, D.J. 1990. Basic local alignment search tool. Journal of Molecular Biology. 215: 403–10.

Bentley, D.R., Deloukas, P., Dunham, A., French, L., Gregory, S.G., Humphray, S.J., Mungall, A.J., Ross, M.T., Carter, N.P., Dunham, I., Scott, C.E., Ashcroft, K.J., Atkinson, A.L., Aubin, K., Beare, D.M., Bethel, G., Brady, N., Brook, J.C., Burford, D.C., Burrill, W.D., Burrows, C., Butler, A.P., Carder, C., Catanese, J.J., Clee, C.M., Clegg, S.M., Cobley, V., Coffey, A.J., Cole, C.G., Collins, J.E., Conquer, J.S., Cooper, R.A., Culley, K.M., Dawson, E., Dearden, F.L., Durbin, R.M., de_Jong, P.J., Dhami, P.D., Earthrowl, M.E., Edwards, C.A., Evans, R.S., Gillson, C.J., Ghori, J., Green, L., Gwilliam, R., Halls, K.S., Hammond, S., Harper, G.L., Heathcott, R.W., Holden, J.L., Holloway, E., Hopkins, B.L., Howard, P.J., Howell, G.R., Huckle, E.J., Hughes, J., Hunt, P.J., Hunt, S.E., Izmajlowicz, M., Jones, C.A., Joseph, S.S., Laird, G., Langford, C.F., Lehvaslaiho, M.H., Leversha, M.A., McCann, O.T., McDonald, L.M., McDowall, J., Maslen, G.L., Mistry, D., Moschonas, N.K., Neocleous, V., Pearson, D.M., Phillips, K.J., Porter, K.M., Prathalingam, S.R., Ramsey, Y.H., Ranby, S.A., Rice, C.M., Rogers, J., Rogers, L.J., Sarafidou, T., Scott, D.J., Sharp, G.J., Shaw_Smith, C.J., Smink, L.J., Soderlund, C., Sotheran, E.C., Steingruber, H.E., Sulston, J.E., Taylor, A., Taylor, R.G., Thorpe, A.A., Tinsley, E., Warry, G.L., Whittaker, A., Whittaker, P., Williams, S.H., Wilmer, T.E., Wooster, R., and Wright, C.L. 2001. The physical maps for sequencing human chromosomes 1, 6, 9, 10, 13, 20 and X. Nature. 409: 942–3.

Coulson, A., Sulston, J., Brenner, S., and Karn, J. 1986. Towards a physical map of the genome of the nematode *Caenorhabditis elegans*. Proc. Natl. Acad. Sci. USA. 83: 7821–7825.

Dear, P.H. 1997. HAPPY mapping. In Genome mapping: A practical approach (ed. P.H. Dear). 95–124. IRL Press, Oxford, UK.

Dear, P.H., and Cook, P.R. 1993. Happy mapping: linkage mapping using a physical analogue of meiosis. Nucleic Acids Research (Online). 21: 13–20.

Deloukas, P., Matthews, L.H., Ashurst, J., Burton, J., Gilbert, J.G., Jones, M., Stavrides, G., Almeida, J.P., Babbage, A.K., Bagguley, C.L., Bailey, J., Barlow, K.F., Bates, K.N., Beard, L.M., Beare, D.M., Beasley, O.P., Bird, C.P., Blakey, S.E., Bridgeman, A.M., Brown, A.J., Buck, D., Burrill, W., Butler, A.P., Carder, C., Carter, N.P., Chapman, J.C., Clamp,

M., Clark, G., Clark, L.N., Clark, S.Y., Clee, C.M., Clegg, S., Cobley, V.E., Collier, R.E., Connor, R., Corby, N.R., Coulson, A., Coville, G.J., Deadman, R., Dhami, P., Dunn, M., Ellington, A.G., Frankland, J.A., Fraser, A., French, L., Garner, P., Grafham, D.V., Griffiths, C., Griffiths, M.N., Gwilliam, R., Hall, R.E., Hammond, S., Harley, J.L., Heath, P.D., Ho, S., Holden, J.L., Howden, P.J., Huckle, E., Hunt, A.R., Hunt, S.E., Jekosch, K., Johnson, C.M., Johnson, D., Kay, M.P., Kimberley, A.M., King, A., Knights, A., Laird, G.K., Lawlor, S., Lehvaslaiho, M.H., Leversha, M., Lloyd, C., Lloyd, D.M., Lovell, J.D., Marsh, V.L., Martin, S.L., McConnachie, L.J., McLay, K., McMurray, A.A., Milne, S., Mistry, D., Moore, M.J., Mullikin, J.C., Nickerson, T., Oliver, K., Parker, A., Patel, R., Pearce, T.A., Peck, A.I., Phillimore, B.J., Prathalingam, S.R., Plumb, R.W., Ramsay, H., Rice, C.M., Ross, M.T., Scott, C.E., Sehra, H.K., Shownkeen, R., Sims, S., Skuce, C.D., Smith, M.L., Soderlund, C., Steward, C.A., Sulston, J.E., Swann, M., Sycamore, N., Taylor, R., Tee. L., Thomas, D.W., Thorpe, A., Tracey, A., Tromans, A.C., Vaudin, M., Wall, M., Wallis, J.M., Whitehead, S.L., Whittaker, P., Willey, D.L., Williams, L., Williams, S.A., Wilming, L., Wray, P.W., Hubbard, T., Durbin, R.M., Bentley, D.R., Beck, S., and Rogers, J. 2001. The DNA sequence and comparative analysis of human chromosome 20. Nature. 414: 865–71.

Deloukas, P., Schuler, G.D., Gyapay, G., Beasley, E.M., Soderlund, C., Rodriguez_Tome, P., Hui, L., Matise, T.C., McKusick, K.B., Beckmann, J.S., Bentolila, S., Bihoreau, M., Birren, B.B., Browne, J., Butler, A., Castle, A.B., Chiannilkulchai, N., Clee, C., Day, P.J., Dehejia, A., Dibling, T., Drouot, N., Duprat, S., Fizames, C., and Bentley, D.R. 1998. A physical map of 30,000 human genes. Science. 282: 744–6.

Dunham, I., Hunt, A.R., Collins, J.E., Bruskiewich, R., Beare, D.M., Clamp, M., Smink, L.J., Ainscough, R., Almeida, J.P., Babbage, A., Bagguley, C., Bailey, J., Barlow, K., Bates, K.N., Beasley, O., Bird, C.P., Blakey, S., Bridgeman, A.M., Buck, D., Burgess, J., Burrill, W.D. et al. 1999. The DNA sequence of human chromosome 22. Nature. 402: 489–95.

Gregory, S.G., Howell, G.R., and Bentley, D.R. 1997. Genome mapping by fluorescent fingerprinting. Genome Res. 7: 1162–8.

Gyapay, G., Schmitt, K., Fizames, C., Jones, H., Vega_Czarny, N., Spillett, D., Muselet, D., Prud_Homme, J.F., Dib, C., Auffray, C., Morissette, J., Weissenbach, J., and Goodfellow, P.N. 1996. A radiation hybrid map of the human genome. Human Molecular Genetics. 5: 339–46.

Hall, N. et al. 2002. Sequence of *Plasmodium falciparum* chromosomes 1, 3-9 and 13. Nature. 419: 527–31.

Hudson, T.J., Stein, L.D., Gerety, S.S., Ma, J., Castle, A.B., Silva, J., Slonim, D.K., Baptista, R., Kruglyak, L., and Xu, S.H. 1995. An STS-based map of the human genome. Science. 270: 1945–54.

Humphray, S.J., Knaggs, S.J., and Ragoussis, I. 2001. Contiguation of bacterial clones. Methods in Molecular Biology. 175: 69–108.

Ioannou, P.A., Amemiya, C.T., Garnes, J., Kroisel, P.M., Shizuya, H., Chen, C., Batzer, M.A., and de_Jong, P.J. 1994. A new bacteriophage P1-derived vector for the propagation of large human DNA fragments. Nature Genetics. 6: 84–9.

Konfortov, B.A., Cohen, H.M., Bankier, A.T., and Dear, P.H. 2000. A high-resolution HAPPY map of Dictyostelium discoideum chromosome 6. Genome Res. 10: 1737–42.

International Human Genome Sequencing Consortium. 2001. Initial sequencing and analysis of the human genome. Nature. 409: 860–921.

Marra, M.A., Kucaba, T.A., Dietrich, N.L., Green, E.D., Brownstein, B., Wilson, R.K., McDonald, K.M., Hillier, L.W., McPherson, J.D., and Waterston, R.H. 1997. High throughput fingerprint analysis of large-insert clones. Genome Res. 7: 1072–84.

The International Human Genome Mapping Consortium. 2001. A physical map of the human genome. Nature. 409: 934–41.

Morton, N.E. 1991. Parameters of the human genome. Proceedings of the National Academy of Sciences of the United States of America. 88: 7474–6.

Mungall, A.J., Edwards, C.A., Ranby, S.A., Humphray, S.J., Heathcott, R.W., Clee, C.M., East, C.L., Holloway, E., Butler, A.P., Langford, C.F., Gwilliam, R., Rice, K.M., Maslen, G.L., Carter, N.P., Ross, M.T., Deloukas, P., Bentley, D.R., and Dunham, I. 1996. Physical mapping of chromosome 6: a strategy for the rapid generation of sequence-ready contigs. DNA Sequence. 7: 47–9.

Mungall, A.J., Humphray, S.J., Ranby, S.A., Edwards, C.A., Heathcott, R.W., Clee, C.M., Holloway, E., Peck, A.I., Harrison, P., Green, L.D., Butler, A.P., Langford, C.F., William, R.G., Huckle, E.J., Baron, L., Smith, A., Leversha, M.A., Ramsey, Y.H., Clegg, S.M., Rice, C.M., Maslen, G.L., Hunt, S.E., Scott, C.E., Soderlund, C.A., and Dunham, I. 1997. From long range mapping to sequence-ready contigs on human chromosome 6. DNA Sequence. 8: 151–4.

Olson, M., Hood, L., Cantor, C., and Botstein, D. 1989. A common language for physical mapping of the human genome. Science. 245: 1434–5.

Piper, M.B., Bankier, A.T., and Dear, P.H. 1998. A HAPPY map of Cryptosporidium parvum. Genome Res. 8: 1299–307.

Riley, J., Butler, R., Ogilvie, D., Finniear, R., Jenner, D., Powell, S., Anand, R., Smith, J.C., and Markham, A.F. 1990. A novel, rapid method for the isolation of terminal sequences from yeast artificial chromosome (YAC) clones. Nucleic Acids Research (Online). 18: 2887–90.

Ross, M.T., LaBrie, S., McPherson, J., and Stanton, V.P. 1999. in Current Protocols in Human Genetics (eds Dracopoli *et al.*) 5.6.1–5.6.5 (Wiley, New York, 1999).

Schuler, G.D. 1997. Sequence mapping by electronic PCR. Genome Res. 7: 541–50.

Schuler, G.D., Boguski, M.S., Stewart, E.A., Stein, L.D., Gyapay, G., Rice, K., White, R.E., Rodriguez_Tome, P., Aggarwal, A., Bajorek, E., Bentolila, S., Birren, B.B., Butler, A., Castle, A.B., Chiannilkulchai, N., Chu, A., Clee, C., Cowles, S., Day, P.J., Dibling, T., Drouot, N., Dunham, I., Duprat, S., East, C. *et al.* 1996. A gene map of the human genome. Science. 274: 540–6.

Shizuya, H., Birren, B., Kim, U.J., Mancino, V., Slepak, T., Tachiiri, Y., and Simon, M. 1992. Cloning and stable maintenance of 300-kilobase-pair fragments of human DNA in Escherichia coli using an F-factor-based vector. Proceedings of the National Academy of Sciences of the United States of America. 89: 8794–7.

Soderlund, C., Humphray, S., Dunham, A., and French, L. 2000. Contigs built with fingerprints, markers, and FPC V4.7. Genome Res. 10: 1772–87.

Soderlund, C., Longden, I., and Mott, R. 1997. FPC: a system for building contigs from restriction fingerprinted clones. Computer Applications in the Biosciences : Cabios. 13: 523–35.

Thomas, J.W., Prasad, A.B., Summers, T.J., Lee_Lin, S.Q., Maduro, V.V., Idol, J.R., Ryan, J.F., Thomas, P.J., McDowell, J.C., and Green, E.D. 2002. Parallel construction of orthologous sequence-ready clone contig maps in multiple species. Genome Res. 12: 1277–85.

Venter, J.C., Adams, M.D., Myers, E.W., Li, P.W., Mural, R.J., Sutton, G.G., Smith, H.O., Yandell, M., Evans, C.A., Holt, R.A., Gocayne, J.D., Amanatides, P., Ballew, R.M., Huson, D.H., Wortman, J.R., Zhang, Q., Kodira, C.D., Zheng, X.H., Chen, L., Skupski, M., Subramanian, G., Thomas, P.D., Zhang, J., Gabor_Miklos, G.L., Nelson, C., Broder, S., Clark, A.G., Nadeau, J., McKusick, V.A., Zinder, N., Levine, A.J., Roberts, R.J., Simon, M., Slayman, C., Hunkapiller, M., Bolanos, R., Delcher, A., Dew, I., Fasulo, D., Flanigan, M., Florea, L., Halpern, A., Hannenhalli, S., Kravitz, S., Levy, S., Mobarry, C., Reinert, K., Remington, K., Abu_Threideh, J., Beasley, E., Biddick, K., Bonazzi, V., Brandon, R., Cargill, M., Chandramouliswaran, I., Charlab, R., Chaturvedi, K., Deng, Z., Di_Francesco, V., Dunn, P., Eilbeck, K., Evangelista, C., Gabrielian, A.E., Gan, W., Ge, W., Gong, F., Gu, Z., Guan, P., Heiman, T.J., Higgins, M.E., Ji, R.R., Ke, Z., Ketchum, K.A., Lai, Z., Lei, Y., Li, Z., Li, J., Liang, Y., Lin, X., Lu, F., Merkulov, G.V., Milshina, N., Moore, H.M., Naik, A.K., Narayan, V.A., Neelam, B.,

Nusskern, D., Rusch, D.B., Salzberg, S., Shao, W., Shue, B., Sun, J., Wang, Z., Wang, A., Wang, X., Wang, J., Wei, M., Wides, R., Xiao, C., Yan, C., Yao, A., Ye, J., Zhan, M., Zhang, W., Zhang, H., Zhao, Q., Zheng, L., Zhong, F., Zhong, W., Zhu, S., Zhao, S., Gilbert, D., Baumhueter, S., Spier, G., Carter, C., Cravchik, A., Woodage, T., Ali, F., An, H., Awe, A., Baldwin, D., Baden, H., Barnstead, M., Barrow, I., Beeson, K., Busam, D., Carver, A., Center, A., Cheng, M.L., Curry, L., Danaher, S., Davenport, L., Desilets, R., Dietz, S., Dodson, K., Doup, L., Ferriera, S., Garg, N., Gluecksmann, A., Hart, B., Haynes, J., Haynes, C., Heiner, C., Hladun, S., Hostin, D., Houck, J., Howland, T., Ibegwam, C., Johnson, J., Kalush, F., Kline, L., Koduru, S., Love, A., Mann, F., May, D., McCawley, S., McIntosh, T., McMullen, I., Moy, M., Moy, L., Murphy, B., Nelson, K., Pfannkoch, C., Pratts, E., Puri, V., Qureshi, H., Reardon, M., Rodriguez, R., Rogers, Y.H., Romblad, D., Ruhfel, B., Scott, R., Sitter, C., Smallwood, M., Stewart, E., Strong, R., Suh, E., Thomas, R., Tint, N.N., Tse, S., Vech, C., Wang, G., Wetter, J., Williams, S., Williams, M., Windsor, S., Winn_Deen, E., Wolfe, K., Zaveri, J., Zaveri, K., Abril, J.F., Guigo, R., Campbell, M.J., Sjolander, K.V., Karlak, B., Kejariwal, A., Mi, H., Lazareva, B., and Hatton, T. 2001. The sequence of the human genome. Science. 291: 1304–1351.

Watson, J.D., and Jordan, E. 1989. The Human Genome Program at the National Institutes of Health. Genomics. 5: 654–6.

From: *Genome Mapping and Sequencing*
© 2003 Horizon Scientific Press, Wymondham, UK

7

FPC: A Software Package for Physical Maps

Fred Engler and Cari Soderlund

Abstract

FPC (FingerPrinted Contigs) builds contigs from marker data and fingerprinted clones. Contigs are ordered by framework markers. The initial versions of FPC mainly supported interactive assembly of maps, which would not scale up efficiently to whole genome maps. Subsequent versions of FPC added increased automation for assembling maps. In this chapter, we will explore how to efficiently build physical maps with FPC using the most recent features. We begin with a brief history of physical mapping as it relates to FPC, and then analyze the concepts and parameters incorporated by the software. Finally, we present a tutorial that guides you through the most useful features of FPC. The demo files used in the tutorial are available online.

1. Introduction

FPC (FingerPrinted Contigs, Soderlund *et al.,* 1997a) is a program initially developed at the Sanger Centre, Cambridge, UK to aid in the construction of physical maps of human chromosomes from restriction fingerprint data. FPC replaced the `contigC` (Sulston *et al.,* 1988) package that was written to map the nematode *C.elegans*, a genome of approximately 100 MB (million bases). FPC was written as a temporary software package because it was thought that restriction fingerprint maps would not be used much longer as automatic techniques for building maps had not been successful; hence, it was time consuming for the user to build maps, and the technique obviously would not scale up to larger genomes. During the development of FPC, automatic assembly routines were successfully written; additionally, there was increased utility for interactively building maps, incorporating markers and frameworks, and support for manually selecting a minimal tiling path (MTP; Soderlund *et al.*, 1997b). Marra *et al.*, (1997) improved fingerprinting techniques in order to reduce the error and uncertainty in the data. Due to the increased quality of the data and increased automation of the assembly, restriction digest fingerprinting is now routinely used for building maps in order to select clones for sequencing and determine the structure of a genome (Soderlund *et al.*, 2000). There are many regional, chromosomal, and whole genome FPC projects. The following are some of the major whole genome sequence ready maps: Marra *et al.*, (1999) assembled an *Arabidopsis* map, which was used to select the MTP for sequencing (The Arabidopsis Genome Initiative, 2000). Hoskins *et al.*, (2000) assembled a *Drosophila melanogaster* map for 81% of the genome, which was used to verify and locate sequenced contigs from a whole genome shotgun approach (Adams *et al.*, 2000). The International Human Genome Mapping Consortium (2001) assembled a whole genome map of human, which was used as the basis for sequencing the human genome (The International Human Genome Sequencing Consortium, 2001). Chen *et al.*, (2002) assembled a whole genome map of rice, which is used for the basis for the International Rice Genome Sequencing Project (IRGSP; Wing *et al.*, 2001). FPC was written to use restriction fragments from one digest, but Ding *et al.*, (1999) demonstrated that it could be used with three separate sets of fingerprints to increase the sensitivity of overlap calculation; with some slight modification to FPC, this technique is now being used by Dupont to map maize (Tingey, 2000).

Currently, whole genome maps are being constructed for a number of organisms including mouse, rat, cow, zebrafish, sorghum, maize, and tomato. Most whole genome maps are accessible from a web-based display (see http://www.genome.arizona.edu/fpc for links to the corresponding web sites). The maps of sorghum, maize, and rice are displayed at AGCoL (Arizona Genomics Computational Laboratory) with a Java program called WebFPC, developed at AGCoL. The maps of human, mouse, rat, and cow are accessible at the BCGSC (British Columbia Cancer Agency, Genome Sequencing Center) with a Java program called ICE (Internet Contig Explorer), developed at the BCGSC. A fingerprint comparison tool is available for the human map at the GSC in St. Louis.

We have recently developed a FPC based routine called BSS (BLAST Some Sequence), which uses the program BLAST (Altschul *et al.*, 1997). It takes as input any sequence (*e.g.* draft or complete genomic, marker sequence such as ESTs) and blasts it against any sequence associated with clones in the FPC database, e.g. genomic or STCs (Sequence Tagged Connectors, also called BES for BAC End Sequences). BSS consolidates the output into an interactive report and adds the results to FPC as markers or remarks. This has been extremely valuable in mapping genetic markers and draft sequence to the rice FPC map, which helps anchor contigs, close gaps, and select a minimal tiling path. We have also developed a function called FSD (FPC Simulated Digest) to perform a simulated digest on a sequenced clone; the resulting *in silico* fingerprint can be automatically assembled into FPC. This also allows us to close gaps and anchor contigs. FSD has a synergistic relation with the BSS: as more sequence is added, more electronic markers can be mapped. These features are covered in tutorials available at http://www.genome.arizona.edu/software/fpc/fac.html and in Soderlund *et al.*, (2002). A parallel version of FPC (Ness *et al.*, 2002) greatly decreases the time to perform a complete assembly of a FPC database, which has been incorporated into the V6 release.

1.1 Background

As mentioned, FPC was written as a temporary package. The initial development was to provide interactive tools for building maps. This included comparing two clones based on the number of shared fragments (alias bands), where two fragments are shared if they are within a tolerance of each other. Using the number of shared fragments, FPC computes the probability of overlap between the two clones; this equation is taken directly from the `contigC` program (Sulston *et al.*, 1988), and is referred to as the Sulston score. The following salient features were added:

- Contig display and Edit Contig menu: The contig display shows the layout of markers, clones, remarks, and frameworks; this display was taken from aceDB (Durbin and Thierry-Meig, 1994). A whole set of editing functions were added that allows the user to move or remove one or more clones from the contig.
- Clone Text and Marker Text: Clones and markers are clickable and bring up a text box showing information about the entity (*i.e.* clone or marker). From this text box, an Edit window can be initiated and the entity edited; for example, a clone can be given a remark.
- Fingerprint and Gel window: The fingerprint bands from a set of clones can be viewed. Additionally, the raw gel image of clones from different gels can be displayed as if they were in adjacent lanes.

- The Main Analysis (from the main window) and Contig Analysis (from the contig window): These provide a set of functions to compare clones based on the Sulston score for a keyset (*i.e.* group) of clones against all the clones in a contig or in the database, compare the clones at the end of one contig with those at the ends of all other contigs, or compare a clone against all clones in a contig.
- Merge (from the contig window): Merges two contigs interactively.

Given this set of features, the user can interactively build a contig, evaluate it, and select a minimal tiling path. This is fine for small datasets, but does not scale up to ordering 1000's of clones.

During the development of the above features, automatic assembly routines were developed (Soderlund *et al.*, 1997a; Soderlund *et al.*, 2000). We feel it is now possible to build sequence ready maps with a minimal amount of human interaction. This thesis is being tested as part of the Maize Mapping Project (Coe *et al.*, 2002). We are developing a physical map of the 2700 MB genome of maize using fingerprints and markers. 10,000 ESTs have been hybridized to a 10x set of BACs in collaboration with Dupont and Inctye Genomics. The map will have 4800 genetic framework markers. A 24x coverage of fingerprinted clones will come from three libraries cut with different enzymes. We are incrementally assembling in new clones and markers, and displaying the results on the web monthly (see http://www.genome.arizona.edu/fpc/maize). We will wait until all data is in the maize FPC database before manually merging contigs. This is the first time a genome has been assembled in this way, and should drastically reduce the time spent in manual editing. A new feature that is described in this manuscript, the DQer, has been extremely important for this project: it allows us to assemble with a relatively low stringency so as to minimize the number of contigs, yet reduce the number of bad contigs automatically.

The User's Manual (Soderlund, 1999) was kept up-to-date until 2000, and can be downloaded from http://www.genome.arizona.edu/software/fpc/download. Since then, all new features and changes are documented in the User's Guide, which is an on-line document at http://www.genome.arizona.edu/software/fpc/faq.html. Therefore, all features are documented. But users never found this approach easy. So we are trying a new approach, which is to present the features of FPC as a tutorial. The remainder of this chapter is a tutorial that covers the features we are using to build the maize physical map – that is, incrementally assembling the map, ordering contigs based on framework markers, adding markers and remarks, and searching. Other than merging and adding remarks, we will not cover any editing functions as these are nearly obsolete. If needed, they are covered in the

User's Manual. We will briefly describe comparing multiple gel images using the Gel Image window. This feature is used in selecting a MTP, which is covered by Chapter 6 in this book. The displays in this tutorial are from FPC V6.2; the only new feature covered in this chapter that is not in V5 is the DQer.

2. Analysis

We will start out describing the major aspects of analysis. You may want to skip this and come back to the various sections when they are referenced during the tutorial. We assume that the reader is familiar with the fingerprinting technique by restriction digest (Marra *et al.*, 1997).

2.1 Tolerance and Cutoff

The bands of two clones are compared to determine the probability that the two clones overlap. The FPC assembly algorithm uses two user-defined variables for measuring clone overlap: tolerance and cutoff. The tolerance determines how closely two bands must match to consider them the same band. If you are using migration rates, a fixed tolerance is used; that is, the same tolerance is used regardless of the value. If you are using sizes, a variable tolerance is generally used; see Soderlund *et al.*, (1997a). The probability that the matching bands are just a coincidence is computed, and the cutoff value is a threshold on the probability score. If the result of the equation is *below* the cutoff, the two clones are said to overlap, *i.e.* the matching bands are less likely to be a coincidence. The cutoff is expressed in scientific notation: a 1e-03 is the same as 0.001 and 1e-05 is 0.00001. A higher exponent is a lower score; a lower score is a higher stringency. We will usually refer to a high or low stringency when discussing the cutoff value. The equation that is used for comparing two clones is stated as follows:

$$\sum_{m=M}^{nL} \left[\binom{nL}{m} \left((1-p)^m p^{nL-m} \right) \right]$$

where $p = (1-b)^{nH}$, $b = 2t/gellen$, t is the tolerance, *gellen* is the number of possible values for bands, nL and nH are the minimum and maximum number of bands for the two clones ($nL<nH$), and M is the number of shared bands. Since the tolerance is used in the equation, it is desirable to set it at the beginning of your analysis and never change it; a change requires reassembly of the entire database. Making the tolerance more stringent can make the coincidence score more or LESS stringent (see Table I). Since the number of bands is also used in the equation, two clone pairs with the same number of matching bands may have two very different probabilities of coincidence (see Table II).

Table I. Tolerance Varies Independent of the Probability of Coincidence. In Set A, the lower tolerance produces a lower probability of coincidence. In Set B, the higher tolerance produces a lower probability of coincidence.

		Number of Bands			
Set	Tol	Clone 1	Clone 2	Matching Bands	Prob. of Coin.
A	7	22	15	11	3e-09
	5	22	15	11	3e-11
B	7	22	36	18	3e-12
	5	22	36	14	3e-09

Table II. Number of Matching Bands Versus the Probability of Coincidence. Even though the two clone pairs have the same number of matching bands, they have different probabilities of coincidence.

Number of Bands			
Clone 1	Clone 2	Matching Bands	Prob. of Coin.
52	38	12	3e-02
52	16	12	3e-06

2.2 CB Maps and CB Units

FPC orders overlapping clones and puts them into contigs based on the probability of coincidence scores. As it orders the clones, it tries to order the bands to provide a more precise definition of the endpoints of clones. As shown in Soderlund *et al.,* (2000), better data yields more precise endpoints. Even with the high quality data being produced today, much ambiguity remains: (1) two bands may have the same length, but be different, (2) two bands may have values where the difference is outside the tolerance, but be the same, (3) bands may be missing, and (4) there may be extra bands, for example, many digests result in end bands. Therefore, slippage occurs in the endpoints; but unless the data is of especially low quality or contains Q clones, clones that are supposed to overlap based on the cutoff do overlap. Also, the algorithm is greedy – that is, to save time, it does not try all possible combinations. Consequently, it cannot guarantee the best solution; it tries ten different solutions, each time starting with a different clone that generates a different solution, and takes the best of the ten.

The ordering of clones and their fragments is called a Consensus Bands (CB) map, as shown in Figure 1. The coordinate system used in the contig display is in *CB units*: each distinct band is one unit of measurement. The length of each clone is equal to the number of bands in the clone. Endpoint coordinates are assigned as follows: N is equal to the number of bands in the clone divided by 2, M is the midpoint of the location of the clone in the CB map, $M-N$ is the left coordinate of the clone and $M+N$ is the right coordinate of the clone. The left endpoint of the contig is set to zero, but

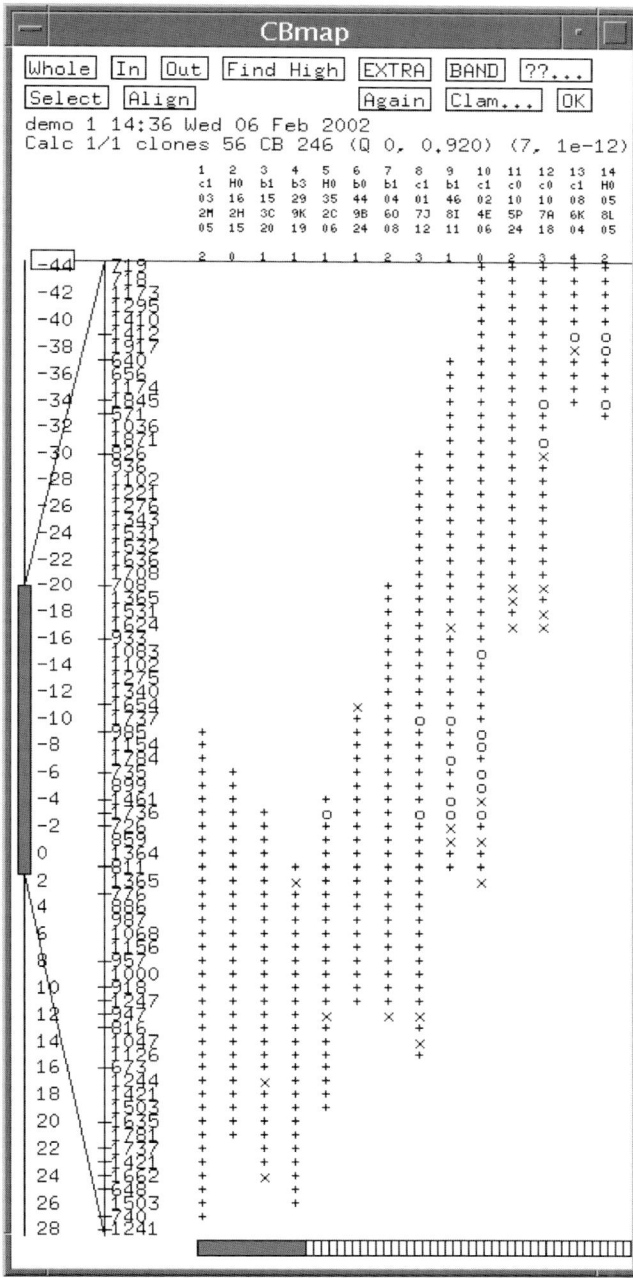

Figure 1. A CB map displayed in FPC. The consensus bands are shown along the left. The tick marks represent partially ordered groups. The {+,x,o} character columns represent the clones. A '+' indicates a match with the band to the left within the tolerance, a 'x' indicates a match within twice the tolerance, and a 'o' indicates no match. The number of extra bands for each clone is listed under the clone name.

can go negative. Note that the coordinates do not have any meaning relative to the chromosome until they are mapped by a framework marker.

2.3 Q Clones

A large number of Q clones generally result from one or more false positive overlaps. Say clone x from contig A falsely overlaps with a clone y from contig B. As clones are being added to the CB map from contig A, when clone x is added, it brings in clone y, which in turn brings in all of contig B. Since there is no way to provide a linear order for two contigs in the same space, the clones in the second contig end up in a stack (see Figure 2). The CB map software recognizes that it cannot order the bands for these clones, and consequently marks them as Q clones. In assembly, a low stringency cutoff results in contigs with many Q clones (*i.e.* many false positives); a high stringency cutoff results in too many contigs (*i.e.* many false negatives). Empirical evidence shows that for BAC clones with an average of 28-35 bands, a 1e-12 cutoff works well to minimize the number of contigs and contigs with many Q clones. NB It is not unusual to have a few Q clones in a contig due to poor fingerprints or as a result of the greedy nature of the assembly algorithm.

3. Getting Started

3.1 Some Unix Basics

FPC runs under Unix and Linux. An extensive knowledge of Unix is not necessary to use FPC effectively. Users must know how to logon to a Unix terminal, and perhaps have a basic knowledge of the directory structure. In this tutorial, any necessary commands are given as they are used. Two basic commands to know are cd, which changes a directory, and ls, which lists all files in the current directory.

3.2 Installing FPC

If FPC is not installed on your system, ask your system administrator to download a FPC executable from http://www.genome.arizona.edu/fpc/ and place it in a shared area for all users to access. Currently, executables for SPARC Solaris, x86 Solaris, Linux, and Dec Alpha are provided. If none of these match your machine type, you will need to have your system administrator download the source code and compile an executable in order to run FPC.

(a)

(b)

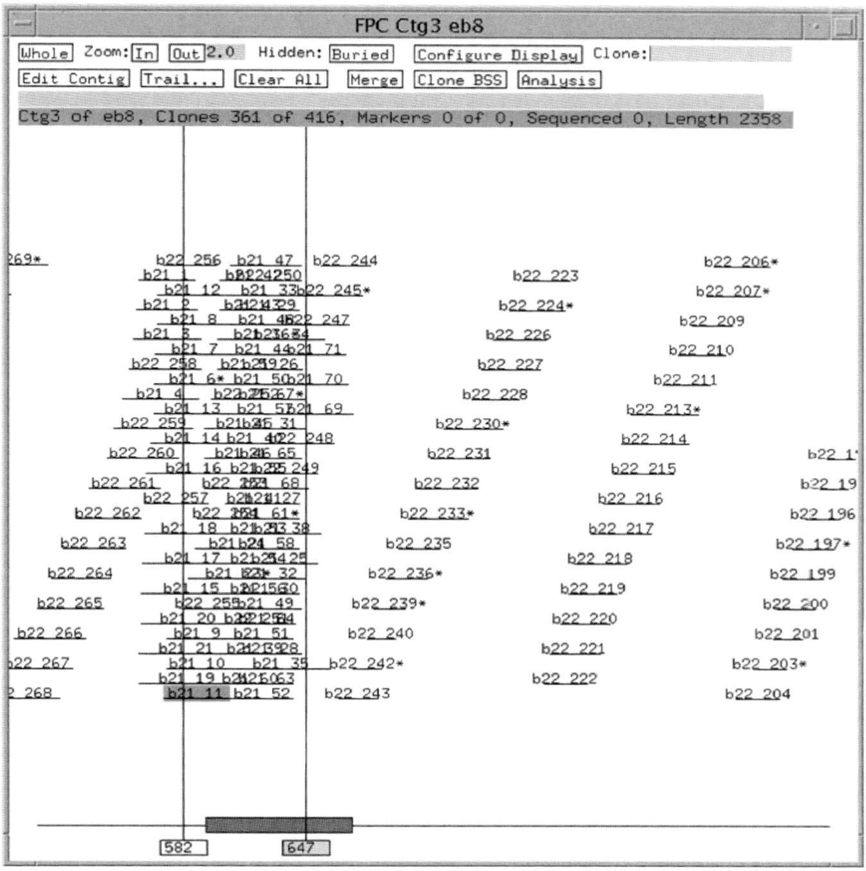

Figure 2. Q clones in FPC. (a) Shows how Q clones get created. A clone from one contig overlaps with a clone from another contig. The circle shows where the overlap occurs. Since no other clones overlap, they cannot accurately be placed on a two-dimensional map. Hence they are placed in a stack. (b) Shows a contig that has a lot of Q clones. These are visually obvious as a thick stack of clones on the display.

3.3 Downloading the Demo Files

Download `demo.tar` from http://www.genome.arizona.edu/software/
fpc/download. When this file has finished downloading, type `tar  xvf`
`demo.tar` on the command line and press return. This action creates a
directory called `demo` in your current directory. Type `cd  demo` to move
into that directory. Then type `ls`. The following files should be listed:

```
copyNew.pl   files    Image       Sizes
cleanup.pl   Gel      Newbands
```

`Image,  Sizes`, and `Gel` are directories, and the files in them are
generated from the Image program (see www.sanger.ac.uk/Software/
Image). If at any time during the tutorial you wish to bring the demo back
to its initial condition, type `cleanup.pl` on the command line while in
the `demo` directory. This will restore all files and directories back to their
original condition so you can restart the demo.

4. Building a Physical Map with FPC

4.1 Creating a New Project

The commands covered in this section:
- **Create new project (File...)**
- **Update .cor**
- **Build Contigs** (from Main Analysis window)
- Buried clones and contig navigation
- **Clean up**

From the `demo` directory, start FPC by typing `fpc` on the command line.
The **Main Menu** window appears (see Figure 3). Right-click on the button
labeled **File**... and a menu appears. Select **Create new project** from
the menu and a window appears as is shown in Figure 4. Choose a name
for your project and type it in the **File**: text entry. For this demo, type the
name "demo". **Click OK**; a `demo.fpc` file is created, and the following is
written to the terminal window:

```
Adding Bands Directory
New project is initalized.
```

Now click on the **Update .cor** button on the Main Menu window. This
function moves all migration rate files from the `Image` directory to a newly

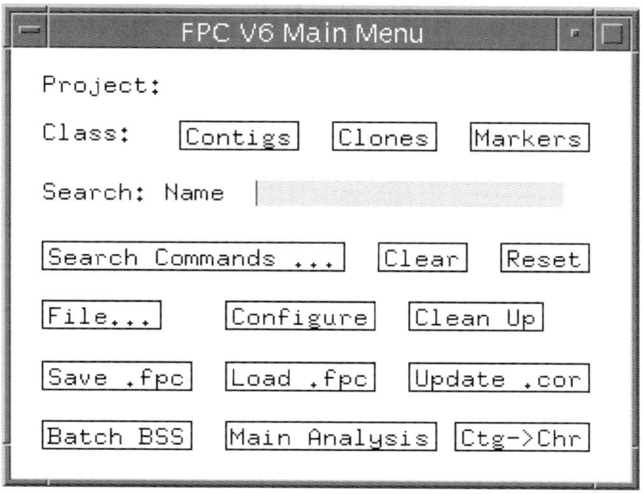

Figure 3. FPC's Main Menu window. This is the first window you see when you start FPC.

Figure 4. The File Chooser. This window is used to select files read or write. In this window, the new file demo.fpc is being created.

created Bands directory. It also creates the file demo.cor, which is the file FPC uses to read the migration rates of clones. When this function has completed, the last few lines written on your terminal window will be (with your path name substituted for /u/efriedr):

```
Read 311 files. Add 345 gel entries and 9456 bands.
Cor file has 9456 bands.
Saving File /u/efriedr/demo/demo.fpc ......Done
```

Click on the **Main Analysis** button on the Main Menu window and the Main Analysis window opens (see Figure 5). Change the cutoff to 1e-12 in the Cutoff text box. Leave all other values unchanged. Next, click on the **Build Contigs (Kill/Calc/OkAll)** button. This starts the map-building process. This process may take several hours for large clone libraries, but for the demo, it should only take a few seconds. When this process completes, the last few lines on your terminal window will be:

```
Ignore 5 AvgOverlap 3.8 AvgScore 0.868 Qs 30(1)
Create 4 contigs (from ctg1 to ctg4)
Contig sizes: Max 121, 4 (>25), 0 (25:10), 0
(9:3), 0 (=2), 5 Singles
```

Note, the number of Q clones may be 30 or 31. The Project window pops up, as shown in Figure 6a. The assembly resulted in four contigs. Double-click on the row of contig 2 and it is displayed. Not all the clones are shown

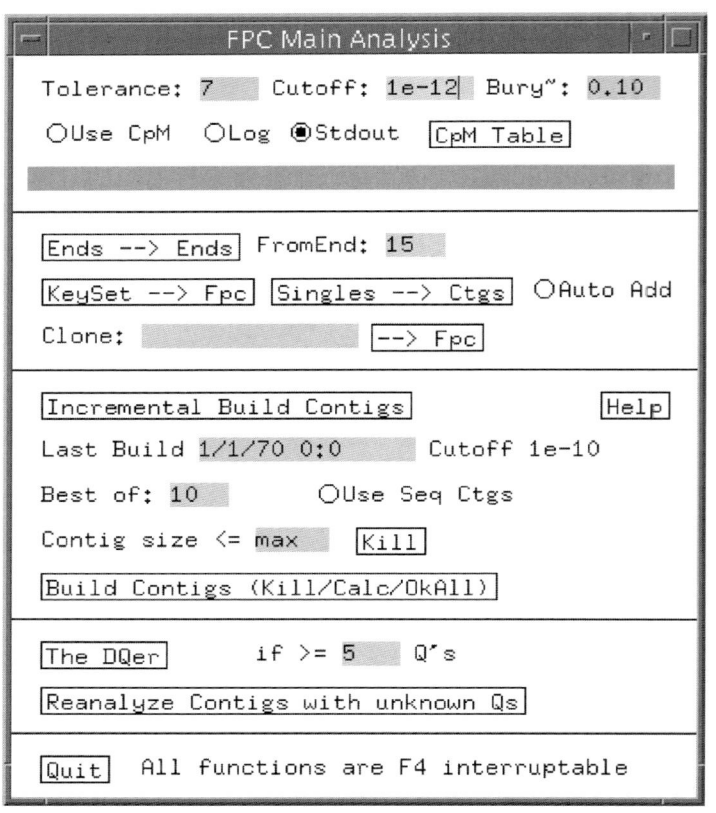

Figure 5. The Main Analysis window. From this window, you build initial maps, or incorporate new clones into an existing map.

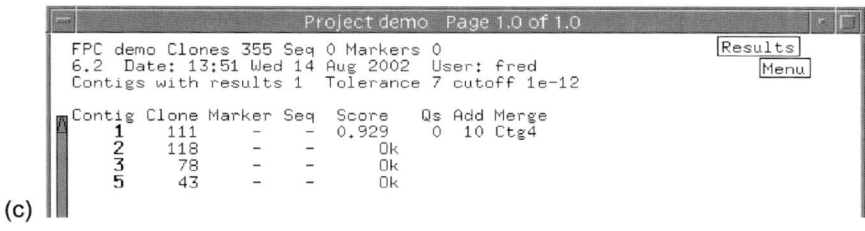

```
                    Project demo   Page 1.0 of 1.0
 FPC demo Clones 345 Seq 0 Markers 0                      Results
 6.2  Date: 13:15 Wed 14 Aug 2002  User: fred             Menu
 Contigs with results 4  Tolerance 7 cutoff 1e-12

 Contig Clone Marker Seq   Score   Qs
    1      61    -    -    0.933    0
    2     118    -    -    0.900    0
    3     121    -    -    0.701   30
    4      40    -    -    0.937    0
(a)
```

```
                    Project demo   Page 1.0 of 1.0
 FPC demo Clones 345 Seq 0 Markers 0                      By ctg..
 6.2  Date: 13:45 Wed 14 Aug 2002  User: fred             Menu
 Contigs 5  Max Contig 5

 Contig Clone Marker Seq        Date      Status  Qs Remark
    0      5     -    -    14aug02 13:15    Ok    -
    1     61     -    -    14aug02 13:15    Ok    0
    2    118     -    -    14aug02 13:15    Ok    0
    3     78     -    -    14aug02 13:44    Ok    0 Split 1e-13.
    4     40     -    -    14aug02 13:15    Ok    0
    5     43     -    -    14aug02 13:44    Ok    0 From Ctg3.
(b)
```

```
                    Project demo   Page 1.0 of 1.0
 FPC demo Clones 355 Seq 0 Markers 0                      Results
 6.2  Date: 13:51 Wed 14 Aug 2002  User: fred             Menu
 Contigs with results 1  Tolerance 7 cutoff 1e-12

 Contig Clone Marker Seq   Score   Qs Add Merge
    1     111    -    -    0.929    0  10 Ctg4
    2     118    -    -           Ok
    3      78    -    -           Ok
    5      43    -    -           Ok
(c)
```

Figure 6. The Project window. (a) Shows the window after an initial build, (b) after the DQer was run, and (c) after new clones were incorporated after an initial build. Note the change in Q clones from (a) to (b), and the merging of contigs from (b) to (c).

as redundant clones are *buried*. Click the button called **Buried** beside the label **Hidden**, it toggles to **None** and all the clones are shown (see Figure 7). If a clone has a set of bands similar to another, it can be buried in the second clone. Click on a clone that has an '*' at the end of the clone name; the '*' implies that it has buried clones and the buried clones are highlighted. A clone ending with a '=' has all the same bands as the parent clone. A clone ending with a '~' has approximately the same set of bands as the parent clone. Clicking on the **None** button changes it to **Pseudo**, which has no effect in this demo. We will not cover pseudo clones in this tutorial. Click on the button again to switch back to the buried state. You can zoom in and out by clicking on the appropriate buttons at the top. The text entry beside the buttons determines the zoom factor. Select **In** a couple of times. To scroll around the contig display, move the mouse pointer

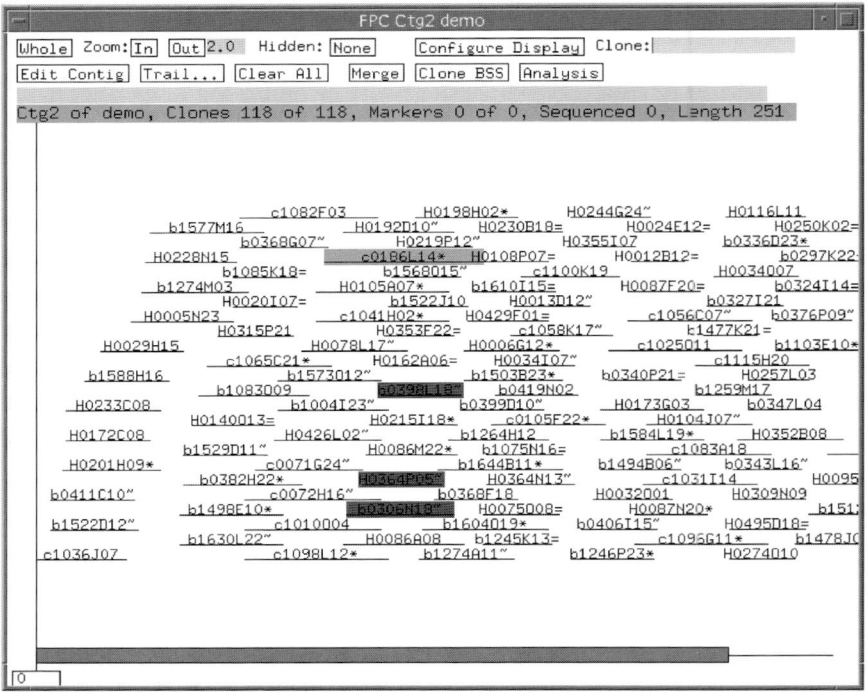

Figure 7. The Contig display. The Hidden state has been toggled from **Buried** to **None**, so all clones are shown. Selecting a clone shows the buried clones highlighted green.

towards the right of the map and click on the middle mouse button (feature not available on two-button mouse). The map scrolls to the left. To move back, position the pointer towards the left of the map and click. You can alternatively use the green scrollbar at the bottom of the display.

Experiment running **Build Contigs** with cutoffs above and below this value (*e.g.*. 1e-8, 1e-16) and examine the results to see the effect on the number of contigs and contigs with Q clones (as discussed in the *Analysis* section). When you are done experimenting, reassemble at 1e-12 before going on with this demo. To close any window in FPC, right-click on an empty space in the window; from the pop-up menu, select **Quit**. To close all windows at once except the Main Menu window, select **Clean Up** on the Main Menu.

4.2 The DQer

Commands covered in this section:
- **The DQer** (from Main Analysis)
- **Save .fpc**

Open the Project window by double-clicking on the bold-faced project name (**demo**) at the top of the main menu. Look at the column with the heading 'Qs' on the Project window. Notice that three contigs have a 0 in this column, while contig 3 has a 30. (Contig 0 contains all clones that could not be placed in the map; hence, Q clones do not apply.) After an initial build with a moderate cutoff, we need to take the contigs with many Q's and re-run them at a lower cutoff. The DQer performs this function automatically. Click the **Main Analysis** button from the Main Menu. Towards the bottom you will see a button labeled **The DQer** with a text entry to the right of it. The value in the text entry determines how many Q's a contig must have in order to be re-evaluated. Empirical evidence has shown that a value around 5 yields good results. Click on the **The DQer** button and the reanalyzation starts. The contigs with Q's above the cutoff are reassembled up to three times, where each time the cutoff is lowered by a factor of 10 (*e.g.* 1e-13, 1e-14, 1e-15). The assembly may result in one or more CB maps (see *Analysis* section). The software tries to merge the CB maps by comparing the end clones at a lower stringency. If the CB maps cannot be merged, one or more new contigs are created. When the DQer is done, the Project window pops to the front (see Figure 6b). Contig 3 has been split into contigs 3 and 5. NB Sometimes a contig is split into multiple contigs, yet there are still Q clones; in this case, the number of Qs for each resulting contig will be '-' as it is not known which contigs have Qs. Also, a contig with many Qs may not change if lowering the cutoff 3-fold does not make a difference; that is, when all clones remain in the same contig and the number of Qs remains high. This indicates a very repetitive fingerprint.

Save the current contigs by clicking on the **Save .fpc** button on the Main Menu window. The **ONLY** time the FPC project is automatically saved is after an **Update .cor.** Therefore, whenever you have made some changes that you want saved, do so immediately. You can save any number of times during an FPC session. The benefit of saving often is that if you make a mistake (*e.g.* merge two contigs) and then decide you did not really want to do that, you can quit and restart FPC from your last save. Right-click in an empty space on the Main Menu and select **Exit** from the pop-up menu to exit FPC. Type **ls** on the command line to see the new files created by FPC. The following should now be listed:

```
Bands           demo.cor.backup    files    Sizes
cleanup.pl      demo.fpc           Gel
copyNew.pl      demo.fpc.backup    Image
demo.cor        demo.fpp           Newbands
```

Incremental Builds

Commands covered in this section:
- Adding additional clones to an existing FPC project.
- **Incremental Build Contigs** (IBC, on Main Analysis window)

We will now add some additional clones to our FPC project. Generally, as new gels are band-called, Image places the files in the Image, Sizes, and Gel directories. For this demo, a new set of files was temporarily put into the Newbands directory, and the files can be moved to the correct locations using the copyNew.pl perl script. From the demo directory, type copyNew.pl on the command line to copy the files from the Newbands directory into their respective Image, Gel, and Sizes directories. Thereafter, launch FPC with the previously created project (we called it "demo") by typing fpc demo on the command line. When the Main Menu window appears, click on the **Update .cor** button. This will copy the files from the Image directory to the Bands directory, and it updates the demo.cor file with the new migration rates. We are now ready to add the new clones to our map. Open the Main Analysis window. The cutoff should still be set at 1e-12. Click on the **Incremental Build Contigs** button. This adds the new clones to our map and merges contigs if the new information allows us to do so. When the build is done, the Project window will pop to the front (see Figure 6c). Contigs 1 and 4 have been merged indicating that one or more of the new clones hits both. Save the new map by clicking on the **Save .fpc** button on the Main Menu.

5. Adding Remarks and Markers

5.1 Manually Adding Remarks

Commands covered in this section:
- Finding a clone in a contig
- Viewing the Clone text window
- Editing the clone

Open the Contig display for contig 2 (via the Project window). Towards the top right, in the text entry labeled **Clone:**, type *b0297K22* and hit Return. The clone is found in the contig, so it will be highlighted as shown in Figure 8a. Click on the highlighted clone and the Clone window opens (see Figure 8b). Click on the **Edit** button in the top left corner and the Edit Clone window opens (see Figure 8c). Here we can change the attributes of the clone, including attaching remarks. In the text entry titled **Remarks:**, type *test_remark* to add that remark to the clone. Click on **Accept**. The Clone Edit window closes, and our newly added remark is shown in the contig display and in the Clone text box. Clicking on the remark highlights the clone and vice versa. Select **Clean up** on the Main Menu before going on to the next section.

Adding remarks and markers from a file

Commands covered in this section:

- **Merge remarks** (from Main Menu/File menu)
- **Replace markers (fw and seq)** (from Main Menu/File menu)
- **Framework** (from Project window)
- **By framework order** (from Project window/Menu window)

Often remarks can be generated from an external file, in which case it is faster to automatically add them all at once. Hence, FPC provides features for adding a list of remarks from an external file. The text file is a list with entries such as the following:

```
BAC : "b1046D08"
Remark "new_clone"
```

Look at `remarks.ace` in the `files` directory for an example. On the Main Menu, right-click on the **File...** button and select **Merge remarks** from the drop down menu. This opens the File Chooser window. Double-click on `files/` in the left-hand column. Then double-click on the `remarks.ace` file that appears in the right-hand column. This action adds all remarks in the `remarks.ace` file to our project. These particular remarks note which clones were added after the initial build by adding the remark *new_add* to those clones. Open the contig display for contig 1 to see where the remarks were placed.

A typical scenario for markers is that the clone/marker results are entered into a simple marker database or spreadsheet. A perl script is written to convert the format to the FPC marker file format. As markers are incrementally added to the spreadsheet, all markers are periodically dumped and input into FPC using the **Replace markers** function. The advantage of replacement is that any deleted markers or markers removed from clones from the external marker database are also deleted in FPC. Each time the marker file is read, if a framework file exists, it is also read. This is a file of ordered markers, generally from a genetic or radiation hybrid map, which orders the contigs. As new markers are added, this file is re-read to see if any new frameworks can be added; a framework marker can go into FPC only if it is attached to a clone. The markers file is

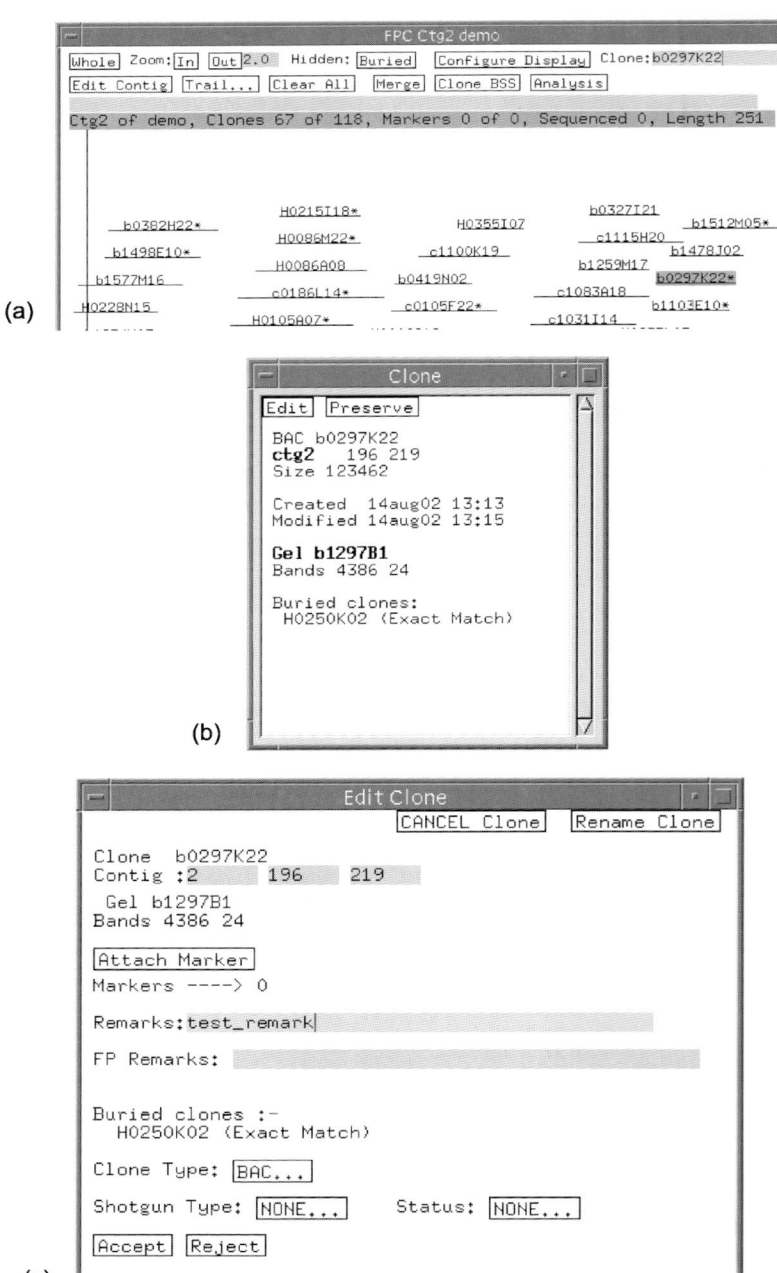

Figure 8. (a) Click the highlighted clone to bring up the Clone text window. (b) The Clone text window. (c) The Edit Clone window.

structured like the remarks file; see `markers.ace` file for an example. Each framework file uses the same name as the marker file, with the `.ace` suffix replaced with `.fw`. Each entry contains three items: 1) marker name, 2) map name, and 3) marker position. To read in the markers and framework file, right-click on **File...** on the Main Menu and select **Replace markers (fw and seq)**. From the File Chooser window, double click `markers.ace`. The markers and framework (called **markers.fw**) are read into our project. When this completes, open the Project window and right-click on the button in the top right corner. Select **Framework** from the menu. The framework markers are shown in order, and the contigs containing these markers are shown in the rightmost column. The contigs are ordered based on the framework. To renumber the contigs based on this new framework ordering, click on the **Menu** button in the top right corner, and then click on the **By framework order** button under **Re-number contigs** from the window that appears. Select **Yes** when the **Move remaining contigs up?** dialog appears.

Select the **B13** marker from the framework project window and select **ctg2** from the Marker text window (see Figure 9a). The framework markers are shown along the bottom of the contig display, while all framework and non-framework markers are displayed along the top (see Figure 9b). Even when only a small region of the contig is displayed, all framework markers are always shown along the bottom. You can center on the region of a framework by clicking on it in the bottom part of the display.

NB If you have chromosome numbers for your framework markers, make that the first position of 4 digits; *e.g.* 1001.1 is position 1.1 on chromosome 1, 12001.1 is position 1.1 on chromosome 12. The 'map name' is the same for all entries.

NB The 'seq' part of the 'fw and seq' refers to reading an optional sequence file to update the sequence status of clones. The option is covered in the User's Manual.

6. Searching

Commands covered in this section:
- Select keyset type: Contigs, Clones, or Markers
- Search by substring of name
- Search by remark
- Search by date
- Add remark to all clones in the keyset

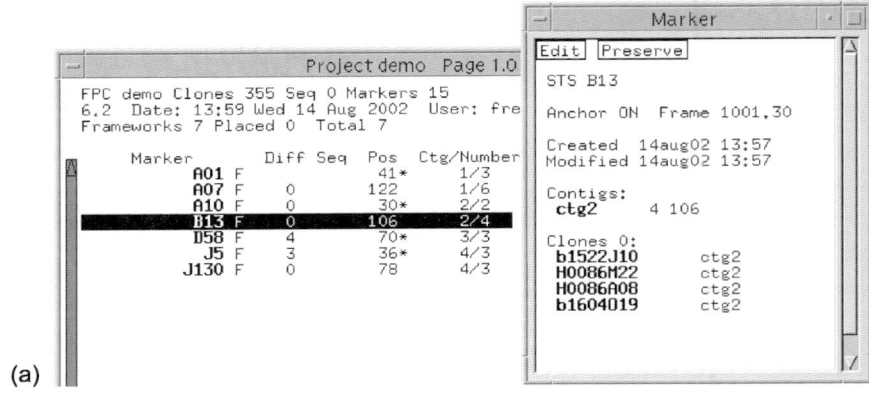

(a)

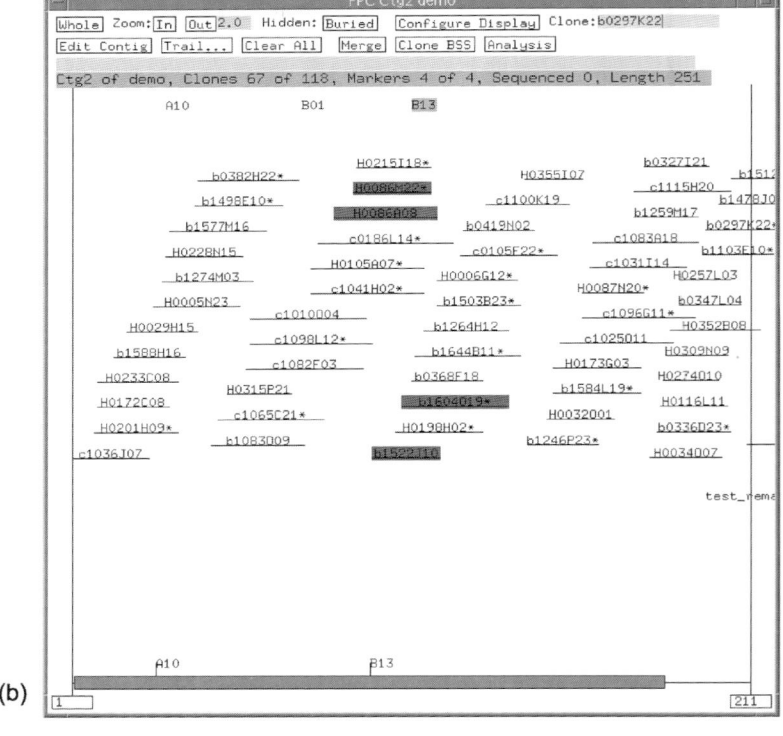

(b)

Figure 9. (a) The Framework window and a Marker window for a framework marker. (b) Contig display after adding markers and frameworks. The markers are displayed towards the top of the window, while the framework is displayed along the bottom.

- Keyset (from Edit Contig window) - show clones from keyset in contig.
- Configure Display (from Contig display) - turn on Fp_remarks.

Using aceDB terminology, there are three *classes* of data in FPC: contigs, clones, and markers. A subset of a class can be shown as a *keyset* of items. The three buttons on the Main Menu, labeled **Contigs, Clones,** and **Markers**, determine which class is searched. A class is selected for searching by clicking on the corresponding button, which highlights it in blue. Once a keyset is displayed, the next search of that class is performed on the existing keyset. Consequently, you can search for multiple conditions; *e.g.* first search for all clones added after a given date, and then search that keyset for all clones in a given contig.

Select the **Contigs** class. On the Main Menu is the label Search:, next to this is the search type, by default **Name**, and next to this is a text box. Type a 1 in the text entry and press return (or click on the Contig button). Contig 1 is shown. Select **Clear** and the text in the Search text box will disappear. Double click the **Contigs** class; all the contigs will be shown in the keyset.

Select the **Markers** class and type *A07* in the text box. The marker window for that marker pops up. Double click on the bold-faced contig number (**ctg1**) to see where the marker is positioned. Next, select **Clear** and then type *A** in the Search text entry. This brings up the Keyset window containing a list of all markers starting with the letter "A." Double-clicking on any marker name will bring up the Marker window for that marker.

Select the **Clones** class and search for clone *c1086K04*, which will open the Clone window. Once again, double-click on the contig number to see the position of the clone in the contig. Next, select **Clear** then type *b**. This brings up a keyset containing all clones starting with "b." To see the distribution of "b" clones among all contigs, open the Project window, right-click on the upper right button (labeled **By ctg...**), select **By keyset**. This gives us the number of "b" clones in each contig sorted in descending order. Now, suppose that we want to see all clones in contig 3 that start with "b." Open contig 3 from the Project window. Click on the **Edit Contig** button in the top left corner. From the window that opens, under the label **Select:**, click the **Keyset** button. All "b" clones are selected, *i.e.* shown in blue. Figure 10 shows us the relation between the Keyset, Project, and Contig windows.

Our next search involves searching for clones containing the *new_add* remark that we attached to all clones added after the initial build. First, reset the keyset to all clones in the contig by selecting **Reset** from the Main Menu. Right-click on the **Search Commands...** button and select **Remark** from the menu. In the text entry, type *new_add*. The

```
Clones 150 (page 1/5)

    b0297K22   b0325O10   b0340P21   b0368G07

    b0303H17   b0326L07   b0341J19   b0372I14

    b0306N07   b0327I21   b0341K22   b0376P09

    b0306N18   b0332M08   b0341O09   b0377M02

    b0307M15   b0333F08   b0343C02   b0379F02

    b0316M15   b0334C23   b0343L16   b0380N03

    b0319B18   b0336D23   b0347L04   b0382H22

    b0324I14   b0336J22   b0368F18   b0382N02
```

(a)

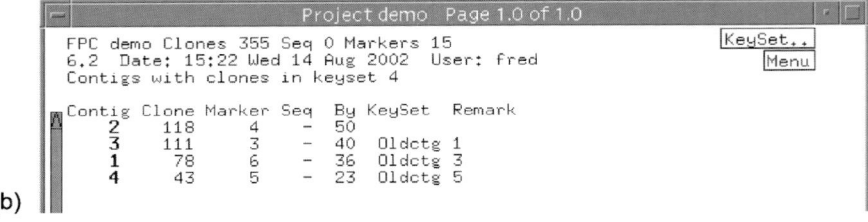

```
                    Project demo   Page 1.0 of 1.0              KeySet..
FPC demo Clones 355 Seq 0 Markers 15                            Menu
6.2  Date: 15:22 Wed 14 Aug 2002  User: fred
Contigs with clones in keyset 4

Contig Clone Marker Seq  By KeySet  Remark
  2     118    4    -  50
  3     111    3    -  40  Oldctg 1
  1     78     6    -  36  Oldctg 3
  4     43     5    -  23  Oldctg 5
```

(b)

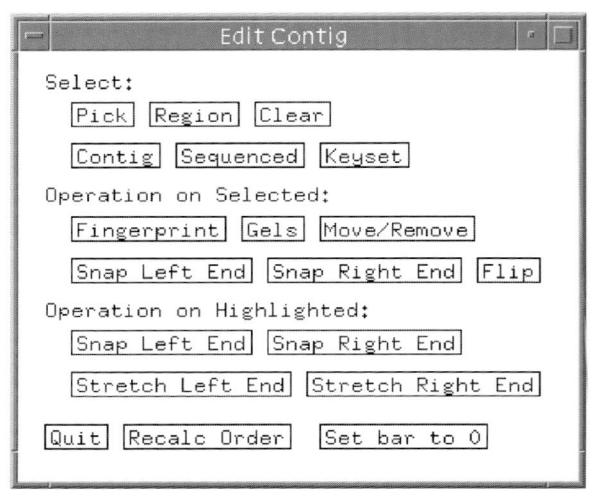

```
                    Edit Contig

   Select:
       Pick  Region  Clear

       Contig  Sequenced  Keyset

   Operation on Selected:
       Fingerprint  Gels  Move/Remove

       Snap Left End  Snap Right End  Flip

   Operation on Highlighted:
       Snap Left End  Snap Right End

       Stretch Left End  Stretch Right End

   Quit  Recalc Order   Set bar to 0
```

(c)

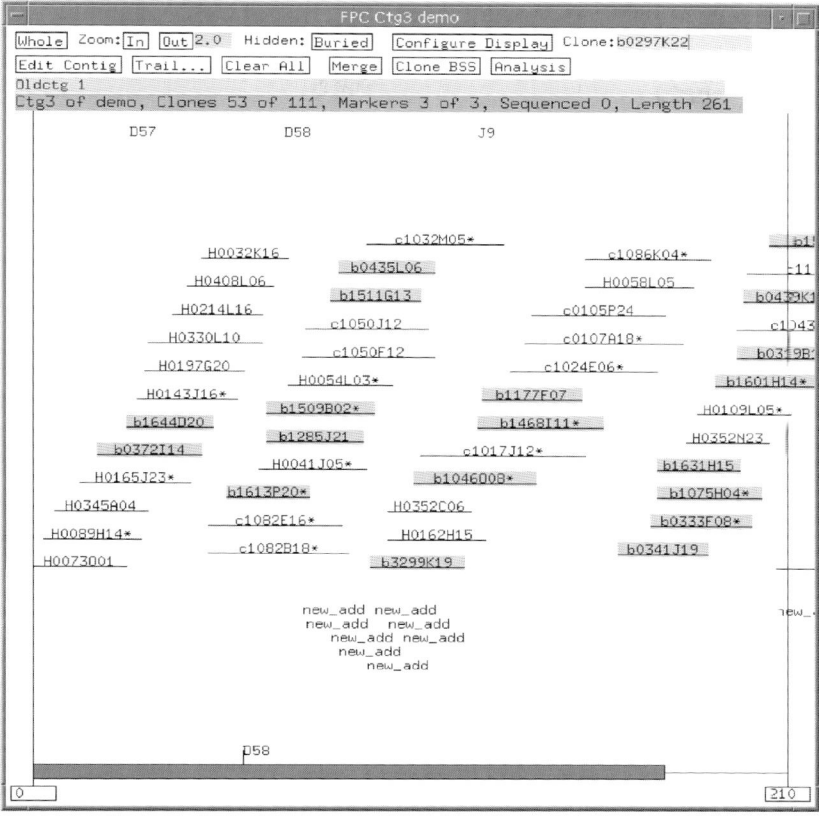

(d)

Figure 10. Viewing keysets in the Contig display. (a) The keyset shown is a subset of all the clones. (b) The Project window displays the number of clones from the keyset that are in each contig. Double-clicking on the row for contig 1 brings up the Contig display. (c) From the Edit Contig window, click on **Keyset** to select all clones from the keyset that are in contig 1. (d) All clones from the keyset in contig 1 are selected (blue).

Keyset window gives us all ten clones containing that remark. Looking at the keyset values in the Project window shows us that all of our new clones were added to contig 3. Open that Contig display and select all newly added clones via Edit Contig window.

Our final search involves searching for clones by date and time. We will find all clones that were added after the initial build by searching for all clones that were created after a specified time. Since the creation time of the added clones is very close to the creation time of the initial clones (unless you added the initial clones one day, and the additional clones on a subsequent day), we need to include the time when specifying the date.

Reset the clone keyset to show all clones. Double click the first one. It does not have the remark *new_add*; note the creation date and time of this clone. On the Main Menu, select **After Create Date** from the **Search Commands...** menu. In the Search text entry, type in the date and time such that the time is at least one minute **AFTER** the creation time of the clone. Type the date and time in the format dd/mm/yy hh:mm. (NB this is the European date format - the day comes first.) Press return. For example, in Figure 8b, the created day and time is '14aug02 13:13' so we would enter the time as '14/8/02 13:14'. The Keyset window opens, listing all clones created after this date and time. Select a clone and make sure it has the *new_add* remark. Now, suppose that we want to add a remark to the two clones in this set that start with an "H." Without closing the Keyset window, select **Name** from **Search Commands...** on the Main Menu. In the Search text entry, type *H**. Now, only those clones from the former set that started with an "H" are shown in the keyset. Right-click anywhere in the white space on the Keyset window, and select **Add Fp Remark** from the pull-down menu. The **Add Fp remark** window pops up. Type in any remark up to ten characters long and click on **Add fp_remark**. Next, double-click on one of the clones in the keyset, and then double-click on the bold contig number in the Clone window to open the Contig display. Note that the new remarks are not present; this is because there are two types of clone remarks: Remarks and Fp_remarks (see Figure 11a). Fp_remarks are typically less important; *i.e.* notes about the clones that are not typically viewed. Click on **Configure Display** at the top, and turn on the **Fp Remarks** switch (see Figure 11b). Then click **Redraw Map**. The newly added Fp_remarks are shown along the bottom of the Contig display along with the other remarks.

With the *left* button, select **Search Commands....** The Clone Commands window containing various types of searches is shown. Many of these are intuitive. The ones we use most are **Multiple Fingerprints**, which show all the clones that have multiple gels (none in this set). The **Selected** option makes a keyset of all the selected clones in the current contig, where a clone can be selected by clicking on its name with the right button and then clicking **Selected**. The selected set can be cleared by clicking **Clear All** on the contig window.

7. Finishing a Project

7.1 Merging Contigs

Commands covered in this section:
- Ends→Ends (Main Analysis window)

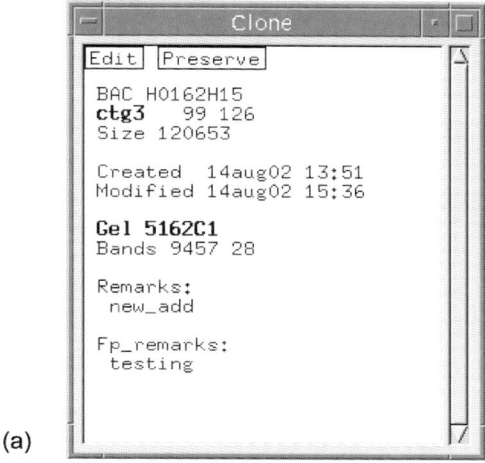

(a)

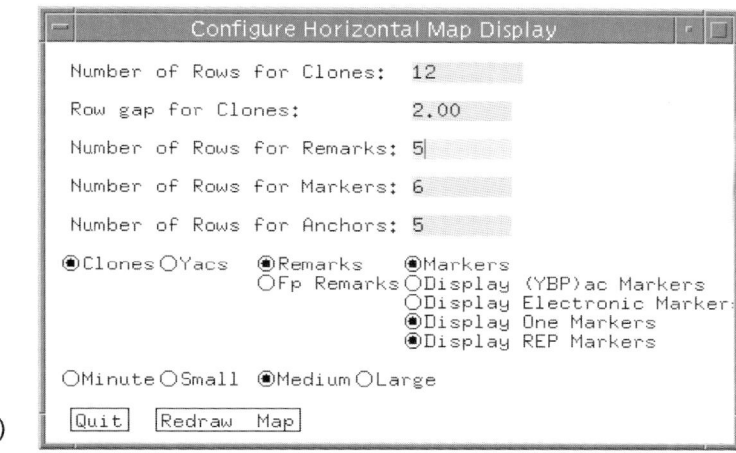

(b)

Figure 11. (a) Clone window showning the two types of remarks. (b) The Contig Configure display window allows the display of different entity types to be turned on and off.

- CpM table (Cutoff plus Markers for scoring overlap, Main Analysis window)
- **Merge** (Contig display)

After the majority of the data is entered into FPC, it is advantageous to find contigs that can be merged. An easy way of finding candidates for merging is to lower the stringency and only compare clones close to the ends of contigs. Lower the stringency by setting the cutoff to a 1e-10. We will use the CpM table to help us identify contigs to merge. When this table is used, clones that share one or more markers can have a less stringent cutoff and still be considered overlapping. On the Main Analysis window, turn on the CpM table by clicking on the radio button labeled Use **CpM**. On the terminal window, you will see:

```
Cutoff 1e-10 CpM (1 1e-09)(2 1e-08)(3 1e-07)
```

This information is also shown on the CpM window, which you can view by clicking **CpM Table**. Click on the **Ends→Ends** button. When this finishes, the Project window pops up showing us that contigs 1 and 4 can be merged (see Figure 12a). The *RR-1* means that one clone from the right end of contig 4 overlaps with one clone from the right end of contig 1. Open the Contig display for contig 1 and click on the **Merge** button at the top. The Merge contig window opens (Figure 12b). Click on the **Do Not Flip** button, which toggles it to **Flip**. Enter a '4' next to **Contig** and click on **Start merge**. Contig 4 is appended to contig 1 as is shown in Figure 13. Notice the clones from contig 4 are in a smaller font indicating that they are not permanently part of contig 1. Select marker F100 and you will see that it is in a clone in contig 1 and a clone in contig 2. Also, a window appears (see Figure 12c) allowing us to move the two contigs closer together or farther apart. Click on the arrows to move the contigs. A value of -20 gives a reasonable merge. This can be checked by either of the methods described in the *Verify overlap* section. When you are done, click on **Accept merge**. The two contigs permanently become one.

NB We generally turn the CpM table on from the beginning of the project. As new clones and markers are added to FPC, the IBC (Incremental Build Contigs) takes into consideration new markers and reanalyzes those clones with new markers for joins.

7.2 Adding Singletons

Commands covered in this section:
- **Singles→Ctgs** (from Main Analysis window)
- **Show additions** (from Contig Analysis window)

There are times when we want to add clones to our map, but we do not want to merge any contigs. For example, after all data is added and manual merges have been completed, we do not want merges to occur automatically anymore; therefore, we cannot run the IBC. Another time this is used is towards the end of a project when *singletons* are added at a lower stringency and should not be used for merges, where a singleton is a clone that has not been placed in a contig. Hence, FPC provides the capability to

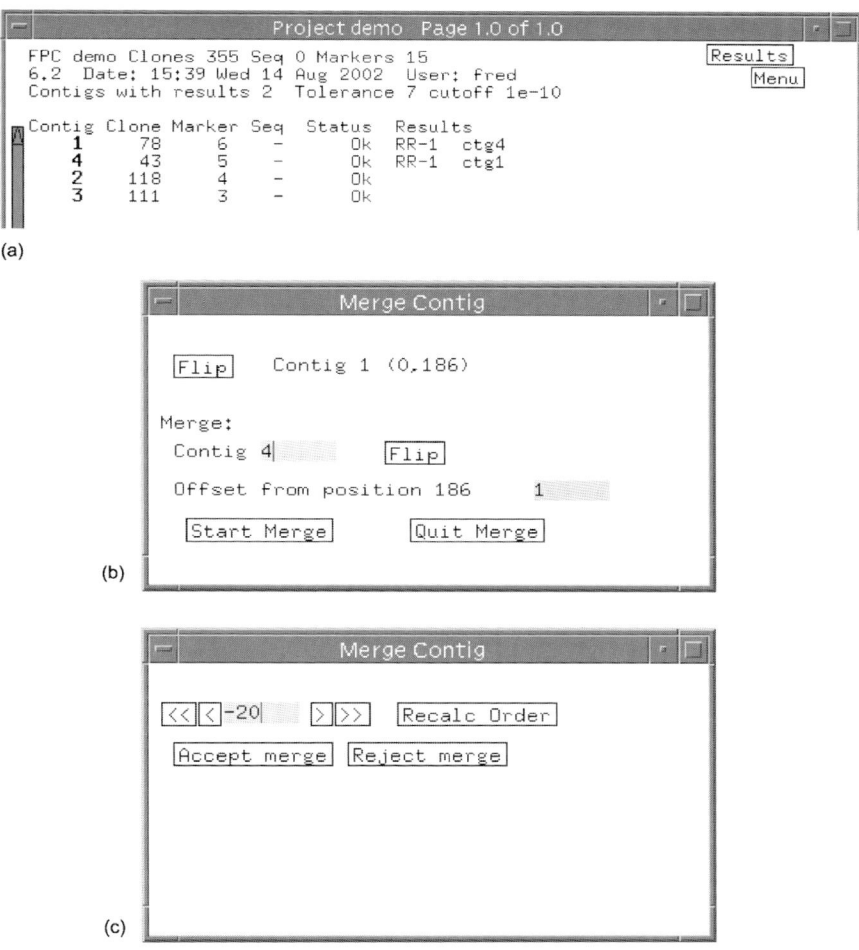

Figure 12. (a) The Project window after an **Ends→Ends** comparison. (b) Choose which contig to merge and whether or not to flip the current contig. (c) Move the newly merged contigs closer together or farther apart.

add singletons to contigs if there exists an overlap with one or more clones in a contig, without doing any merges. The clone is positioned where the best overlap occurs. Users should be warned not to use this function inappropriately; a low stringency cutoff value can add many clones to the map, but they could be positioned incorrectly as there is no global analysis taking place. On the Main Analysis window, set the cutoff to 1e-10, turn on the **Auto Add** radio button, and click on the **Singles→Ctgs** button. All singletons that overlap clones already in a contig at this lower stringency are automatically added to the contig. When the function completes,

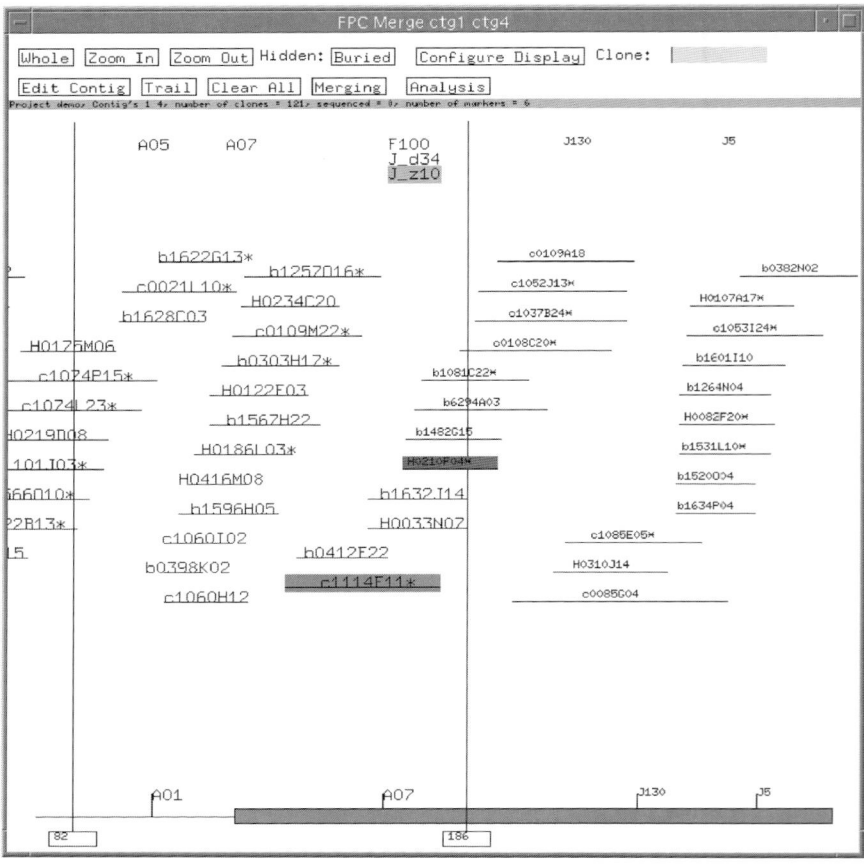

Figure 13. Contig 1 after a merge with contig 4.

the Project window pops up, showing us in which contigs our singletons found overlapping clones. The number in the **Results** column gives us the number of clones that overlap our singletons at the lower stringency. Open the Contig display for contig 1. Click on the **Analysis** button at the top of the window. The Contig Analysis window opens (see Figure 14). Click on the **Show Additions** button. All newly added clones are highlighted in dark blue, so clones *H0278N12* and *H0125D19* should be highlighted in dark blue. NB The added clones may be buried; in which case, toggle the buried state to see them.

Bring up the Project window and select By contig number from the pull-down menu in the top right corner. Note that the number of Qs for contig 1 and contig 3 are set to '-'; this is because the number of Q clones

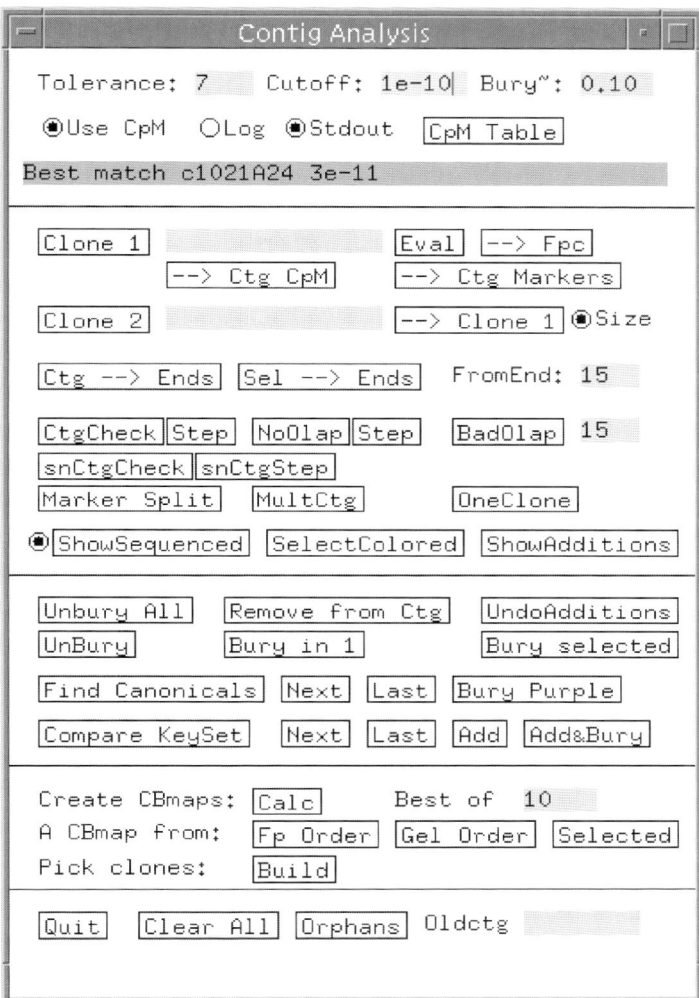

Figure 14. The Contig Analysis window.

is no longer known since the algorithm for building a CB map has not been run with the added clones.

7.3 Verify Overlap

Commands covered in this section:

- **Clone 1→Ctg CpM** (from Contig Analysis, compare clone 1 against contig)

- From Clone text window, bring up Gel Image window
- Gel Image window, add and move clones
- **Clone 1→Clone 2** (from Contig Analysis, compare two clones)

When clones are added at a lower stringency and are not positioned with the automatic analysis, there is a greater risk of a false positive or poorly positioned clone. Therefore, it is worth verifying the addition by looking at the raw fingerprints. It takes a while to learn to be an expert at looking at gel images, but we will take you through one to get you started. First, we will compare the fingerprint of clone *H0278N12* with its immediate neighbors by clicking on the clone and then clicking on the **Clone 1** button in the Contig Analysis window. The clone name will be copied to the text entry. Next, click on the **→Ctg CpM** button. The overlapping clones are highlighted in purple and the following text is displayed on your terminal window:

```
>> - -> Ctg1  H0278N12  28b  32  60  (Tol  7  Cutoff
       1e-10 CpM NoBuried)
    c1021A24 ( 27, 64)         38b 20 3e-11
    b1537B07 ( 31, 65)         35b 19 9e-11      canon
```

Right-click on clone *H0278N12* and select **Gel image** from the pull-down menu that appears. This brings up a window containing the gel image of our clone, with its 'called bands' marked as lines along the side. The numbers along the left side are the scale for migration rates. We need to determine if this banding pattern is similar to that of its neighbors. Bring up the images of the clone's immediate neighbors (*b1537B07* and *c1021A24*) by turning on the **Add** button in the Gel Image window by clicking on it, and then clicking on the neighbor clones in the contig display. The images for the clicked clones will appear in the Gel Image window. Turn off **Add** by clicking on the button again. Now, we will position the clones in the Gel Image window such that our selected clone is flanked by its two neighbors. Turn on **Move** and arrange the gels by dragging and dropping the images such that clone *H0278N12* is in the middle. Then turn off **Move**. Next, click on the name of our newly added clone in the Gel Image window. All bands that have matches with the neighbors are colored in blue, while bands that are not matched remain black. On the neighbors, any bands that match are shown in red. You can also zoom in and out as desired by clicking on the corresponding buttons at the top of the Gel Image window. Zoom out a couple of times to view a similar image as is shown in Figure 15. The GreyRamp tool adjusts the contrast of the gels. If the banding pattern seems plausible, we can keep this clone. The method of analyzing the fingerprints is discussed in greater detail by Chapter 6 in this book.

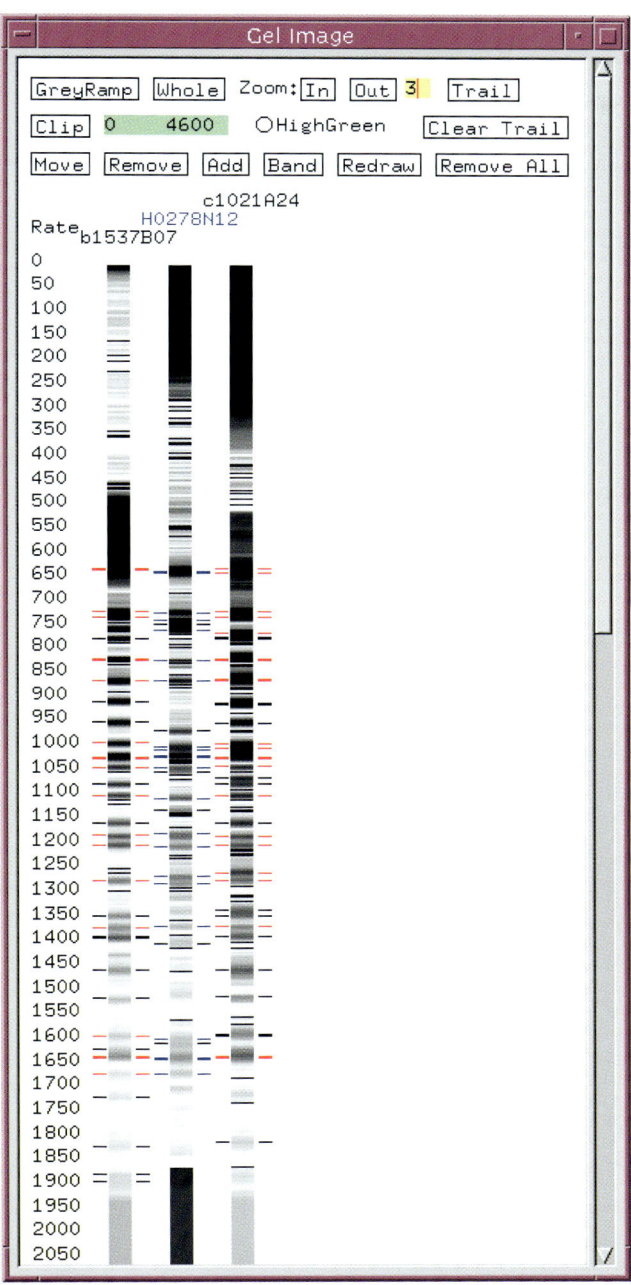

Figure 15. The Gel Image window. The center clone was added by the **Singles→Ctgs** function, and the blue bands show which bands overlap with the immediate neighbors. A red band on a neighbor clone indicates a match with a band from the newly added clone.

Quit out of the Gel Image window. On your terminal window are the results of the previous clone comparison. Put the cursor beside the text c1021A24, hold down the left button, and drag the cursor over the name. Put the cursor in the yellow box beside **Clone 2** on the Contig Analysis window and click the middle button and the text will be dropped into the yellow box. Using this technique, text can be dragged and dropped into any text box in FPC. Select → **Clone 1**. On your terminal window, the last few lines will say:

```
Olap 28 Match 11 1e-03 (Nsizes 28 38)
Total size overlap 50239 Shared markers 0
```

Olap 28 means that the two clones overlap by 28 CB units. Note that it says they have an overlap of 1e-03, but previously it said these two clones matched based on a cutoff of 1e-10. The following is happening: FPC notes that there is a Size directory, so it reads the size value for these clones and compares them with a variable 0.007 tolerance. It also determines the physical overlap based on the variable tolerance, which is approximately 50239 bases. Furthermore, it gives the total sizes for each clone at the beginning of this printout (*i.e.* 126163 and 168382). Click on the **Size** radio button to turn it off. Hit →**Clone 1** again. Now the output is:

```
Olap 28 Match 20 3e-11 (Nbands 28 38)
Total band overlap 20893 Shared markers 0
```

Now the output is using bands. The fixed tolerance 7 on bands finds more shared ones than a variable tolerance 0.007 on sizes. The Size option is there so that you can see the sizes and is not practical for matching unless the variable tolerance value has been optimized.

7.4 Summary

From the Project window, you can view the results in different ways. We have already looked at the Framework window. To see the length of the contigs in CB units, go to the Project window, pull down on the upper right button, and select **By Contig Length**. Using the same pull-down button, select **Summary;** this shows you the average number of bands per clone and other statistics (see Figure 16).

8. What We Didn't Talk About

We did not discuss most of the functions on the Edit Contig window; these allow editing, setting clones to selected, viewing the fingerprints of the

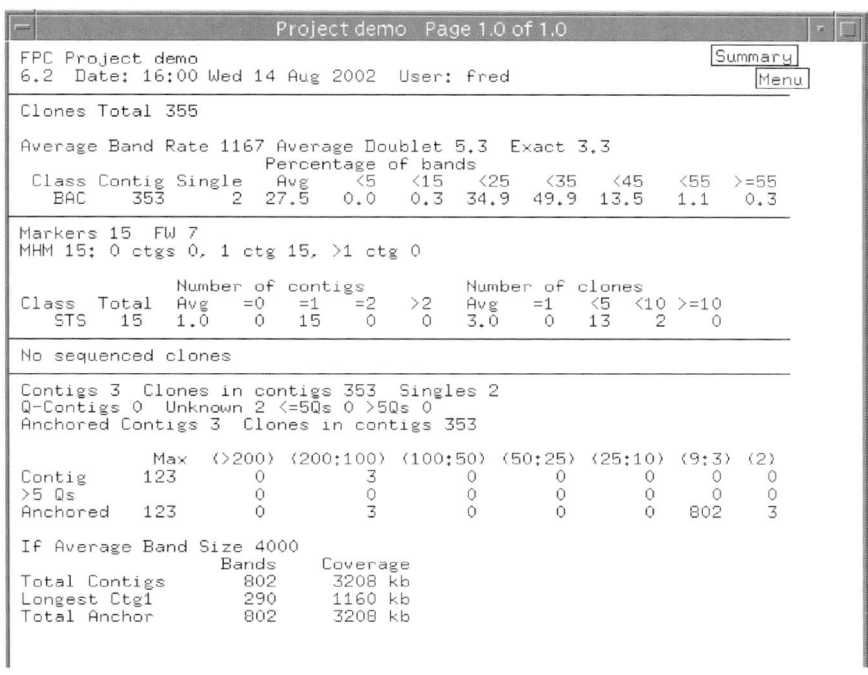

Figure 16. The Summary window after completion of our demo.

selected set, viewing the gel of the selected set, and moving clones around.
Experiment with these functions; many are intuitive. Also, they are
documented in the User's Manual. We did not discuss many of the functions
on the Contig Analysis window, such as **Find Canonical**, which looks
for potential clones to bury. This would be useful to bury at a low stringency for a contig with too much coverage. But most of these functions are
obsolete, as they were in support of manual creation of contigs before the
automatic routines took over. Still, it is worth playing with them to find out
what they do, as they may come in handy for special cases. Most of the
functions on the ProjectMenu window were not discussed; these are
discussed in the User's Guide, but may not be clear. Please feel free to send
us email asking if there is a way to do 'such-and-such', or how to use a
function; it could save you a lot of time and frustration. Finally, no mention
of BSS (BLAST Some Sequence), a tool for locating sequence on the map,
has been made. A separate tutorial on BSS is available at http://www.
genome.arizona.edu/software/fpc/faq.html.

Acknowledgements

F. Engler is supported by USDA/IFAFS grant #11180. We would like to thank James Hatfield for his valuable feedback on this manuscript.

References

Adams, M.D., Celniker, S.E., Holt, R.A., Evans, C.A., Gocayne, J.D., Amanatides, P.G., Scherer, S.E., Li, P.W., Hoskins, R.A., Galle, R.F., George, R.A., Lewis, S.E., Richards, S., Ashburner, M., Henderson, S.N., Sutton, G.G., Wortman, J.R., Yandell, M.D., Zhang, Q., Chen, L.X., Brandon, R.C., Rogers, Y.H., Blazej, R.G., Champe, M., Pfeiffer, B.D., Wan, K.H., Doyle, C., Baxter, E.G., Helt, G., Nelson, C.R., Gabor Miklos, G.L., Abril, J.F., Agbayani, A., An, H.J., Andrews-Pfannkoch, C., Baldwin, D., Ballew, R.M., Basu, A., Baxendale, J., Bayraktaroglu, L., Beasley, E.M., Beeson, K.Y., Benos, P.V., Berman, B.P., Bhandari, D., Bolshakov, S., Borkova, D., Botchan, M.R., Bouck, J., Brokstein, P., Brottier, P., Burtis, K.C., Busam, D.A., Butler, H., Cadieu, E., Center, A., Chandra, I., Cherry, J.M., Cawley, S., Dahlke, C., Davenport, L.B., Davies, P., de Pablos, B., Delcher, A., Deng, Z., Mays, A.D., Dew, I., Dietz, S.M., Dodson, K., Doup, L.E., Downes, M., Dugan-Rocha, S., Dunkov, B.C., Dunn, P., Durbin, K.J., Evangelista, C.C., Ferraz, C., Ferriera, S., Fleischmann, W., Fosler, C., Gabrielian, A.E., Garg, N.S., Gelbart, W.M., Glasser, K., Glodek, A., Gong, F., Gorrell, J.H., Gu, Z., Guan, P., Harris, M., Harris, N.L., Harvey, D., Heiman, T.J., Hernandez, J.R., Houck, J., Hostin, D., Houston, K.A., Howland, T.J., Wei, M.H., Ibegwam, C., Jalali, M., Kalush, F., Karpen, G.H., Ke, Z., Kennison, J.A., Ketchum, K.A., Kimmel, B.E., Kodira, C.D., Kraft, C., Kravitz, S., Kulp, D., Lai, Z., Lasko, P., Lei, Y., Levitsky, A.A., Li, J., Li, Z., Liang, Y., Lin, X., Liu, X., Mattei, B., McIntosh, T.C., McLeod, M.P., McPherson, D., Merkulov, G., Milshina, N.V., Mobarry, C., Morris, J., Moshrefi, A., Mount, S.M., Moy, M., Murphy, B., Murphy, L., Muzny, D.M., Nelson, D.L., Nelson, D.R., Nelson, K.A., Nixon, K., Nusskern, D.R., Pacleb, J.M., Palazzolo, M., Pittman, G.S., Pan, S., Pollard, J., Puri, V., Reese, M.G., Reinert, K., Remington, K., Saunders, R.D., Scheeler, F., Shen, H., Shue, B.C., Siden-Kiamos, I., Simpson, M., Skupski, M.P., Smith, T., Spier, E., Spradling, A.C., Stapleton, M., Strong, R., Sun, E., Svirskas, R., Tector, C., Turner, R., Venter, E., Wang, A.H., Wang, X., Wang, Z.Y., Wassarman, D.A., Weinstock, G.M., Weissenbach, J., Williams, S.M., Woodage, T., Worley, K.C., Wu, D., Yang, S., Yao, Q.A., Ye, J., Yeh, R.F., Zaveri, J.S., Zhan, M., Zhang, G., Zhao, Q., Zheng, L., Zheng, X.H., Zhong, F.N., Zhong, W., Zhou, X.,

Zhu, S., Zhu, X., Smith, H.O., Gibbs, R.A., Myers, E.W., Rubin, G.M., and Venter, J.C. 2000. The genome sequence of Drosophila melanogaster. Science. 287: 2185–95.

Altschul, S.F., Madden, T.L., Schaffer, A.A., Zhang, J., Zhang, Z., Miller, W., and Lipman, D.J. 1997. Gapped BLAST and PSI-BLAST: a new generation of protein database search programs. Nucleic Acids Research. 25: 3389–3402.

Chen, M., Presting, G., Barbazuk, W., Goicoechea, J., Blackmon, B., Fang, G., Kim, H., Frisch, D., Yu, Y., Higingbottom, S., Phimphilai, J., Phimphilai, D., Thurmond, S., Gaudette, B., Li, P., Liu, J., Hatfield, J., Sun, S., Farrar, K., Henderson, C., Barnett, L., Costa, R., Williams, B., Walser, S., Atkins, M., Hall, C., Bancroft, I., Salse, J., Regad, F., Mohapatra, T., Singh, N., Tyagi, A., Soderlund, C., Dean, R., and Wing, R. 2002. An integrated physical and genetic map of the rice genome. Plant Cell. 14:537–545.

Coe, E., Cone, K., McMullen, M., Chen, S., Davis, G., Gardiner, J., Liscum, E., Polacco, M., Paterson, A., Sanchez-Villeda, H., Soderlund, C., Wing, R. 2002. Access to the maize genome: an integrated physical and genetic map. Plant Physiology. 128: 9–12.

Ding, Y., Johnson, M., Colayco, R., Chen, Y., Melnyk, J., Schmitt, H., and Shizuya, H. 1999. Contig assembly of bacterial artificial chromosome clones through multiplexed fluorescent-labeled fingerprinting. Genomics. 56: 237–246.

Durbin, R., and Thierry-Mieg, J. 1994. The AceDB Genome Database. (ed. Suhai, S.), Computational Methods in Genome Research. Plenum Press, New York, pp. 7821–7825.

Hoskins, R., Nelson, C., Berman, B., Laverty, T., George, R., Ciesiolka, L., Naeemuddin, M., Arenson, A., Durbin, J., David, R., Tabor, P., Bailey, M., DeShazo, D., Catanese, J., Mammoser, A., Osoegawa, K., de Jong, P., Celniker, S., Gibbs, R., Rubin, G., and Scherer, S. 2000. A BAC-based physical map of the major autosomes of Drosophila melanogaster. Science. 287: 2271–2274.

Marra, M., Kucaba, T., Dietrich, N., Green, E., Brownstein, B., Wilson, R., McDonald, K., Hillier, L., McPherson, J., and Waterston, R. 1997. High throughput fingerprint analysis of large-insert clones. Genome Research. 7: 1072–1084.

Marra, M., Kucaba, T., Sakhon, M., Hillier, L., Martienssen, R., Chinwalla, A., Crockett, J., Fedele, J., Grover, H., Gund, C., McCombie, W., McDonald, K., McPherson, J., Mudd, N., Parnell, L., Schein, J., Seim, R., Shelby, P., Waterston, R., and Wilson, R. 1999. A map for sequence analysis of the Arabidopsis thaliana genome. Nature Genetics. 22: 265–275.

Ness, S.R., Terpstra, W., Krzywinski, M., Marra M., Jones, A., and Ness, S.J.M. 2002. Assembly of fingerprinted contigs: parallelized FPC. BioInformatics 18:484–484.

Soderlund, C., Engler, F., Hatfield, J., Blundy, S., Chen, M., Yu, Y., and Wing, R. 2002. Mapping sequence to Rice FPC. In C. Wu, P. Wang, and J. Wang (ed). Computational Biology and Genome Informatics. Selected papers from CBGI 2001. World Scientific Publishing.

Soderlund, C., Gregory, S., and Dunham, I. 1997b. Sequence ready clones. (ed. M. Bishop) Guide to Human Genome Computing. Academic Press. pp. 151–177.

Soderlund, C., Humphrey, S., Dunham, A., and French, L. 2000. Contigs built with fingerprints, markers and FPC V4.7. Genome Research. 10: 1772–1787.

Soderlund, C., Longden, I., and Mott, R. 1997a. FPC: a system for building contigs from restriction fingerprinted clones. CABIOS 13: 523–535.

Soderlund, C. 1999. FPC V4.0: User's Manual. Technical Report SC-01 -99. The Sanger Centre, Hinxton Hall, Cambridge UK.

Sulston, J., Mallet, F., Staden, R., Durbin, R., Horsnell, T., and Coulson, A. 1988. Software for genome mapping by fingerprinting techniques. CABIOS. 4: 125–132.

The Arabidopsis Genome Initiative. 2000. Analysis of the genome sequence of the flowering plant Arabidopsis thaliana. Nature. 408: 769–815.

The International Human Genome Mapping Consortium. 2001. A physical map of the human genome. Nature. 409: 934-941.

The International Human Sequencing Consortium. 2001. Initial sequencing and analysis of the human genome. Nature. 409: 860–920.

Tingey, Scott. 2000. The International Conference on the Status of Plant & Animal Genome Research. San Diego, CA.

Wing, R.A., Yu, Y., Presting, G., Frisch, D., Wood, T., Woo, S-S., Budiman, M.A., Mao, L., Kim, H.R., Rambo, T., Fang, E., Blackmon, B., Goicoechea, J.L., Higingbottom, S., Sasinowski, M., Tomkins, J., Dean, R.A., and Soderlund, C. 2001. The CUGI Rice Genome Framework Project and Application to Sequence Rice Chromosomes 10 and 3. The International Rice Research Institute, International Rice Genome Sequencing Meeting, October 2000.

From: *Genome Mapping and Sequencing*
© 2003 Horizon Scientific Press, Wymondham, UK

8

Mapping Pericentromeric Regions

Devin P. Locke, Julie E. Horvath and Evan E. Eichler

Abstract

The genomes of most organisms, from the simplest unicellular organism to more complex species, consist of a variety of genomic "landscapes." Each landscape has a unique profile of high-copy repeats, low-copy repeats, genic content, GC-richness and so forth. Particularly near regions of structurally important sequences, such as the centromere, the genomic landscape can become quite problematic. The primary factor contributing to the additional difficulty in studying these areas within the human genome is that clusters of genomic duplications and unusual repeat structures often lie in close proximity (1-2 Mb) to the centromeres. Consequently, sequence similarity-based methods of global genome assembly fail to properly assign the correct positions of duplicated sequences. Because of this, artificial overlaps form, significant "warping" of working draft sequences occurs, and numerous gaps appear in the assembly, reducing the overall quality and relevance of the assembly in these regions. These effects are further compounded by the absence of "unique" STS within such regions and a general under-representation of such areas in clone-by-clone sequencing projects. Thus, the presence of large spans of duplicated sequence near centromeres (and telomeres) interferes with a generalized approach to genome analysis. To combat this problem, specialized computational/ experimental approaches have been developed to accurately map and assemble these difficult, but biologically relevant, genomic landscapes.

Herein we will explore recently developed techniques aimed at building sequence-ready maps of duplicated regions and "solving" the structure of pericentromeric regions.

1. Introduction

Alongside the significant strides in understanding the nature of our own genome, the Human Genome Project (HGP) has shed light on a host of practical issues concerning the sequencing and assembly of complex metazoan genomes. One facet of this discovery process has been the relative inability of sequence assembly algorithms to accurately distinguish allelic sequence overlap from paralogous sequence overlap, which preferentially causes misassemblies within pericentromeric regions. In light of this, the pericentromeric regions of human chromosomes may be considered one of the "serious and unanticipated challenges" of the Human Genome Project (Collins *et al.*, 1998). For the purposes of this discussion, the term pericentromeric refers to a large transition zone which begins immediately distal to the alpha-satellite repeat and extends into the first distinguishable cytogenetic Giemsa-staining light band on either side of the centromere. Numerous investigations into the genomic organization of several of these regions have revealed an unprecedented degree of recent gene duplication (Borden *et al.*, 1990; Tomlinson *et al.*, 1994; Wohr *et al.*, 1996). Furthermore, the presence of both intra- and interchromosomal duplications with a high degree of sequence similarity (95-99%) complicates traditional mapping and sequencing strategies which have been almost exclusively designed on the concept of "unique" sequence. Sequence assembly algorithms are not solely responsible for this problem, however, as the underlying fingerprint maps upon which assembly decisions are based inherently break down in regions of high sequence similarity as well as in the transition from unique to duplicated sequence. One factor thought to account for the relative paucity of finished pericentromeric sequence is a cloning bias against these often highly duplicated sequences, however recent technical developments aimed at accurately assigning duplicated sequences to their chromosomal positions demonstrates that currently utilized genomic libraries are in fact much better than expected. It is our inability to properly differentiate duplicated sequences from each other and position them with respect to adjacent, possibly also duplicated, sequences that causes confusion.

Adding further complexity to the issue at hand, pericentromeric regions are not all structured in a similar fashion. There is significant variation in the type and copy number of sequences within pericentromeric regions, as well as variation in the extent of structural similarity to other pericentromeric regions, depending on the chromosome being analyzed. The pericentromeric regions of many human chromosomes such as 16p11, 15q11, 22q11, 10q11 (Stallings *et al.*, 1993; Eichler *et al.*, 1996; Eichler *et al.*, 1997; Ritchie *et al.*, 1998; Jackson *et al.*, 1999; Guy *et al.*, 2000; Horvath *et al.*, 2000b) are composed of a mosaic or natural chimera of many different gene segments which have been concatenated over recent evolutionary time. The individual genomic segments are not arrayed in tandem, but originate from a variety of non-homologous chromosomes or from different cytogenetic bands within the same chromosome. In contrast, other pericentromeres such as those in 22p11 and 19p12 are composed of large arrays of tandemly duplicated genes which share a high degree of sequence homology (Wohr *et al.*, 1996; Eichler *et al.*, 1998). In both cases, the high degree of sequence similarity, the large size of the duplication events (10-150 kb) and the fact that many of the so-called "repeat" sequences can not be distinguished as such, represent a serious dilemma for both clone selection and the assignment of overlap between two clones during sequence assembly (Eichler, 1998).

Recent sequence analyses indicate that pericentromeric duplications can exhibit extreme structural complexity (Jackson *et al.*, 1999; Horvath *et al.*, 2000a; Horvath *et al.*, 2000b). Simplistically, pericentromeric regions are organized as duplications within duplications, in that duplicated copies of genic and non-genic sequences are the building blocks of larger more complex pericentromeric duplications (Figure 1). The arrangement of these different simple duplication units may be dramatically scrambled within different pericentromeric regions resulting in mosaic patterns of organization that extend up to 2 Mb in size (Eichler, 1999; Guy *et al.*, 2000; Horvath *et al.*, 2000a). Despite the disruptions in the co-linearity of modular units, the individual duplication units distributed among different chromosomes are remarkably consistent in size, sharing in many instances identical breakpoint sequences among non-homologous chromosome (Eichler *et al.*, 1996; Eichler *et al.*, 1997; Eichler *et al.*, 1999). This has led to the development of a two-step model for their origin. The first step involves the transposition of non-pericentromeric sequences (various intron-exon containing genic segments) toward an ancestral pericentromeric region (5-10 mya) (Eichler *et al.*, 1999; Horvath *et al.*, 2000b). The second event entails the duplication of larger segments (often made up of the originally transposed material) among non-homologous pericentromeric regions (~5 mya). Subsequent deletion, inversion, and further duplication events may then be responsible for additional disruptions in the

Duplications in Human Chromosome 22

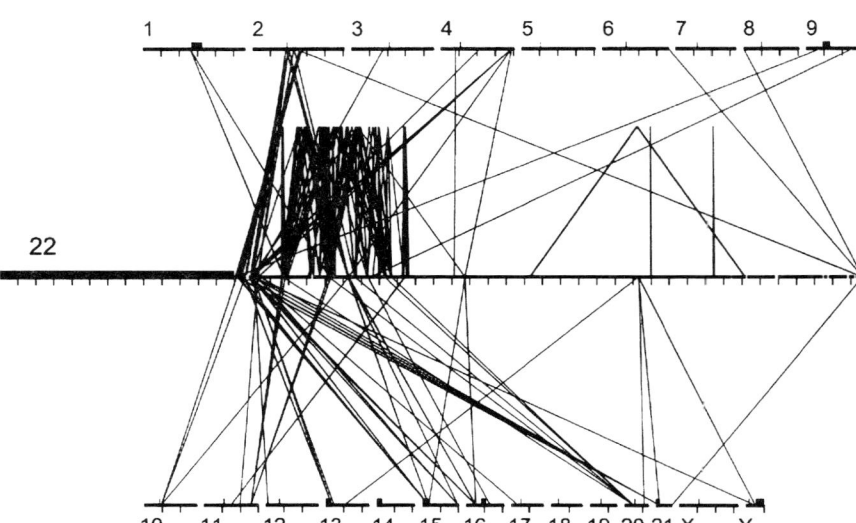

Figure 1. Chromosome 22 Duplications. A view of chromosome 22 is presented with a line drawn between duplicated segments >10 kb in length and 90-98% sequence similarity (adapted from Bailey *et al.*, 2001).

co-linearity of these segments. Studies of a few of these regions (15q11 and 16p11) indicate extensive structural polymorphism within the human population suggesting that such large-scale rearrangement events may still be ongoing (Ritchie *et al.*, 1998; Barber *et al.*, 1999). Despite this extreme genomic plasticity, it may be noteworthy that several genes have been recently described embedded within these regions (Iyer *et al.*, 1996; Chen *et al.*, 1999; Dunham *et al.*, 1999; Guy *et al.*, 2000; Johnson *et al.*, 2001). The function of these "genes" remains largely unknown.

1.1 Why Bother?

Given the difficulty in determining the sequence of the pericentromeric regions, one might wonder what the benefits may be of undertaking such a task. There are several important factors that have driven the development of these methods. Currently, the gap-laden nature of these areas of the human genome may cause genic content to be overlooked. Additionally, evolutionarily important regions that may be prone to rearrangement remain not only unexplored but refractory to further analysis for lack of an intact sequence foundation in these regions. Also, clusters of duplicated sequences in these regions have medical relevance, as rearrangements between highly similar duplications have been shown to mediate numerous genomic diseases. Thus, we believe there are many reasons to delve into highly duplicated regions and present two approaches derived from practical experience.

2. Methodology and Strategy

2.1 Overview

The main obstacle to complete and accurate sequence assembly within pericentromeric regions is their repetitive nature. Two types of repeats may be distinguished: An enrichment of high-copy repetitive elements within the euchromatic sequences flanking the centromeric alpha-satellite (e.g. 19p12) and/or the presence of large-blocks of highly homologous segmental duplications (e.g. 2p11). These properties create gaps within the assembly of regions, not because these areas are particularly recalcitrant to subcloning but rather due to their unusual repeat content which creates biases in clone selection and sequence assembly overlap (Eichler, 1998). There are two important steps in resolving the complexity of these regions. First, the identification of clones that map to pericentromeric regions, for which there is little reliable mapping information. Second, the distinction of true sequence overlaps from those that may be induced by the repeat content. Two different models of pericentromeric structure are considered and an approach to resolve each is outlined below: In each case, a specific working example is illustrated.

2.2 The Mosaic Structure of Chromosome 2p11

Detailed analyses of pericentromeric DNA structure from 16p11, 10q11, 2p11 and 15q11 (Eichler et al., 1999; Jackson et al., 1999; Guy et al., 2000; Horvath et al., 2000a; Horvath et al., 2000b) indicated that the degree of sequence similarity among various pericentromeric regions ranges over a relatively narrow interval (95-99%). It was reasoned that this high degree of sequence identity while problematic at many levels, could be exploited to readily distinguish and identify other pericentromeric clones. Central to this analysis was the characterization of paralogous sequence variants (PSVs) that could be used to tag each duplicated copy at the sequence level. The paralogous sequence variant is analogous to the single nucleotide polymorphism (SNP) with the exception that the sequence variation accumulates after a duplication event as opposed to descent from a common founding allele. Our approach entailed using a reference or seed pericentromeric clone sequence as a template to develop a series of STS, called paralogous or pSTS, that (unlike traditional STS) cross-amplified multiple loci. pSTS primers are designed to hybridize within regions of higher sequence conservation than the intervening region, to maximize the cross-amplification potential as well as the number of putative diagnostic nucleotide changes (*i.e.* PSVs) between duplicated sequences. A typical pSTS is ~400 bp in length, to allow forward and reverse sequencing reactions to cover the entire pSTS sequence, reducing ambiguities that could otherwise be induced by sequencing. Direct sequencing of the PCR products from monochromosomal resources (monochromosomal somatic cell hybrids or chromosome-specific libraries, such as cosmid libraries) allowed us to distinguish and assign the paralogous segments. These paralogous sequence tags, then, could be utilized to identify corresponding large insert BAC clones (from the RPCI-11, Caltech D BAC libraries) and integrate the sequence of these clones into the appropriate assembly in order to provide genome sequence continuity within these problematic regions. In effect, PSV-verified coverage is increased for all duplicated loci simultaneously, which aids in assembling multiple pericentromeric regions through the analysis of a single duplicated sequence.

As a verification of this methodology, we selected a completely sequenced clone from the pericentromeric region of 2p11. FISH analysis using the 162 kb insert as a probe had indicated the presence of multiple pericen-tromeric signals (1p12, 2p11, 7p11, 9p11, 9q13, 10p11, 15q11, 16p11, 22q11, and Yp11) (Horvath *et al.*, 2000a). Database searches using this 2p11 sequence revealed the presence of various duplicated genomic fragments including segments that contain duplicated exon-intron structure of genes from 4q24 (Hs13584), Xq28 (adrenoleukodystrophy gene) and 2p12 (Vk immunglobulin gene). A total of 24 PCR assays were developed within "unique" sequence (as determined by RepeatMasker) within the 2p11 reference sequence. These assays were subsequently used to screen a panel of monochromosomal hybrid DNAs (NIGMS, Human Genetic Mutant Cell Repository Mapping Panel 2) and a paralogy map was generated (Figure 2). In this manner, the monochromosomal hybrid panel PCR results were used to measure the extent of duplication for each pSTS and uncovered the mosaic structure of duplications within this 2p11 seed sequence. The sequence of the monochromosomal hybrid panel PCR products was then used to identify chromosome-specific PSVs for each pSTS, an example of which is shown in Figure 3.

2p11 Paralogy Map

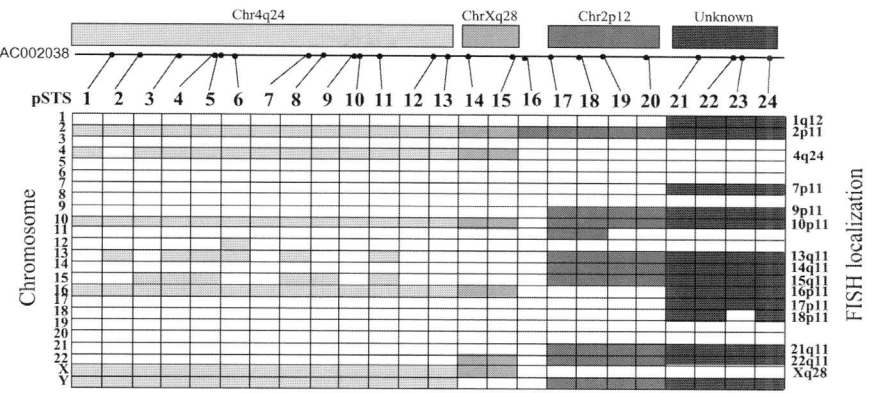

Figure 2. Paralogy Map of AC002038. The position of the 24 pSTS with respect to a schematic organization of duplications for the 162 kb of 2p11 sequence is shown. Shaded cells indicate monochromosomal hybrids that scored positive by PCR. FISH localizations were determined using long-range PCR products from and subclones of AC002038 as probes. The originating chromosomal position of each duplicon is indicated above the horizontal bars above the pSTS matrix (adapted from Horvath *et al.*, 2000a).

A total of 685 paralogous sequence variants from the RPCI-11 human BAC library were identified by examination of 13 different pSTS (a total of 34.7 kb of paralogous sequence) based on our original 2p11 reference sequence. Subsequent FISH mapping of long-range PCR probes and monochromosomal hybrid analysis allowed us to assign 642/685 paralogous sequence variants to the pericentromeric region of 14 different chromosomes, and the remaining 23 sequence variants mapped outside of pericentromeric DNA (Xq28, 2p12, 4q24). Correlation of the FISH results, using long-range PCR FISH probes derived from the 2p11 clone, with the monochromosomal hybrid panel PCR results increased our confidence in the assignment of PSV sequences to specific chromosomes. Overall, this analysis revealed that the 2p11 (AC002038) reference sequence was virtually devoid of any unique sequence, as each pSTS primer pair amplified on average 6 different chromosomal locations. Sequencing of each PCR product identified an average of 4-5 diagnostic differences. Furthermore the resulting paralogy map indicated a complex arrangement of duplications of various lengths. Some pericentromeric regions, such as 10q11, 22q11 and 16p11, shared sequence similarity over the entire length of the 2p11 clone while others shared homology over a much smaller domain (Figure 2).

2p11 pSTS #1 Example Paralogous Sequence Variant

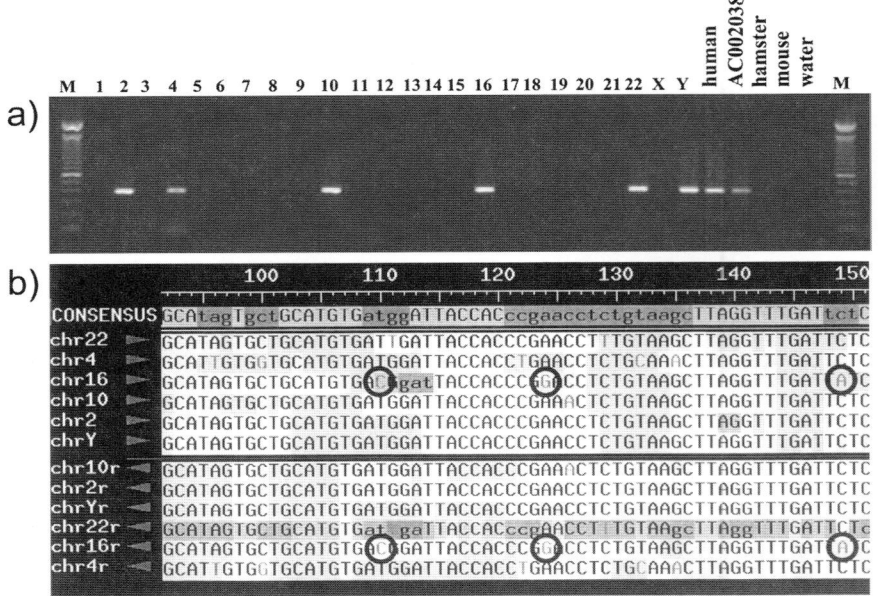

Figure 3. Hybrid PCR and Paralogous Sequence Variants. a) An example of monochromosomal hybrid panel PCR results with pSTS #1 from Figure 1. Marker lanes are marked with an M, the chromosome of origin of the template in the panel is marked above the corresponding lane. b) Consed view of forward and reverse ('r') sequence reads of products (Gordon, 1998). Variant sites are circled in black (Adapted from Horvath et al 2000a).

Following this analysis, three of the most informative pSTS sites were selected and used as probes to recover clones from RPCI-11 library. A total of 702 BACs were identified (11.8X) of which a subset was analyzed at the sequence level. The same STS were amplified from the BACs, the products sequenced and the variants compared to those generated from the monochromosomal hybrids (Figure 4). This approach allowed us to assign nearly 70 of the BACs to the pericentromeric regions of 9 different chromosomes (the remaining 37 required further validation and were found to represent intrachromosomal duplications within pericentromeric regions). Subsequent analysis revealed that nearly ~46% of these BACs were not represented by sequence within GenBank. A total of 35 of these BACs were submitted to various sequencing centers to increase coverage in these areas. In so doing, our analysis of a single 162 kb pericentromeric clone that was highly duplicated allowed us to effectively identify and

Using PSVs to Assign Clones to Chromosomes

Reference base position within AC002038

8272	8263	8238	8236	8230	8228	8227	8225	8186	8177	8174	8171	8166	8116	8095	8081	8078	8075	8072	8070	8068	8058	8057	8056	8046	8042	8015	7987	CLONE/ HYBRID
T	C	G	G	A	G	T	A	T	T	A	A	A	C	C	G	T	C	C	A	C	G	G	T	C	A	G	A	**101B6**
.	.	.	.	.	.	A	.	.	.	.	.	.	.	.	.	.	.	.	.	.	.	.	.	.	.	.	.	**MCH2**
.	.	.	.	.	.	.	.	.	.	.	.	.	.	.	.	.	.	.	.	.	.	.	.	.	.	.	.	13P10
.	.	.	.	.	.	.	.	.	.	.	.	.	.	.	.	.	.	.	.	.	.	.	.	.	.	.	.	95M16
.	.	.	A	.	.	.	C	.	C	C	.	G	.	.	.	A	C	.	.	.	T	.	.	G	T	T	G	**MCH4**
.	.	.	A	.	.	.	C	.	C	C	.	G	.	.	.	A	C	.	.	.	T	.	.	G	T	T	G	134C6
.	.	.	A	.	.	.	C	.	C	C	.	G	.	.	.	A	C	.	.	.	T	.	.	G	T	T	G	231M21
.	.	.	A	.	.	.	C	.	C	C	.	G	.	.	.	A	C	.	.	.	T	.	.	G	T	T	G	255F14
.	.	.	A	.	.	.	C	.	C	C	.	G	.	.	.	A	C	.	.	.	T	.	.	G	T	T	G	289C17
C	.	.	A	.	.	.	.	.	.	.	.	.	.	.	C	.	.	.	.	.	A	.	.	.	.	.	.	**MCH10**
C	.	.	A	.	.	.	.	.	.	.	.	.	.	.	C	.	.	.	.	.	A	.	.	.	.	.	.	109B16
C	.	.	A	.	.	.	.	.	.	.	.	.	.	.	C	.	.	.	.	.	A	.	.	.	.	.	.	43M19
C	.	.	A	.	.	.	.	.	.	.	.	.	.	.	C	.	.	.	.	.	A	.	.	.	.	.	.	453N3
.	.	.	A	.	.	.	.	.	.	.	.	.	.	.	C	.	.	.	.	.	A	.	.	.	.	.	.	464B16
T	.	.	A	.	.	.	C	.	C	T	.	.	A	.	.	.	.	G	.	.	.	C	.	.	.	.	.	**MCH16**
.	T	.	A	.	.	.	C	.	C	T	.	.	A	.	.	.	.	G	.	.	.	C	.	.	.	.	.	169A5
.	T	.	A	.	.	.	C	.	C	T	.	.	A	.	.	.	.	G	.	.	.	C	.	.	.	.	.	370N13
.	T	.	A	.	.	.	C	.	C	T	.	.	A	.	.	.	.	G	.	.	.	C	.	.	.	.	.	69H20
.	.	A	G	.	.	.	.	.	.	C	.	.	T	.	.	.	T	.	.	.	.	T	.	.	.	.	.	**MCH22**
.	.	.	A	.	.	.	.	.	.	C	.	.	T	.	.	.	T	.	.	.	.	T	.	.	.	.	.	25N14
.	.	A	G	.	.	.	.	.	.	C	.	.	T	.	.	.	T	.	.	.	.	T	.	.	.	.	.	276I7
.	.	A	G	.	.	.	.	.	.	C	.	.	T	.	.	.	T	.	.	.	.	T	.	.	.	.	.	394J3
.	.	A	A	.	.	C	.	.	.	C	.	.	.	.	.	.	.	.	.	.	.	.	.	.	.	.	.	**MCHY**

Figure 4. Paralogous Sequence Variants and Corresponding BACs. A Findpoly output (Kashuk et al., 2002) showing paralogous sequence variants with respect to the AC002038 (101B6) reference sequence (coordinates shown along top) that anchor RPCI-11 BAC sequence signatures to monochromosomal hybrid sequence data (*e.g.* MCH2 = monochromosomal hybrid 2) (Adapted from Horvath et al., 2000a).

characterize an estimated 6-8 Mb of pericentromeric DNA of which more than half mapped to sequence gaps within the Human Genome Project. Furthermore, our detailed analysis of the paralogous sequence variant map assignments allowed us to correct at least two misassignments within the HGP.

A three-tiered approach is recommended for gap closure and sequence assembly within pericentromeric DNA associated with highly duplicated region of the genome (Figure 5; Protocol 3.1). The first step involves the

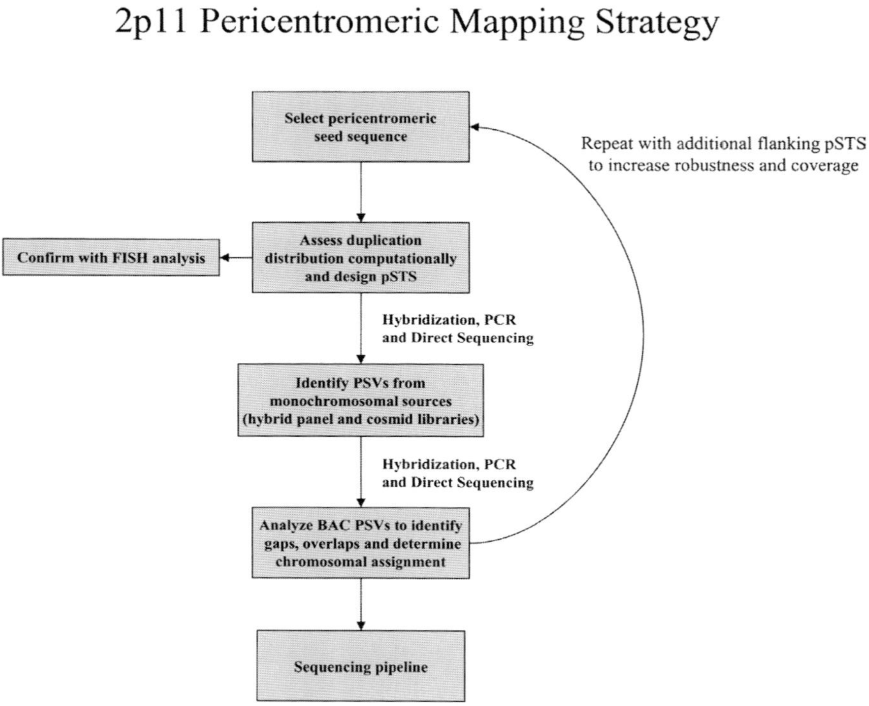

2p11 Pericentromeric Mapping Strategy

Figure 5. 2p11 Pericentromeric Mapping Strategy. See text for details.

selection of pericentromeric seed sequence from GenBank and its detailed computational analysis complemented by a cursory cytogenetic analysis. This is used to assess the distribution of paralogous sequences and the location of potential gaps within the duplicated regions. Next, based on the reference sequence, diagnostic variants, derived from monochromosomal and single haplotype source material, are compared to those from the widely available large-insert genomic libraries (RPCI-11 and Caltech D BAC libraries and the chromosome-specific cosmid libraries). This identifies clones that map to gap regions within the genome and allows one to rapidly identify potential misassemblies within these regions. As clones are identified that map to gaps, these can be incorporated into sequencing pipelines, subsequent to additional rounds of experimental validation and computational analysis. Several cycles of analysis are required in order to increase coverage of these regions significantly. This approach exploits the recent and duplicated nature of sequences within these regions to develop a systematic approach to identify such gaps, recover clones that map to these regions and to validate the integrity of the overlaps. Based on *in silico* analysis of the human genome project (Horvath *et al.*, 2000a; Bailey *et al.*, 2001), it has been suggested that 26/48 pericentromeric regions may share this type of organization. The most critical aspect of this strategy is the distinction between allelic and paralogous sequence variation—discernment at the sequence level provides the greatest sensitivity in this regard.

2.3 Complex Satellite Repeat Structures in Chromosome 19p12

The pericentromeric region of 19p12 harbors one of the largest clusters of tandemly duplicated gene families of the human genome. It has been estimated that there are approximately 30-40 KRAB ZNF (Krueppel-associated box zinc finger) genes and pseudogenes tandemly reiterated throughout the region (Bellefroid *et al.*, 1991; Bellefroid *et al.*, 1993; Bellefroid *et al.*, 1995; Eichler *et al.*, 1998). This cluster of genes spans ~4.5 Mb of the most proximal portion of DNA up to and abutting the alpha-satellite DNA of the 19p11 centromere. The entire gene cluster is situated within variable-staining C-banded chromatin, which is generally associated with the heterochromatic portions of human DNA (ISCN, 1985). Sequence analysis of transcripts from the region indicate a remarkably high degree of sequence and gene structure conservation both among individual members in this cluster (~91%) and other gene members scattered throughout the human genome (85%) (Villa *et al.*, 1993; Derry *et al.*, 1995; Baban *et al.*, 1996; Grondin *et al.*, 1996). This high degree of conservation has led to speculation that the general organization of the pericentromeric region of 19p12 evolved after the divergence of primates from the mammalian radiation (64 mya) (Bellefroid *et al.*, 1995). Despite numerous searches, no orthologues of the genes in the mouse genome have yet been identified.

Our analysis of the region (Eichler *et al.*, 1998) revealed that the ZNF genes are distributed in a head-to-tail fashion throughout the pericentromeric region with an average density of one ZNF gene every 150-180kb of genomic distance. Sequence analysis of a ~165 kb segment from 19p12 identified the presence of large complex blocks of (25 kb) inverted beta-satellite like repeat structures bracketing the functional ZNF (ZNF208) gene (Figure. 6). The beta-satellite repeats showed considerable (65-70%) sequence divergence from the classic beta-satellite 68 bp motif (Eichler *et al.*, 1998). Hybridization studies indicated that the beta-satellite repeat sequences hybridize specifically to the 19p12 region and were a general property for more than 3.0 Mb of the pericentromeric/ZNF region

Organization of the 19p12 Pericentromeric Region

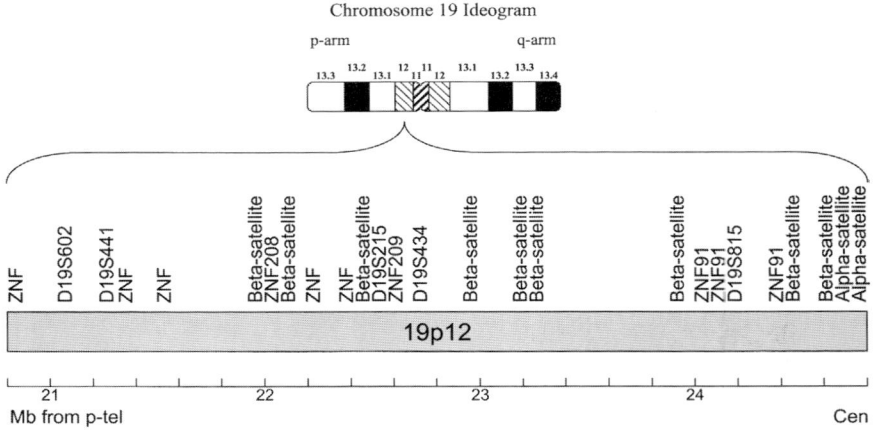

Figure 6. Organization of the 19p12 Pericentromeric Region. Clusters of beta-satellite sequences are found interspersed among ZNF genes, which are themselves duplicated throughout the p12 band.

of 19p12. In addition, multiple probes designed to the p12 beta-satellite repeat structures did not cross-hybridize by FISH to other sites in the genome known to harbor beta-satellites. This led to the formation of a working model in which the region was composed of tandemly duplicated ZNF genes punctuated by large (25-50 kb) beta-satellite repeat clusters. Furthermore, the beta-satellite sequence of 19p12 appeared to be a novel 19p12-specific beta-satellite which had not been previously described. This biological distinction between generic and p12-specific repeat probes offers a considerable mapping advantage. Although the 19p12 region is virtually devoid of classically defined "unique" sequence (<15%), the presence of p12-specific repeat probes, due to the tandem duplication structure of the region, effectively provides a molecular handle for the identification of a large proportion of BAC and cosmid clones which map to the region (a 19p12-enriched sub-genomic resource).

Our approach therefore initially focused on the identification, collection and sequence characterization of cloned reagents which map specifically to 19p12 using 19p12-specific repeats as probes (Figure.7; Protocol 3.2). To test the validity of this approach, we developed a series of 19p12

19p12 Pericentromeric Mapping Strategy

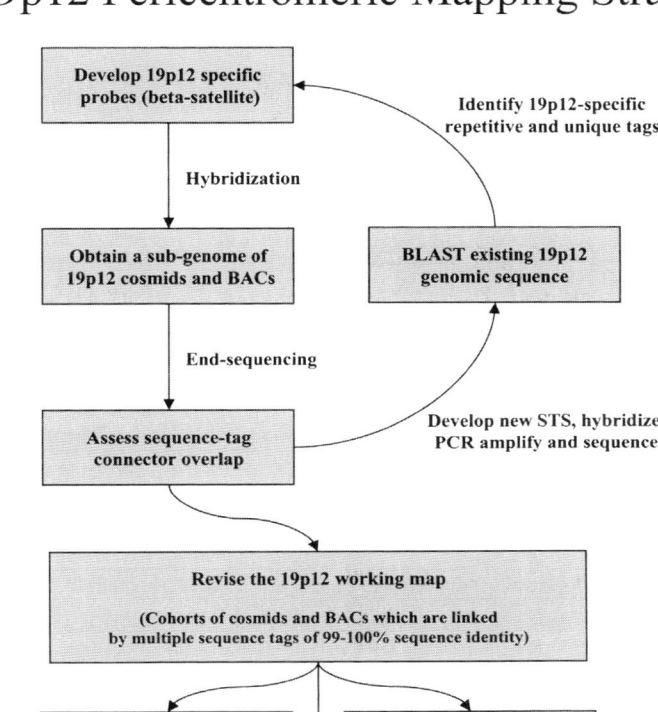

Figure 7. 19p12 Pericentromeric Mapping Strategy. See text for details.

beta-satellite repeat probes based on the consensus of existing sequence and hybridized them to the chromosome 19-specific and whole genome BAC libraries. Utilizing such probes for screening by hybridization facilitated the isolation of a collection of BACs and cosmids which are specific to the 19p12 region. Once a collection of 19p12-specific genomic reagents has been generated, the second step of our analysis involved the generation of end sequence tags from a subset of these p12 clones and anchoring these clones to available seed sequence within the regions. Typical genome sequencing strategies involve a combination of whole-genome shotgun sequence, working draft sequence from BAC clones and sequence-tag connectors to provide clonal overlap during assembly. It is true that targeted approaches of specific regions will effectively double the number of end-sequence reads in such areas. This high density of sequence-tag connectors, however, is essential due to the paralogous nature and the virtual absence of unique sequence in such regions. Once the sequence-anchored overlaps have been established, all clones tagged by identical sequence are subjected to fingerprint analysis to provide additional confidence in the suspected overlap. This analysis not only provides an important check-and-balance to increase the confidence of the 19p12 sequence-ready map, but it more importantly allows the complete integration of the sequence-anchor map with the portions of the 19p12 physical map which have been previously assembled. This approach was successfully used for the sequencing and assembly of the 4 Mb transition region between the centromere and chromosomal arm (Eichler, in preparation; http://www.genome.ucsc.edu).

One of the advantages of the integrated sequence-anchored approach is that it allows rapid identification of discrepancies that may arise due to misassembly or biological heteromorphism. Due to the high density of BAC end sequences which have been placed within the sequence map, one will in effect have generated two different framework maps-one from each chromosome within the diploid BAC library. If appreciable large-scale heteromorphism exists, serious discrepancies will be encountered when comparing these two frameworks. Each of these discrepancies may be analyzed in more detail to evaluate the biological significance (O'Keefe et al., 1996; O'Keefe et al., 1997). High resolution FISH technology, for example, has been used to investigate the genomic architecture of the pericentromeric region of 16p11 where it has proven invaluable in confirming or negating overlaps suggested by molecular data (Eichler, in preparation). This final stage of the analysis provides a means for the biological confirmation of the sequence-ready map.

3. Protocols

3.1 Protocol – 2p11 Pericentromeric Analysis

1) Select pericentromeric seed sequence based upon multiple sequence alignment over a region of sequence masked for high-copy repeats (RepeatMasker, BLAST)
2) Design PCR primers (preferably 24 nucleotides in length) in conserved regions of duplicated sequence flanking potentially informative paralogous sequence variants (PSVs) observed by multiple sequence alignment.

3) Confirm computational results by FISH with the seed sequence.

4) Hybridize genomic BAC libraries, PCR amplify pSTS sequences and identify PSVs.

5) Obtain pSTS sequence from monochromosomal sources (mono-chromosomal cosmid libraries, somatic cell hybrid panel, etc) to link PSVs to specific chromosomes.

6) Analyze BAC PSVs to identify clones in sequence gaps and submit resulting clones for sequencing.

3.2 Protocol – 19p12 Pericentromeric Analysis

1) Identify chromosome-specific sequence elements within the pericentromeric region (*i.e.*, beta-satellite) by sequence comparison (BLAST).

2) Develop a PCR amplicon capable of identifying the sequence element and amplify the probe from a monochromosomal source.

3) Hybridize genomic BAC and cosmid libraries with the chromosome-specific probe.

4) End-sequence the resulting chromosome-specific sub-genome.

5) Compare end sequences against regional unique sequence islands to anchor clones.

6) Identify gap-filling clones and submit for sequencing.

Conclusion

Within the pericentromeric regions of chromosomes, definitions of unique and repetitive DNA are obscure. From a classical perspective, virtually all sequence within the pericentromeric region may be viewed as various gradations of repeat. For sequencing and mapping strategies which have been developed around the concept of unique sequence, such areas pose particular problems. We have developed systematic approaches to construct sequence-ready maps within the pericentromeric regions of 19p12 and 2p11, which we believe are applicable to other regions rich in complex intra and interchromosomal segmental duplications. Our approach has been to uncouple these regions from the pipeline of traditional mapping and sequencing approaches, recognize their unusual biological properties, distinguish the different characteristics of repeats within the region, and exploit these for clone identification, sequencing and assembly. We realize that such approaches require greater upfront effort and initial cost. If the goal of the Human Genome Project, or any genome effort focused on an organism with significant levels of segmental duplication, however, is to provide a true understanding of the structure and organization of the entire genome, these regions can not be avoided. To incisively investigate the interesting biological properties of pericentromeric and subtelomeric regions, a robust sequence assembly is required. A strong sequence foundation is not only essential for comparative evolutionary analysis, but also for the analysis of genomic disease rearrangements involving clusters of duplicated sequences. The strategy we have proposed to develop a sequence-ready map of 19p12 and 2p11 is critical in ensuring that these complex regions of the genome are sequenced and that the end-product has biological meaning. Such approaches allow the euchromatic portions of entire chromosomes to be sequenced. In addition, these strategies should serve as a model for the analysis of other similarly problematic regions and facilitate future research into the structure and function relationship of these regions to disease, evolution and genomic instability.

Acknowledgements

The authors thank Jeffrey A. Bailey for his contribution to the development of the computational techniques discussed herein.

References

Baban, S., Freeman, J., and Mager, D. 1996. Transcripts from a novel human KRAB zinc finger gene contain spliced Alu and endogeneous retroviral segments. Genomics. 33: 463–472.

Bailey, J.A., Yavor, A.M., Massa, H.F., Trask, B.J., and Eichler, E.E. 2001. Segmental duplications: organization and impact within the current human genome project assembly. Genome Res. 11: 1005–17.

Barber, J.C., Reed, C.J., Dahoun, S.P., and Joyce, C.A. 1999. Amplification of a pseudogene cassette underlies euchromatic variation of 16p at the cytogenetic level. Hum Genet. 104: 211–8.

Bellefroid, E., Marine, J., Matera, A., Bourguignon, C., Desai, T., Healy, K., Bray-Ward, P., Martial, J., Ihle, J., and Ward, D. 1995. Emergence of the ZNF91 Kruppel-associated box-containing zinc finger gene family in the last common ancestor of anthropoidea. Proc. Natl. Acad. Sci. USA. 92: 10757–10761.

Bellefroid, E., Marine, J.-C., Ried, T., Lecocq, P., Riviere, M., Amemiya, C., Poncelet, D., Coulie, P., deJong, P., Szpirer, C., Ward, D., and Martial, J. 1993. Clustered organization of homologous KRAB zinc-finger genes with enhanced expression in human T lymphoid cells. EMBO J. 12: 1363–1374.

Bellefroid, E., Poncelet, D., Lecocq, P., Relevant, O., and Martial, J. 1991.

The evolutionarily conserved Kruppel-associated box domain defines a subfamily of eukaryotic multifingered proteins. Proc. Nat. Acad. Sci. USA. 88: 3608–3612.

Borden, P., Jaenichen, R., and Zachau, H. 1990. Structural features of transposed human Vk genes and implications for the mechanism of their transpositions. Nucleic Acids Res. 18: 2101–2107.

Chen, H., Rossier, C., Morris, M.A., Scott, H.S., Gos, A., Bairoch, A., and Antonarakis, S.E. 1999. A testis-specific gene, TPTE, encodes a putative transmembrane tyrosine phosphatase and maps to the pericentromeric region of human chromosomes 21 and 13, and to chromosomes 15, 22, and Y. Hum Genet. 105: 399–409.

Collins, F.S., Patrinos, A., Jordan, E., Chakravarti, A., Gesteland, R., and Walters, L. 1998. New goals for the U.S. Human genome project: 1998-2003. Science. 282: 682–9.

Derry, J., Jess, U., and Francke, U. 1995. Cloning and characterization of a novel zinc finger gene in Xp11.2. Genomics. 30: 361–365.

Dunham, I., Shimizu, N., Roe, B.A., Chissoe, S., Hunt, A.R., Collins, J.E., Bruskiewich, R., Beare, D.M., Clamp, M., Smink, L.J., Ainscough, R., Almeida, J.P., Babbage, A., Bagguley, C., Bailey, J., Barlow, K., Bates, K.N., Beasley, O., Bird, C.P., Blakey, S., Bridgeman, A.M., Buck, D., Burgess, J., Burrill, W.D., O'Brien, K.P., and et al. 1999. The DNA sequence of human chromosome 22. Nature. 402: 489–95.

Eichler, E., Archidiacono, N., and Rocchi, M. 1999. CAGGG repeats and the pericentromeric duplication of the hominoid genome. Genome Res. 9: 1048–1058.

Eichler, E.E. 1998. Masquerading repeats: Paralogous pitfalls of the Human Genome. Genome Res. 8: 758–762.

Eichler, E.E. 1999. Repetitive conundrums of centromere structure and function. Hum Mol Genet. 8: 151–155.

Eichler, E.E., Budarf, M.L., Rocchi, M., Deaven, L.L., Doggett, N.A., Baldini, A., Nelson, D.L., and Mohrenweiser, H.W. 1997. Interchromosomal duplications of the adrenoleukodystrophy locus: a phenomenon of pericentromeric plasticity. Hum Molec Genet. 6: 991–1002.

Eichler, E.E., Hoffman, S.M., Adamson, A.A., Gordon, L.A., McCready, P., Lamerdin, J.E., and Mohrenweiser, H.W. 1998. Complex beta-satellite repeat structures and the expansion of the zinc finger gene cluster in 19p12. Genome Res. 8: 791–808.

Eichler, E.E., Lu, F., Shen, Y., Antonacci, R., Jurecic, V., Doggett, N.A., Moyzis, R.K., Baldini, A., Gibbs, R.A., and Nelson, D.L. 1996. Duplication of a gene-rich cluster between 16p11.1 and Xq28: a novel pericentromeric-directed mechanism for paralogous genome evolution. Hum Molec Genet. 5: 899–912.

Grondin, B., Bazinet, M., and Aubry, M. 1996. The KRAB zinc finger gene ZNF74 encodes an RNA-binding protein tightly associated with the nuclear matrix. J. Biol. Chem. 271: 15458–15467.

Guy, J., Spalluto, C., McMurray, A., Hearn, T., Crosier, M., Viggiano, L., Miolla, V., Archidiacono, N., Rocchi, M., Scott, C., Lee, P.A., Sulston, J., Rogers, J., Bentley, D., and Jackson, M.S. 2000. Genomic sequence and transcriptional profile of the boundary between pericentromeric satellites and genes on human chromosome arm 10q [In Process Citation]. Hum Mol Genet. 9: 2029–42.

Horvath, J., Schwartz, S., and Eichler, E. 2000a. The mosaic structure of a 2p11 pericentromeric segment: A strategy for characterizing complex regions of the human genome. Genome Res. 10: 839–52.

Horvath, J., Viggiano, L., Loftus, B., Adams, M., Rocchi, M., and Eichler, E. 2000b. Molecular structure and evolution of an alpha/non-alpha satellite junction at 16p11. Hum Molec Genet. 9: 113–123.

ISCN 1985. Report of the standing committee on human cytogenetic nomenclature. Birth Defects. 21: 1–117.

Iyer, G., Krahe, R., Goodwin, L., Doggett, N., Siciliano, M., Funanage, V., and Proujansky, R. 1996. Identification of a testis-expressed creatine transporter gene at 16p11.2 and confirmation of the X-linked locus to Xq28. Genomics. 34: 143–146.

Jackson, M.S., Rocchi, M., Thompson, G., Hearn, T., Crosier, M., Guy, J., Kirk, D., Mulligan, L., Ricco, A., Piccininni, S., Marzella, R., Viggiano, L., and Archidiacono, N. 1999. Sequences flanking the centromere of human chromosome 10 are a complex patchwork of arm-specific sequences, stable duplications, and unstable sequences with homologies to telomeric and other centromeric locations. Hum Mol Genet. 8: 205–215.

Johnson, M.E., Viggiano, L., Bailey, J.A., Abdul-Rauf, M., Goodwin, G., Rocchi, M., and Eichler, E.E. 2001. Positive selection of a gene family during the emergence of humans and African apes. Nature. 413: 514–9.

Kashuk, C., SenGupta, S., Eichler, E., and Chakravarti, A. 2002. ViewGene: a graphical tool for polymorphism visualization and characterization. Genome Res. 12: 333–8.

O'Keefe, C., Griffin, D., Bean, C., Matera, A., and Hassold, T. 1997. Alphoid variation-specific FISH probes can distinguish autosomal meiosis I from meiosis II non-disjunction in human sperm. Hum. Genet. 101: 61–66.

O'Keefe, C., Warburton, P., and Matera, A. 1996. Oligonucleotide probes for alpha satellite DNA variants can distinguish homologous chromosomes by FISH. Hum. Molec. Genet. 11: 1793–1799.

Ritchie, R.J., Mattei, M.G., and Lalande, M. 1998. A large polymorphic repeat in the pericentromeric region of human chromosome 15q contains three partial gene duplications. Hum Mol Genet. 7: 1253–60.

Stallings, R., Whitmore, S., Doggett, N., and Callen, D. 1993. Refined physical mapping of chromosome 16-specific low-abundance repetitive DNA sequences. Cytogenet. Cell Genet. 63: 97–101.

Tomlinson, I.M., Cook, G.P., Carter, N.P., Elaswarapu, R., Smith, S., Walter, G., Buluwela, L., Rabbitts, T.H., and Winter, G. 1994. Human immunglobulin VH and D segments on chromosomes 15q11.2 and 16p11.2. Hum Molec Genet. 3: 853–860.

Villa, A., Zucchi, I., Pilia, G., Strina, D., Susani, L., Morali, F., Patrosso, C., Frattini, A., Lucchini, F., and Repetto, M. 1993. ZNF75: isolation of a cDNA clone of the KRAB zinc finger gene subfamily mapped in YACs 1 Mb telomeric of HPRT. Genomics. 18: 223–229.

Wohr, G., Fink, T., and Assum, G. 1996. A palindromic structure in the pericentromeric region of various human chromosomes. Genome Res. 6: 267–279.

From: *Genome Mapping and Sequencing*
© 2003 Horizon Scientific Press, Wymondham, UK

9

Cloning, Mapping, and Sequencing Telomeres

Harold Riethman

Abstract

The DNA sequence organization of human telomeres includes large stretches of highly similar duplicated and low-copy DNA adjacent to terminal telomere repeat sequences. This unusual sequence organization has led to significant complications with respect to mapping and sequencing. Our approach to solving these problems, described in this paper, has been to isolate each telomere region using a specialized yeast artificial chromosome (YAC) system that permits propagation of large telomere-terminal human DNA fragments as linear plasmids in yeast. Each YAC contains a terminal repeat tract, the entire subtelomeric repeat region, and the adjacent single-copy DNA region, physically linked on a single large DNA segment that has been purified from the rest of the human genome. From this starting material, the most distal single-copy segments of each chromosome arm can be identified, analyzed, and used to validate subtelomeric sequence structure.The particular repeat organization and DNA sequence of each subtelomeric region can then be deciphered without interference from duplicons derived from elsewhere in the genome. This basic approach has been used successfully for most human telomeres, and is applicable to all vertebrate and most eukaryotic genomes.

1. Introduction

Human telomeres are complex, dynamic chromosomal structures (Figure 1). The DNA at each chromosome terminus is a simple repeat sequence tract (TTAGGG)n, typically 5 kb to 15 kb in length in somatic cells (Moyzis *et al.*, 1988), that ends with a single-stranded extension of the G-strand of DNA (Griffith *et al.*, 1999). The lengths of the terminal repeat tracts are dynamically modulated by a complex interplay of factors involved in cell cycle regulation, cellular senescence, and cellular immortalization (Blasco *et al.*, 1999; de Lange and Jacks, 1999). Adjacent to this "terminal repeat" is a subtelomeric repeat region comprised of a mosaic patchwork of segmentally duplicated DNA tracts. This class of low-copy repeat DNA is characterized by very high sequence similarity (90 % to >99.5 %) between duplicated tracts, and variably-sized but often very large duplicated tract lengths (1 kb to > 200 kb). Some of the segmental duplications are unique to subtelomeric repeat regions, some are shared with a subset of pericentromeric repeat regions, and a few are shared with one or several

CEN   **TEL**

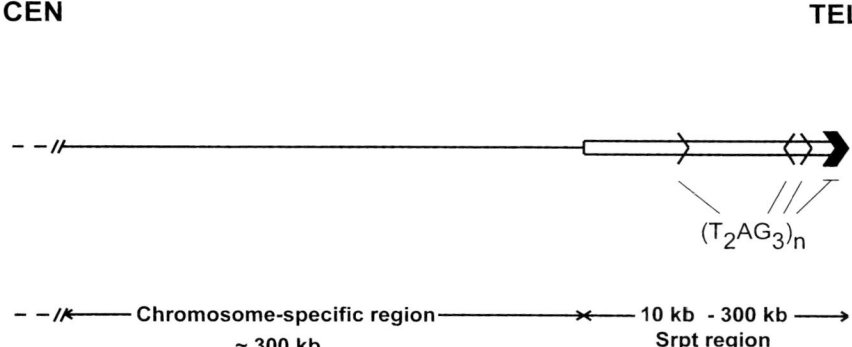

$(T_2AG_3)_n$

– –/⊬⟵———— Chromosome-specific region ————⟶✕⟵— 10 kb - 300 kb ⟶
          ~ 300 kb   Srpt region

Figure 1. Model for human subtelomeric DNA regions. The telomere terminus consists of a 2-50 kb tract of the simple repeat sequence (TTAGGG)n, the length of which is determined by the specific physiology of the cell. Adjacent to this terminal repeat sequence is a subtelomeric repeat (Srpt) region comprised of a mosaic patchwork of segmentally duplicated DNA tracts. The Srpt region can be shorter than 10 kb or longer than 500 kb, depending on the specific telomere. In addition, the size of a Srpt region can vary by hundreds of kb for separate alleles at a single polymorphic telomere in a human population. Centromeric to the Srpt region is conventional single-copy genomic DNA, typically with a high GC content and high gene density. Short (50-250 bp) and often degenerate (TTAGGG) tracts are interspersed within the Srpt region and, occasionally, within the adjacent single-copy subtelomeric DNA. Half-YAC cloning targets the terminal 200 to 400 kb chromosome fragment for isolation, and often but not always spans the entire Srpt region and includes single-copy DNA near the vector end of the cloned fragment.

interstitial chromosomal loci. The duplicated segments can be in either orientation relative to the telomere, and are interspersed with internal (TTAGGG)n homology segments (typically 50 bp to 250 bp tracts of interspered perfect and imperfect (TTAGGG)n motifs), also in either orientation (Azzalin *et al.*, 2001; Riethman *et al.*, 2001). The aggregate size of a subtelomeric repeat region varies according to the specific telomere; the shortest subtelomeric repeat region is 2 kb in length and the longest is greater than 500 kb. At many individual telomeres, allelic differences in the sizes of subtelomeric repeat regions can be quite large, on the order of hundreds of kilobases in length.

The sequence organization of human telomere regions has led to two major complications with respect to mapping and sequencing. The first is underrepresentation of subtelomeric clones in genomic libraries prepared using typical restriction enzymes. The (TTAGGG)n repeat tract at the end of each chromosome lacks restriction sites, preventing the cloning of this sequence and causing a statistical underrepresentation of adjacent sequences for all size-selected clone libraries. There is some speculation that the clone under represention may be exacerbated by clone instability caused by multiple copies of small duplicons and other unstable sequences within clones, but for most telomeres evidence for this is lacking. The second and more severe problem is caused by the large size and close similarity of duplicated subtelomeric DNA segments. The large stretches of nearly identical DNA sequence cause false joins, not only in draft sequence assemblies but even in restriction fragment fingerprint-based clone maps. Whole-genome shotgun sequence assembly of these regions is expected to be especially unreliable. In the latter respects, the problems faced in mapping and sequencing subtelomeric repeat regions are similar in principle (but smaller in scope) than those faced in mapping and sequencing pericentromeric regions (Bailey *et al.*, 2001).

2. Methodology and Strategy

2.1 Overview

Our basic approach to resolving the subtelomeric mapping and sequencing problem has been to isolate each telomere region using a specialized yeast artificial chromosome (YAC) system that permits propagation of large telomere-terminal human DNA fragments as linear plasmids in yeast (Riethman *et al.*, 1989). The result of a successful cloning event is a yeast strain which carries a YAC clone similar to that shown in Figure 2. The

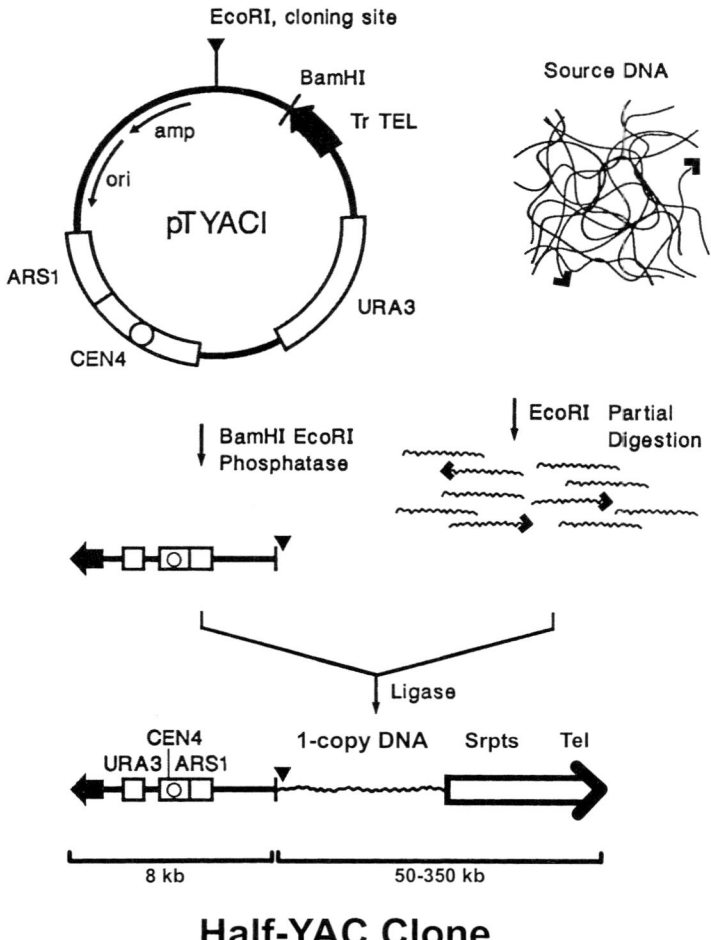

Half-YAC Clone

Figure 2. Half-YAC cloning strategy. The 8 kb vector fragment generated by double-digestion of pTYAC1 with *Bam*HI and *Eco*RI contains one Terahymena (Tr) telomere (which functions as a chromosomal telomere in yeast), a yeast centromere sequence (CEN4), a replication origin (ARS1), and a selectable marker (URA3). The amp and ori sequences used for propagation of the plasmid in *E. coli* are retained on the half-YAC vector fragment. Following phosphatase treatment, the vector fragment is gel-purified to remove the small *Eco*RI-*Bam*HI stuffer fragment. Large source DNA digested partially using *Eco*RI and size-selected to optimize recovery of 200 kb-400kb fragments is then ligated to the vector fragment. Following a second size-selection to remove unligated vector molecules, the ligation mix is introduced into yeast using spheroplast transformation. Large ligated source fragments whose termini are capable of healing into yeast telomeres survive as URA+ transformants, and the subset of these transformants which contain true (TTAGGG)n sequence are selected by DNA hybridization screening with a (TTAGGG)n probe.

terminal repeat tract, the entire subtelomeric repeat region, and the adjacent single-copy DNA region are physically linked on a single large DNA segment that has been purified from the rest of the human genome. From this starting material, the most distal single-copy segments of each chromosome arm can be identified and analyzed, and the particular subtelomeric repeat organization of each cloned fragment can be deciphered without interference from duplicons derived from elsewhere in the genome. The plasmid-rescue method (Figure 3) has proven the most efficient means of rapidly isolating single-copy sequences from the half-YAC clones.

Half-YAC plasmid end-rescue

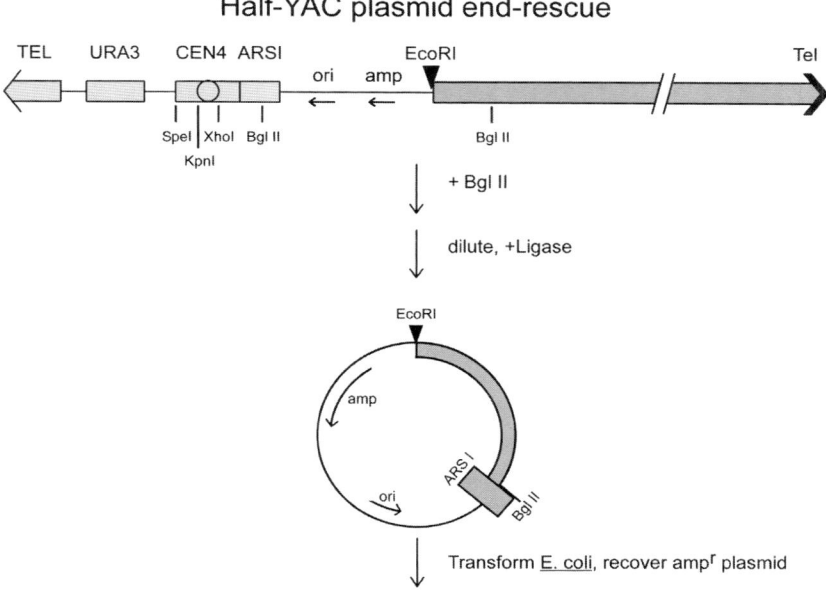

Figure 3. Plasmid end-rescue of single-copy sequences from half-YACs. Digestion of DNA from a half-YAC strain using one of the four enzymes shown (*Bgl*II, *Xho*I, *Kpn*I, or *Spe*I) each generates a single DNA fragment spanning the *Eco*RI cloning site and carrying the ori and amp sequences from the plasmid pBR322. Dilution of the singly-digested DNA (shown for BglII in Figure 3), followed by circularization of the restriction fragments by self-ligation of the fragment ends, generates pBR322-based molecules carrying subtelomeric DNA directly adjacent to the cloning site. Transformation of *E. coli* with this ligation mix results in the selection of only those circularized molecules containing the amp and ori sequences. The half-YAC DNA subcloned in this manner is farthest from the molecular telomere, (most likely to contain chromosome-specific sequences; see Figure 1), and large enough to select only single-copy sequences amidst potential interspersed repeats near the cloning site. These probes are ideal for developing reagents (*e.g.*, STS, hybridization probes) for localization/mapping of the DNA and for isolation of overlapping subtelomeric DNA fragments from large-insert genomic clone libraries.

A major challenge associated with this strategy is to validate the YAC clone structure and ensure that the deduced sequence organization reflects that of native genomic DNA. As with the cloning strategy itself, we took advantage of the unique properties of telomeres to assist with this problem. Because the telomere forms the end of a very large linear DNA molecule, a site-specific cleavage of DNA close to the end of the chromosome is expected to release a DNA fragment whose size corresponds to the distance between the site and the end of the chromosome. A site-specific DNA cleavage method developed in the early 90's (Ferrin and Camerini-Otero, 1991; Koob *et al.*, 1992) was thus applied to this problem in order to validate the sequence organization of individual alleles of many telomeres (Figure 4, Figure 5). Similar mapping methods are likely to be an important tool for analyzing variant subtelomeric regions in human populations.

The elucidation of finished DNA sequence from each of these regions is following several distinct paths, each of which takes advantage of the telomere linkage of half-YAC sequences but which differ in the extent to which DNA sequences are derived from the half-YACs themselves. The phenomenal throughput of the public draft sequencing project allowed connection of most of the telomere regions with the human working draft sequence (Riethman *et al.*, 2001), and for many telomeres relatively small sequence gaps now remain between the end of the working draft sequence and the end of the chromosome (Figure 6). Relatively straightforward strategies, using half-YAC clones as starting material, can be implemented to complete these regions. On the other hand, the sequences of very large subtelomeric regions are not entirely encompassed by single half-YACs, half-YAC clones from the short arm telomeres of the acrocentric chromosome are unstable, and a few human telomeres were not recovered in half-YAC cloning experiments. Significant challenges therefore remain.

2.2 Half-YAC Cloning

Figure 2 shows the basic half-YAC cloning strategy employed in our laboratory. The moderate size and tight size selections targeted by our half-YAC cloning strategy both facilitates library construction (it is technically much less difficult to introduce YACs of this size into yeast as compared to Mega-YACs) and constrains potential chimera and instability problems (clones appearing to be larger than 500 kb are considered likely chimeras, and smaller moderate-sized YACs have a lower probability of carrying specific unstable sequence motifs compared to Mega-YACs). Nonetheless, both chimeric and unstable half-YACs occur in this size range, so any library construction and clone characterization project must take these

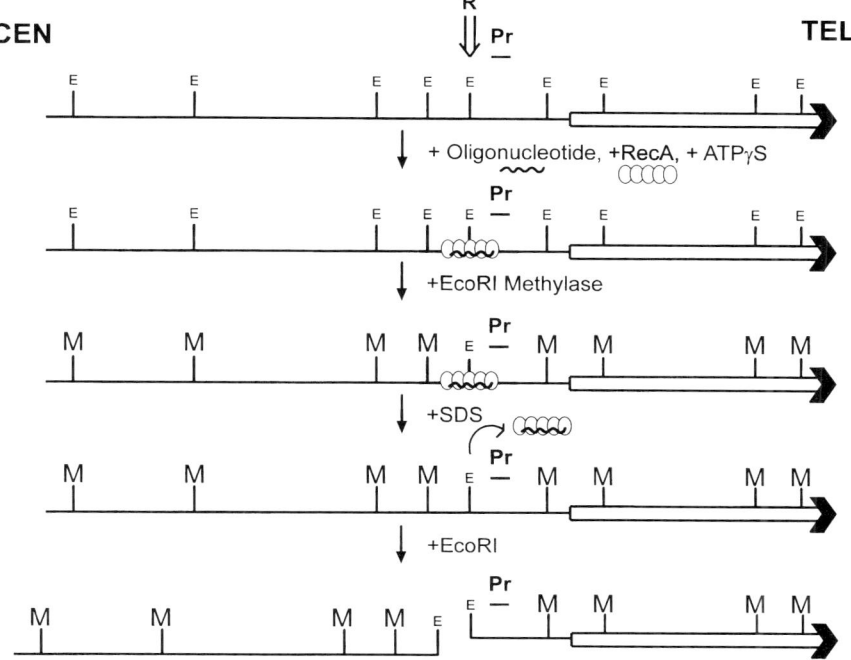

Figure 4. Key steps of *Rec*A-assisted Restriction Endonuclease (RARE) cleavage. A subtelomeric DNA region containing 9 *Eco*RI sites is shown. In the first step, DNA is incubated with an oligonucleotide consisting of unique sequence including and adjacent to the targeted *Eco*RI restriction site, with *Rec*A protein, and with ATP-gamma-S (a non-hydrolyzable analogue of ATP). The product of this reaction is a *Rec*A-protein-coated triple helix at the site that is homologous to the oligonucleotide; this complex protects the targeted *Eco*RI restriction site from the subsequent methylation step. In the second step, all other *Eco*RI restriction sites are methylated by *Eco*RI restriction site methylase. Although only 8 methylated sites are shown in the diagram, for a genomic RARE cleavage reaction this step efficiently methylates the hundreds of thousands of unprotected *Eco*RI sites in the human genome. In the third step, the *Rec*A-protein-oligonucleotide complex is removed by treatment with detergent. Thus, only the targeted site is susceptible to cleavage with restriction endonuclease during the final step, while other *Eco*RI sites recognized by the same endonuclease are protected from cleavage by methylation.

features of YACs into account.

2.3 Isolation of Vector-Insert Junction Probes

The region of the half-YAC clone of most immediate practical value is that portion which is farthest from the human chromosomal telomere. This part of the half-YAC is most likely to contain chromosome-specific sequences (see Figure 1), which are needed to develop mapping/localization reagents either directly (STS and hybridization probes) or indirectly (using these probes to isolate overlapping subtelomeric DNA fragments from separate large-insert genomic clone libraries (*e.g.*, BAC or PAC libraries). DNA

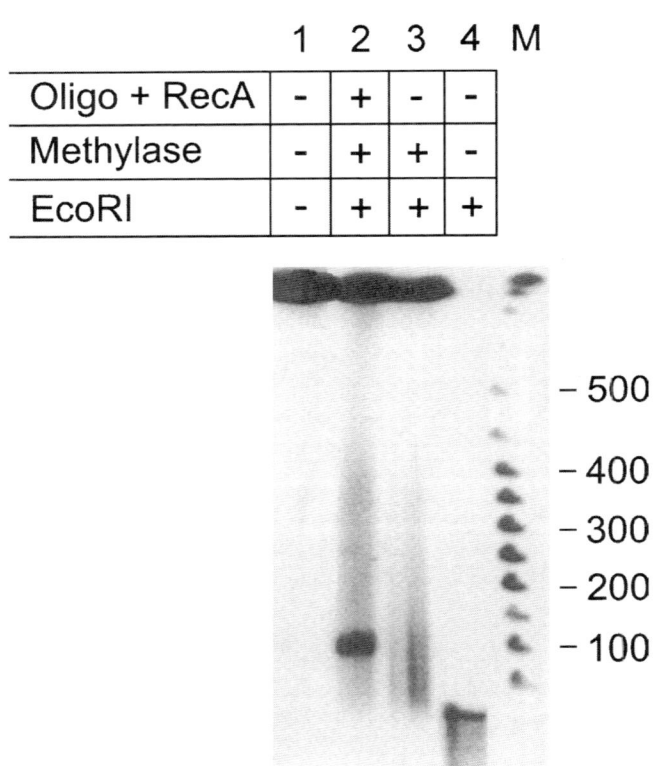

	1	2	3	4	M
Oligo + RecA	-	+	-	-	
Methylase	-	+	+	-	
EcoRI	-	+	+	+	

- 500
- 400
- 300
- 200
- 100

Figure 5. Example of a RARE cleavage experiment of the 13q telomere using the *Eco*RI cloning site from yRM2067 as "R" and a plasmid rescue probe from yRM2067 as "Pr". Genomic DNA was subjected to RARE cleavage at site "R" using an oligonucleotide, then separated on a pulsed-field gel, transferred to a nylon membrane, and hybridized using the plasmid-rescue probe Pr. Lane 1, Genomic DNA taken through the reaction step buffers but not incubated with any of the oligonucleotides or enzymes ("High Molecular Weight DNA quality control"). Lane 2, Genomic DNA sample taken through all of the RARE cleavage reaction steps, producing a single genomic RARE cleavage fragment of the expected size. Lane 3, Genomic DNA subjected to methylation in the absence of oligonucleotide and *Rec*A, then taken through the rest of the protocol ("methylation control"). Lane 4, Genomic DNA taken through the reaction steps in the absence of oligonucleotides, *Rec*A, and methylase, but subjected to *Eco*RI digestion ("*Eco*RI digestion control"). Lane M, lambda DNA concatemer markers (approximate sizes of bands indicated to the right of the figure).

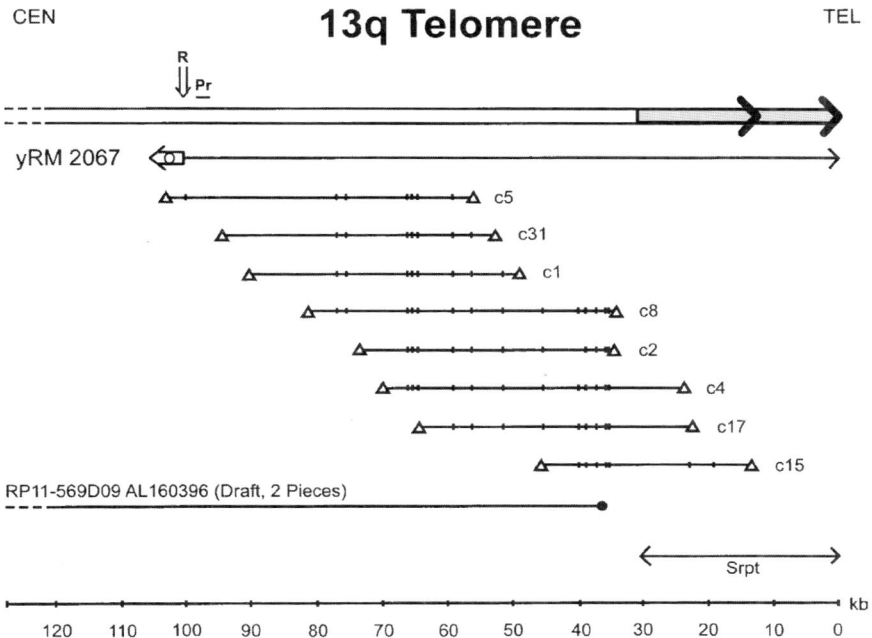

Figure 6. Connecting the 13q telomere to the working draft sequence. The end of the working draft sequence at 13q is represented by the BAC clone RP11-569D09, for which high-quality draft sequence is currently available in two ordered fragments (AL160396). A representation of the chromosomal telomere regions is shown at the top, along with the RARE cleavage site (R) and the single-copy probe isolated from the vector-insert junction region of the half-YAC (Pr) that were used for the mapping experiment shown in Figure 5. The large DNA segment cloned in yRM2067 corresponds in size to the RARE cleavage fragment shown in Figure 5, as predicted if the half-YAC clone carries the telomere-terminal fragment of DNA corresponding to the 13q telomere. Cosmid subclones were derived from the half-YAC clone and the contig of these subclones, which encompasses most of the half-YAC clone, was aligned by *Eco*RI restriction sites (small vertical tics). The cosmid end sequences (triangles) were screened to the working draft sequence to orient the clones relative to the 13q telomere. This physical mapping experiment enabled us to precisely characterize the size and features of the 35kb-telomeric region missing from the 13qtel working draft sequence. The distances (kb) from the 13q (TTAGGG)n telomere are shown on the scale at the bottom. We identified several sequence features, including a 30 kb region of subtelomeric repeat DNA (dark gray) and an internal (TTAGGG)n telomere repeat sequence (internal arrowhead).

isolated from the half-YACs using plasmid end-rescue is the most centromeric segment cloned, and the captured DNA sequences are of sufficient length to permit isolation of probes containing only single-copy sequences amidst potential interspersed repeats near the cloning site.

Alternative PCR-based vector-insert junction isolation strategies, such as vectorette-PCR (Riley *et al.*, 1990) typically produce much smaller fragments than plasmid-rescue; hybridization probe isolation from these short sequences, especially in the presence of interspersed repeat sequences, can be quite problematic. We therefore favor the plasmid-rescue method for isolating vector-insert junction sequences.

2.4 Mapping Telomere Fragments using RARE Cleavage

A RARE cleavage experiment targeting a single genomic site in a subtelomeric region is expected to release a telomere-terminal fragment of genomic DNA. The size of this fragment corresponds to the distance from the cleavage site to the end of the chromosome (Ferrin and Camerini-Otero, 1994). This simple principle makes RARE cleavage mapping an ideal method for physically mapping telomere regions, simplifies validation of half-YAC clone structure, and points the way towards the future analysis of large polymorphisms in subtelomeric regions.

Figure 4 illustrates the key steps of *Rec*A-assisted Restriction Endonuclease (RARE) cleavage. Because the single cleavage is in a subtelomeric region, it liberates a DNA fragment whose size is equal to the distance from the targeted *Eco*RI site to the telomere terminus. Following pulsed-field gel electrophoresis to separate the liberated cleavage fragment from bulk genomic DNA according to size, (and subsequent transfer of the DNA from gel to a membrane by Southern transfer), the fragment can be detected using a single-copy probe (Pr) and its size determined relative to molecular weight markers. When the targeted genomic *Eco*RI site (R) corresponds to the *Eco*RI site adjacent to the vector sequences in the half-YAC cloning experiment, the size of the genomic RARE cleavage fragment should be very similar to the size of the half-YAC. In this instance, single-copy probes derived from half-YAC plasmid end-rescue experiments (Figure 3) provide convenient hybridization probes.

Figure 5 shows an example of a RARE cleavage experiment. A telomere-terminal genomic DNA fragment from the 13q telomere was cleaved using the EcoRI cloning site derived from yRM2067 (designated "R"), and was detected on a Southern blot of the pulsed field gel using a plasmid rescue probe from yRM2067 (designated "Pr"). The size of the genomic RARE cleavage fragment (Lane 2) corresponds to the distance from the yRM2067 cloning site to the telomeric end of the clone (see Figure 6), confirming the structure of the clone and the close physical linkage to the end of 13q. Note the absence of non-specific DNA degradation (Lane 1), and the nearly complete protection of genomic DNA from *Eco*RI digestion by *Eco*RI methylase in the absence of oligonucleotide and *Rec*A (Lane 3). The hybridizing 24 kb *Eco*RI fragment corresponds to the vector-adjacent YAC fragment apparent from the 13qtel cosmid contig (see Figure 6).

2.5 DNA Sequencing of Telomeres

Three basic approaches have been taken in using telomeric half-YACs for sequencing subtelomeric DNA regions. In the first, the half-YAC itself is used to acquire DNA sequence. Typically this is done after subcloning it into molecules more suitable for hierarchical shotgun sequencing (usually cosmids). The cosmid subclones of the half-YAC are aligned using clone fingerprint methods then a tiling path of cosmids across the half-YAC is selected and sequenced (Riethman *et al.*, 1993). The half-YACs can also be gel-purified, shotgun-cloned, and sequenced directly without an intermediate subcloning step; this should be particularly useful for smaller clones, but it is technically more difficult. However, given the success of several sequencing centers in using this approach (Vaudin *et al.*, 1995), as well as the decreasing cost of automated sequencing, it is likely to gain wider acceptance even if the YAC clones overlap significantly with already-sequenced DNA.

In the second approach, the half-YAC is used in combination with more convenient, independently prepared DNA sequencing substrates to identify the telomeric ends of contig maps and to guide clone selection for sequencing particular subtelomeric regions. With this type of approach, the bulk of the sequence data is generated from BACs, PACs, or whole-genome shotgun reads carried out in high-throughput production sequencing mode. The half-YAC clones and their associated reagents are used mainly to identify and correct mislocalized clones and misassembled contigs and sequences (typically caused by subtelomeric repeat sequences in BACs selected for sequencing).

In a third approach, the half-YAC is used both as a mapping reagent and as a source of additional clones to extend DNA sequences aquired in production mode into the most distal telomere regions. In the first stage, vector-insert junction probes are used to identify overlapping BACs and PACs. Then, half-YAC-derived cosmids are mapped to determine the overlap region of the half-YAC with the pre-existing contig. Mapped cosmids are then selected to help cover telomeric gaps in the BAC/PAC coverage, which can be substantial in size (Riethman *et al.*, 2001). Finally, the half-YAC clone is used to directly recover the last, small piece at the telomeric end of a chromsome, which is usually absent from both the BAC/PAC libraries and often from the cosmid subclones of the half-YACs themselves.

Figure 6 illustrates the third approach. The telomeric end of the working draft sequence produced by production sequencing facilities near 13q is represented by the BAC clone RP11-569D09, for which high-quality draft sequence is currently available in two ordered fragments (AL160396). Cosmid subclones were derived from the half-YAC clone yRM2067, and the contig of these cosmid subclones, which encompasses most of the half-YAC clone, was aligned by EcoRI restriction sites (small vertical ticks). The cosmid end sequences (triangles) were ascertained and screened against the working draft sequence to orient the clones relative to the 13q telomere. This physical mapping experiment enabled us to define the terminus of the 13qtel contig in the available working draft sequence, to precisely characterize the size and features of the 35kb-telomeric region missing from the 13qtel working draft sequence, and to identify a half-YAC –derived cosmid (c15) capable of extending the sequence to within about 10 kb of the telomere. We identified several sequence features from the cosmid end-sequences, including a 30 kb region of subtelomeric repeat DNA and an internal (TTAGGG)n telomere repeat sequence.

2.6 Strategies for Recovering and Sequencing Telomeric Gap DNA Fragments

For many telomeres, small segments of DNA remain which are cloned in the half-YACs but are not present in the cosmid subclone libraries of the half-YACs, presumably due to underrepresentation due to restriction site bias or to sequences stable in yeast but unstable in E. coli. Two approaches are currently being taken to recover these sequences and complete the telomere sequences. In the first, primers based upon the subtelomeric repeat content of the missing regions (ascertained by Southern analysis of the half-YACs) are used in Long-range PCR of half-YAC DNA to amplify these sequences. This approach is successful in many cases where the gap size is less than 15 kb.

In the second approach, the existing DNA sequence from the half-YAC is used to target integration of a second half-YAC vector, effectively trimming the large half-YAC clone into a small linear plasmid that contains primarily unsequenced DNA extending to the end of the original half-YAC. This small half-YAC is then maintained under selection, gel-purified, then shotgun cloned and sequenced.

2.7 Missing Telomeres

Several telomeres were not recovered in our half-YAC libraries of human DNA (5p, 19q, 20q), and telomeres from the acrocentric short-arm regions, while represented in the existing half-YAC libraries, were unstable and difficult to work with because of repeated sequences (Riethman, unpublished). All of our libraries were prepared with *Eco*RI partial digests of genomic DNA, and the total depth of half-YAC coverage (after correcting for non-telomeric half-YACs that accounted for approximately 50% of the (TTAGGG)n-hybridizing clones) was roughly three to four-fold, so the absence of these three telomeres might simply be statistical holes, largely remedied by using random shear approaches and/or increasing library depth of coverage. However, it is feasible that some subtelomere sequences, particularly those with extensive tandem repeat structure (Neil *et al.*, 1990), might be refractory to cloning in yeast. Alternative methods for recovering and sequencing these ends might therefore be sought.

3. Protocols

Protocol 1. Half-YAC Cloning

The protocol we have used for half-YAC cloning of telomere fragments avoids preparative agarose gels, using instead high-molecular weight liquid DNA preparations and sucrose density gradients for steps requiring partial digestion and size selection of fragments. The main advantage of this method is that it produces a very high concentration of recombinant YAC molecules in the 200 to 400 kb size range; molecules of this size are relatively easy to introduce into yeast cells at high efficiency, and are of a size range appropriate for spanning most human subtelomeric repeat tracts. Our cloning protocols are based upon those described in Burke and Olson (1991) and elaborated and modified in Macina and Riethman (1992) and Riethman *et al.* (1997). The yeast transformation protocol is based upon Burgers and Percival (1987), and is detailed in Macina and Riethman (1992).

Steps:

1. Preparation of high molecular weight DNA. We prepared bulk high molecular DNA by lysing purified leukocytes or cultured lymphoblasts, then separating the large DNA from cellular debris on a sucrose step gradient. It is important to minimize mechanical

shearing of the large DNA by using wide-bore pipette openings as described in detail (Burke and Olson, 1991; Riethman et al., 1997).

2. Partial digestion of high molecular weight DNA. Light partial digestion is achieved by sampling a range of enzyme concentrations in scaled-up reaction mixes (Riethman et al., 1997). The partially digested DNA is pooled and fractionated on a sucrose gradient as described (Burke and Olson, 1991; Riethman et al., 1997). Fractions containing EcoRI partial digest fragments of the desired size (200 kb to 400 kb) are pooled and concentrated using a vacuum dialysis concentration device.

3. Preparation of vector and ligation to partial digest products. The pTYAC1 vector (Riethman et al., 1989), Figure 2) is doubly digested with EcoRI and BamHI, then the 8 kb linear fragment is gel-purified in bulk. Following Alkaline Phosphatase treatment of vector ends, a large molar excess of the vector fragment is ligated to the concentrated EcoRI partial digest fragments of high molecular weight DNA (Riethman et al., 1989).

4. The ligation products are pooled, and separated from unligated vector using a sucrose density gradient (Riethman et al., 1989; Riethman et al., 1997). This step is critical for half-YAC cloning. Contaminating unligated vector circularizes very efficiently upon transformation into yeast cells, and these contaminating transformants overwhelm true recombinant molecules if they are not physically removed prior to transformation.

5. The sucrose gradient fractions lacking unligated vector contamination are pooled according to size ranges, concentrated by vacuum dialysis, then introduced into the yeast host strain by spheroplast transformation.

6. Our laboratory protocol for spheroplast transformation is detailed in Macina and Riethman (1992). It is based upon that of Burgers and Percival (1987) with modifications to permit direct DNA hybridization screening of primary yeast transformants by colony hybridization using a (TTAGGG)n probe.

7. Transformants selected on the basis of (TTAGGG)n hybridization are analyzed individually on pulsed-field gels, and re-tested by hybridization for the presence of (TTAGGG)n on a YAC molecule.

8. Positive clones are then analyzed by isolation of vector-insert junction probes and FISH localization of YAC DNA. The library is arrayed and prepared for screening by either hybridization to colonies or by PCR of row and column pools of colonies from the arrayed library.

Protocol 2. Isolation of Vector-Adjacent Half-YAC Sequences Using Plasmid-Rescue

Steps:

1. DNA is purified from the YAC-bearing yeast strain. Many protocols are available; we prefer the standard method for purification of yeast DNA in solution of Green *et al.* (1999). Yeast DNA purified in this manner usually cuts well with restriction enzymes and amplifies easily using PCR. The DNA is dissolved in TE8 to a final concentration of 0.2 µg/µl.

2. Two µg of the DNA are digested to completion with 20 units of restriction enzyme, for 2 hrs in a 100 µl reaction volume. Enzymes appropriate for half-YACs constructed using pTYAC1 include *Bgl*II, *Xho*I, *Spe*I, and *Kpn*I. *Bgl*II and *Spe*I typically generate smaller rescued fragments than *Kpn*I or *Xho*I. Ten µl of the digested material is checked on a gel to ensure complete digestion. If digestion is complete, the restriction enzyme is inactivated with heat.

3. Ten µl of the digested material (200 ng DNA) is added to a 200 µl volume ligation reaction. One unit of T4 ligase is added, and ligation is carried out overnight at 16 degrees C. At this DNA concentration, circularization of individual molecules is highly favored over intermolecular ligation.

4. The 200 µl ligation reaction is extracted with 1 volume of phenol, followed by one volume of chloroform. DNA in the aqueous phase is ethanol-precipitated, then dissolved in 5 µl of water.

5. 25 µl of competent DH5alpha cells (subcloning efficiency, Life Technologies) are then electroporated with 2.5 µl of the circularized material and plated onto L + amp.

6. Colonies are grown and screened for inserts using double-digestion with *Eco*RI + the enzyme used for plasmid rescue.

Protocol 3. RARE Cleavage Near Telomeres

Several protocols have been developed for RARE cleavage. A thorough description of the procedure as well as several protocols is provided in Riethman *et al.* (1997). Our current protocol closely resembles that devised by Ferrin and Camerini-Oterro (1994).

RARE cleavage of DNA embedded in agarose requires high-quality agarose blocks of high molecular weight DNA, at the proper DNA concentration (a final embedded cell concentration of 1.5×10^7 cells/ml is optimal for diploid mammalian cells). It is particularly important to avoid falling below this DNA concentration, as this will make eventual detection of RARE cleavage fragments on Southern blots of pulsed field gels extremely difficult. The proteinase K high molecular weight DNA preparation method described in Riethman et al. (1997) works reliably for us – it is important, however, to inactivate all residual protease activity prior to initiating the RARE cleavage experiment (see Riethman et al., 1997 for details). Another key step in the protocol is the penetration of the agarose with a large RecA-oligonucleotide nucleoprotein complex. There are several ways to optimize this step, also discussed in detail in Riethman et al; (1997); we have obtained the best results by chopping agarose blocks into a slurry of small cubes immediately before introducing the nucleoprotein complex. A final key step, described initially by Ferrin and Camerini-Otero (1994), is to set the pH of the RARE buffer to 7.85 rather than 7.5 (the pH of the original protocols). This seemingly small modification has a dramatic effect upon experimental results.

It is important to pre-test both the DNA sample and the hybridization probe by a conventional restriction digest and gel-transfer hybridization. The probe must be able to detect a single-copy fragment in the sample with high sensitivity; the strength of this signal is equivalent to a perfectly efficient RARE cleavage. Actual single-cut RARE cleavage experiments are usually 50 –80 % efficient in our hands.

The size and characteristics of oligonucleotides used for RARE cleavage are discussed in detail elsewhere (Riethman et al., 1997). We recommend using a 60-mer. If the entire sequence is known, position the EcoRI site near the center of the oligonucleotide. If only a half-site is known, add 10 random bases beyond the GAATTC of the EcoRI site, in addition to 44 bases of homology to the known flanking sequence. Either method produces excellent results. We have not had good success with restriction endonuclease/methylase systems other than EcoRI on human genomic DNA samples.

It is recommended to include the following set of control reactions for each RARE experiment: a control without oligonucleotide, RecA protein, and EcoRI restriction endonuclease (i.e., the DNA is methylated but not digested), a control without oligonucleotide, RecA protein, and EcoRI restriction methylase (i.e., not methylated but digested) and a control without oligonucleotide, RecA protein (i.e., methylated and treated with restriction enzyme).

*Rec*A protein is commercially available at concentrations around 2 mg per ml from several suppliers (*e.g.*, New England Biolabs, Epicenter Technologies). ATP-gamma-S at 100 pmol/µl can be purchased from Boehringer Mannheim or can be prepared at this concentration from the powder form (Sigma # A-138; Fluka #01995).

Steps:

1. Purify Leukocytes from peripheral blood using HISTOPAQUE-1077 following instructions from the manufacturer (Sigma) or, for cell lines, simply harvest intact cells using standard methods. Wash cells in phosphate-buffered saline pH 7.0, and suspend the cells to a concentration of 3×10^7 cells/ml PBS. Embed the cells in Low-melting point agarose to a final concentration of 1.5×10^7 cells per ml of 1% agarose, then purify intact DNA embedded in the blocks using the protease K protocol (Riethman *et al.*, 1997).

2. Equilibrate intact agarose blocks for 4×10 minutes in 1 X RARE buffer containing 100 µg/ml acetylated BSA. For each buffer change, use 10 ml of buffer for ten 100-µl plugs in 50-ml culture tube. If done with different DNA samples, use 1 ml of buffer for each plug in a 1.5-ml microcentrifuge tube. Each RARE cleavage reaction requires one 100 µl agarose block containing 1.5×10^7 cells/ml concentration for diploid human cells.

   10 x RARE buffer (without BSA):

   0.25 M Tris-Acetate, pH 7.85

   40 mM Mg-Acetate

   4 mM DTT

   5 mM spermidine-trihydrochloride

   Filter sterilize and store in small aliquots at -20°C

   Add BSA to the diluted RARE buffer immediately before use.

3. After equilibration, dice each plug into small cubes (close to a slurry) using a sharp, sterile, razor blade. Keep the diced agarose from each plug in an individual 1.5-ml microcentrifuge tube.

4. In a separate 1.5-ml microcentrifuge tube, prepare 89 µl of *Rec*A-oligonucleotide mix for each reaction. Prepare by adding, in order:

   49 µl of $2 \times$ RARE buffer containing 200 µg/ml acetylated BSA

   30 µl of *Rec*A protein (2.0 mg/ml stock, NEB)

   4 µl of water plus 60-mer oligonucleotide at the desired concentration

   This mix is incubated for 60 seconds at 37°C prior to adding:

   0.89 µl of 0.25 M ADP

3.9 µl of 5 mM EGTA

1.2 µl of 50 mM ATP-gamma-S (Fluka catalog #01995)

5. Incubate the 89 µl of *Rec*A-oligonucleotide mix and the diced agarose plugs separately for 10 minutes at 37°C.

6. Add the 89 µl of *Rec*A-oligonucleotide mix to the tubes containing the diced plugs. Mix gently and completely by tapping or low-speed vortexing and incubate for 30 minutes at 37°C.

7. Add 1 µl of 32 mM S-adenosyl methionine and 10 µl (400 units) *Eco*RI methylase (New England Biolabs, catalog number 211L). Continue the incubation for an additional 90 minutes at 37°C.

8. Add 200 µl of 2% SDS and incubate for a final 30 minutes at 37°C.

9. Collect the diced plug by centrifugation (5 minutes, room temperature, 4500 rpm in a microcentrifuge).

10. Wash the diced plug 4 × 10 minutes at room temperature with 1 ml of TE buffer, pH 8.0, collecting the agarose as indicated in step 9 after every wash.

11. Equilibrate each diced plug 5 × 10 minutes with 1 ml *Eco*RI RARE-cleavage buffer containing 500 µg/ml BSA. Collect the diced plugs after each step by centrifugation as indicated in step 9.

    10 X *Eco*RI RARE-cleavage buffer (without BSA):

    0.5 M Hepes-NaOH, pH 7.0

    1.5 M NaCl

    100 mM $MgCl_2$

    10 mM DTT

    Filter sterilize and store in small aliquots at -20°C.

    Add BSA to 1x buffer immediately before use.

    *Eco*RI RARE-cleavage buffer is a modified restriction buffer that has neutral pH and a high NaCl concentration to minimize *Eco*RI star activity.

12. After the last equilibration step, suspend the diced plug in 200 µl of *Eco*RI RARE-cleavage buffer with 500 µg/ml BSA and 80 units of *Eco*RI restriction endonuclease. Use a high concentration stock (100 units/ µl) of *Eco*RI so the glycerol conc of the reaction mix stays acceptably low. Let the reaction proceed for 2 hours at 37°C.

13. Load the diced plug pieces into a single well of a 1% agarose pulsed-field gel using a spatula, then overlay with molten 1% low-melting-point agarose to seal in the well. Alternately, equilibrate the diced plug for 10 minutes in 1 ml of 225 mM NaCl in TE buffer, pH 8.0. Collect the diced plug by centrifugation as indicated in step 8, melt the agarose at 65°C and load into a well using a pipette with a cut-off plastic tip.

14. Run the pulsed field gel using conditions to optimize separation and sizing of the expected RARE cleavage fragment (Riethman *et al.*, 1997). Prepare a Southern blot of the pulsed-field gel, and hybridize using conditions designed to optimize detection of single-copy DNA fragments.

Conclusion

The half-YAC cloning strategy has been very successful in isolating human subtelomeric DNA regions, and in assisting with their mapping and sequencing. Many variant telomere alleles have been discovered in the human population by RARE cleavage analysis and by FISH methods (Trask *et al.*, 1998). New half-YAC libraries with a much greater depth of coverage, or perhaps the use of targeted cloning approaches, will be required to recover and analyze these variant telomeres from the appropriate source materials. In addition, each new eukaryotic genome project, whether following a whole-genome shotgun approach or a more conventional clone-based approach, will leave gaps near telomeres that can be effectively identified and closed using half-YAC cloning approaches. Because of the near-universality of the eukaryotic telomere sequence motif, this unique approach for closing the maps and sequences of telomeric regions of genomes should be applicable to almost every eukaryotic genome.

Acknowledgements

This work was supported by the NIH

References

Azzalin, C.M., Nergadze, S.G., and Giulotto, E. 2001. Human intrachromosomal telomeric-like repeats: sequence organization and mechanisms of origin. Chromosoma. 110: 75–82.

Bailey, J.A., Yavor, A.M., Massa, H.F., Trask, B.J., and Eichler, E.E. 2001. Segmental duplications: organization and impact within the current human genome project assembly. Genome Res. 11: 1005–17.

Blasco, M.A., Gasser, S.M., and Lingner, J. 1999. Telomeres and telomerase. Genes Dev. 13: 2353–9.

Burgers, P.M., and Percival, K.J. 1987. Transformation of yeast spheroplasts without cell fusion. Anal Biochem. 163: 391–7.

Burke, D.T., and Olson, M.V. 1991. Preparation of clone libraries in yeast artificial-chromosome vectors. Methods Enzymol. 194: 251–70.

de Lange, T., and Jacks, T. 1999. For better or worse? Telomerase inhibition and cancer. Cell. 98: 273–5.

Ferrin, L.J., and Camerini-Otero, R.D. 1991. Selective cleavage of human DNA: RecA-assisted restriction endonuclease (RARE) cleavage. Science. 254: 1494–7.

Ferrin, L.J., and Camerini-Otero, R.D. 1994. Long-range mapping of gaps and telomeres with RecA-assisted restriction endonuclease (RARE) cleavage. Nat Genet. 6: 379–83.

Green, E.D., Hieter, P., and Spencer, F.A. 1999. Yeast Artificial Chromosomes. In: Genome Analysis: A Laboratory Manual Series, Volume 3: Cloning Systems. B. Birren, E.D. Green, S. Klapholzet al ed. Cold Spring Harbor Laboratory Press, Cold Spring Harbor, NY. 3 p. 297–565.

Griffith, J.D., Comeau, L., Rosenfield, S., Stansel, R.M., Bianchi, A., Moss, H., and de Lange, T. 1999. Mammalian telomeres end in a large duplex loop. Cell. 97: 503–14.

Koob, M., Burkiewicz, A., Kur, J., and Szybalski, W. 1992. RecA-AC: single-site cleavage of plasmids and chromosomes at any predetermined restriction site. Nucleic Acids Res. 20: 5831–6.

Macina, R.A., and Riethman, H.C. 1992. Direct DNA hybridization screening of primary yeast transformants in the construction of targeted yeast artificial chromosome (YAC) libraries. Genet Anal Tech Appl. 9: 58–63.

Moyzis, R.K., Buckingham, J.M., Cram, L.S., Dani, M., Deaven, L.L., Jones, M.D., Meyne, J., Ratliff, R.L., and Wu, J.R. 1988. A highly conserved repetitive DNA sequence, (TTAGGG)n, present at the telomeres of human chromosomes. Proc Natl Acad Sci U S A. 85: 6622–6.

Neil, D.L., Villasante, A., Fisher, R.B., Vetrie, D., Cox, B., and Tyler-Smith, C. 1990. Structural instability of human tandemly repeated DNA sequences cloned in yeast artificial chromosome vectors. Nucleic Acids Res. 18: 1421–8.

Riethman, H.C., Birren, B., and Gnirke, A. 1997. Preparation, manipulation, and mapping of high molecular weight DNA. In: Genome Analysis: A Laboratory Manual Series, Volume 1: Analyzing DNA. B. Birren, E.D. Green, S. Klapholz, R.M. Myers and J. Roskams ed. Cold Spring Harbor Laboratory Press, Cold Spring Harbor, NY. 1 p. 83–248.

Riethman, H.C., Moyzis, R.K., Meyne, J., Burke, D.T., and Olson, M.V. 1989. Cloning human telomeric DNA fragments into Saccharomyces cerevisiae using a yeast-artificial-chromosome vector. Proc Natl Acad Sci U S A. 86: 6240–4.

Riethman, H.C., Spais, C., Buckingham, J., Grady, D., and Moyzis, R.K. 1993. Physical analysis of the terminal 240 kb of DNA from human chromosome 7q. Genomics. 17: 25–32.

Riethman, H.C., Xiang, Z., Paul, S., Morse, E., Hu, X.L., Flint, J., Chi, H.C., Grady, D.L., and Moyzis, R.K. 2001. Integration of telomere sequences with the draft human genome sequence. Nature. 409: 948–51.

Riley, J., Butler, R., Ogilvie, D., Finniear, R., Jenner, D., Powell, S., Anand, R., Smith, J.C., and Markham, A.F. 1990. A novel, rapid method for the isolation of terminal sequences from yeast artificial chromosome (YAC) clones. Nucleic Acids Res. 18: 2887–90.

Trask, B.J., Friedman, C., Martin-Gallardo, A., Rowen, L., Akinbami, C., Blankenship, J., Collins, C., Giorgi, D., Iadonato, S., Johnson, F., Kuo, W.L., Massa, H., Morrish, T., Naylor, S., Nguyen, O.T., Rouquier, S., Smith, T., Wong, D.J., Youngblom, J., and van den Engh, G. 1998. Members of the olfactory receptor gene family are contained in large blocks of DNA duplicated polymorphically near the ends of human chromosomes. Hum Mol Genet. 7: 13–26.

Vaudin, M., Roopra, A., Hillier, L., Brinkman, R., Sulston, J., Wilson, R.K., and Waterston, R.H. 1995. The construction and analysis of M13 libraries prepared from YAC DNA. Nucleic Acids Res. 23: 670–4.

From: *Genome Mapping and Sequencing*
© 2003 Horizon Scientific Press, Wymondham, UK

10

Shotgun Sequencing

M.C. Jones and S.K. Sims

Abstract

Shotgun sequencing is the strategy of choice for large scale sequencing projects. In addition to comparing shotgun sequencing with alternative strategies, this chapter details the methodologies of producing high quality random subclone libraries, template production and sequencing in use at the Wellcome Trust Sanger Institute.

1. Introduction

DNA sequence is obtained by copying a template strand starting from a specific priming site (at which a primer oligonucleotide has annealed), under conditions that allow the inference of the sequence of bases in the template strand. The length of DNA sequence obtained in a single sequencing reaction is constrained. Consequently smaller overlapping sequences have to be built up in order to obtain DNA sequence from large pieces of DNA. The strategy chosen to obtain these overlapping sequences depends upon a number of factors. Directed sequencing (in which specific subclones of the original DNA are generated by restriction endonuclease digestion followed by ligation into sequencing vectors) requires relatively few sequencing reactions, but requires many molecular manipulations that are difficult to automate. Primer walking, in which new primers are generated using information from the preceding round of sequencing reactions, is a process in which progress is limited by the rate of sequence acquisition and primer synthesis. Shotgun sequencing requires more sequencing reactions than either directed sequencing or primer walking, but it is very cost-effective for obtaining sequence from large DNA molecules. This cost-effectiveness derives from the economies of scale that apply when treating different subclones in an identical fashion, yet obtaining their unique DNA sequences. In shotgun sequencing, multiple copies of the DNA molecule from which sequence is to be obtained are fragmented randomly into smaller pieces that are then inserted into the cloning site of a sequencing vector. Near to the cloning site of the sequencing vector is a priming site for the initiation of DNA sequencing. The primer oligonucleotide and other sequencing reagents for different randomly generated subclones are the same, but the sequence generated will be unique to individual subclones. The sequences generated from different templates can be compared one with another and where sequences are identical they can be considered to overlap. A series of overlapping DNA sequences can be used to generate a contiguous consensus sequence, which replicates the sequence of the original DNA molecule.

Shotgun sequencing was first described in the early 1980s (Gardener *et al.*, 1981; Anderson, 1981; Sanger *et al.*, 1982) initially for relatively small sequencing projects. It is possible to generate genomic DNA sequence from a whole genome shotgun, where the entire genome is subjected to shotgun sequencing. However, many large genome DNA sequencing projects depend upon an initial phase of generating large subclones of genomic DNA. Such large initial subclones (referred to subsequently as primary subclones) are assayed at relatively low resolution, for instance by *Hind*III agarose gel fingerprinting, and arrayed into overlapping contigs. A minimal set of overlapping primary subclones (the minimal tiling path) are chosen, and these are subjected to a further round of fractionation and subcloning from which templates and ultimately DNA sequence are obtained. The advancements in sequencing technology and strategies using the shotgun approach have had a dramatic effect on genome projects in the last 5 years.

A wide variety of primary subclone systems have been used in genome sequencing projects. These include lambda-phage clones, cosmids, fosmids, PACs (P1 artificial chromosomes), BACs (bacterial artificial chromosomes) and YACs (yeast artificial chromosomes). Each primary subclone consists of the cloning vector, which is common to all the primary subclones within a library, and the insert DNA derived from the genome of interest. Primary subclone systems have differing capacities in terms of the size of insert DNA that they can accept. It is possible to subclone each type of primary subclone in its entirety without first purifying the insert DNA but the subclone library will contain subclones of the cloning vector. However for systems accommodating large inserts, the cost of wasted effort involved in obtaining cloning vector sequence will be relatively small, and will normally outweigh the problems associated with insert purification. PACs and BACs with inserts ranging up to 200 kbp or more are the best reagent for this type of sequencing currently available.

In this chapter we consider the production and fractionation of primary subclones (see Chapters 3, 4 and 6), ligation of those fragments into sequencing vectors, and subsequent preparation and sequencing of templates derived from those fragments.

There are two main sequencing methods that were developed over 20 years ago, the chain termination method using dideoxynucleotides (ddNTP's) (Sanger *et al.*, 1977) and the Maxim-Gilbert method that used chemical degradation of end-radiolabelled DNA fragments (Maxim, and Gilbert, 1977). In genome sequencing projects the Sanger method is more commonly used as it is technically easy, automatable and produces cost effective high quality sequence. Original sequencing technologies relied on using radioisotopes where the sequence was manually determined by reading autoradiographs. The current sequencing chemistry labels DNA with fluorescent dyes, these are excited by a laser and detected in automated systems. The accuracy and speed by which sequence can be read has increased considerably with each advancement in technology, the most recent advance being the move from gel to capillary-based sequencing instruments allowing 96 samples to be anaylsed in parallel.

2. Theory and Strategies

2.1 Large Scale DNA Preparation

Primary subclones need to be prepared in sufficient quantity to confirm their identity, generate subclone libraries and assist in obtaining and confirming finished DNA sequence. Most DNA preparation methods are based on the alkaline lysis method described by Birnboim and Doly (1979). Following cellular lysis, these methods rely upon the host chromosomal DNA being trapped within precipitated proteinaceous materials (including the remnants of the peptidoglycan bacterial cell walls), whilst the primary subclone DNA is released into the supernatant. Recovery of the primary subclone DNA from the supernatant (the "cleared lysate") is the subject of a large variety of methodologies. The method described in protocol 1 depends upon a series of differential precipitation steps combined with enzymatic removal of RNA and matrix binding of residual proteins. Other methods include ion-exchange and gel-exclusion chromatography, both of which are incorporated into a number of commercially available products designed to make the process quicker or more convenient. Levels of contaminating host DNA in the cleared lysate are critical for all of these methods. All the DNA present at the end of the preparation will be available for subsequent steps, and the level of contaminating DNA will be reflected in the presence of host DNA sequences in the final sequence. Experience with manual large scale DNA preparations suggests that it is essential to be very gentle during the lysis and neutralization steps of the process (see Protocol 1) in order to minimize levels of host DNA contamination.

An alternative if somewhat dated strategy is to employ CsCl isopycnic density ultracentrifugation of the cleared lysate in the presence of ethidium bromide (Clewell and Helinski 1969). This is able to differentiate between primary subclone DNA and contaminating host DNA on the basis of physical conformation. The primary subclone DNA (which is closed circular) physically limits the amount of ethidium bromide which can intercalate. There is no such constraint on contaminating host DNA, and the CsCl gradient exploits this difference to separate the types of DNA. Unfortunately there is a heavy penalty to pay both in terms of time (gradients will typically run in overnight) and loss of potentially useful material (as the band containing host DNA also contains any nicked primary subclone DNA). It is also possible to use an enzymatic method to digest contaminating host DNA. Contaminating host DNA should be a substrate for ATP dependant DNase (available from Epicenter Technologies as Plasmid -Safe).

2.2 Coverage

The Poisson model of Lander and Waterman (1988) used primarily for DNA mapping techniques can be applied to shotgun clone sequencing (Lander and Waterman 1988; Roach, 1995). This calculation can be used to determine the amount of random sequence required to cover a specific percentage of the clone.

For instance to randomly shotgun sequence a 100Kb clone at 1x coverage, would result in approximately 63% of the clone having sequence coverage. This type of random sequencing, results in some regions being sequenced several times, whilst others not at all (Table 1). However given that upwards of 3x coverage is required to give 95% sequence coverage across a clone, it is usual to stipulate the number of reads required by proportion.

Table 1. Shows fold coverage for a clone compared to the percentage of the clone sequenced.

Fold Coverage	% clone sequenced
0.25X	22%
0.5X	39%
1.0X	63%
2.0X	88%
3.0X	95%
4.0X	98%
5.0X	99.4%
8.0X	99.97%
10.0X	99.995%

To determine the number of reads required to cover a clone, there are a number of factors that need to be taken into consideration these include:

1. The size of the clone
2. The coverage (depth of reads) required across the clone
3. The expected average length of the reads

The equation below gives an estimate of the number of reads required:

(clone size (bp) / average read length) x coverage required = no. of reads
e.g. Clone = 150,000bp
Average read length = 450bp
Coverage required = 8 depth

$(150,000/450) \times 8 = 2667$ reads required

A percentage failure rate from template production and sequencing should also be built into the equation. To ensure that the clone has adequate coverage, around 10-20% extra should be added to the final number of reads required.

2.3 Fragmentation of Primary Subclones

Generally in clone-by-clone sequencing, the primary subclone DNA will not be limiting, and so any fragmentation method producing fragments within the optimum range will be suitable. However, it will be most efficient to select a method which produces fragments with a small distribution about the optimum size, since this should minimize the amount of starting primary subclone DNA required which affords a greater range of DNA preparation methodologies.

It is unlikely that any method of fragmentation of DNA is entirely independent of base sequence, and so cannot be truly random. However the methods described here tend towards the greatest levels of sequence-independence, and when combined with gap-closing strategies (see Chapter 11) should enable sequence data to be obtained for most genomes.

Random fragmentation can be enzymatic or rely on physical breakage of the DNA. Typically enzymatic methods would employ mixtures of restriction endonucleases (Flint *et al.*, 1988), or DNaseI in the presence of manganese ions (Anderson 1981). Physical methods include sonication (Deininger 1983), passing through a fine-gauge needle, use of a French pressure cell (Schriefer *et al.*, 1990), use of a recirculating point-sink flow system (Oefner *et al.*, 1996) (Hydroshear), and nebulising (Bodenteich *et al.*, 1993). In the context of clone-by-clone sequencing, factors such as capacity (number of clones which can be fragmented in a unit time), prevention of cross contamination between clones and reliability have to be taken into account alongside issues of optimizing fragment size and distribution. Methods that employ disposable equipment (enzymatic methods, sonication, needle-shear and nebulisation), and have adjustable parameters allowing optimization of fragment sizes are likely to be most suitable for large scale projects.

Protocol 3 details the sonication method that has been used at the Sanger Institute during the sequencing of yeast, nematode worm, human, mouse, zebrafish, and most small genome projects. This produces fragments in the range 1-4kb. Once the system has been calibrated it is very reproducible, especially for homogenous DNA suspensions.

2.4 End Repair and Cloning Strategy

Once fragments have been generated the ends have to be made suitable for ligation into the sequencing vector. Enzymatic methods leave defined ends. If a mixture of restriction endonucleases is chosen, these should leave identical ends, or a mixture of vector molecules cut with suitable enzymes will be required. For physical methods of disruption it will be necessary to make fragment ends physically identical so as to be compatible with the ends of the sequencing vector. Normally this involves making the ends blunt. Cloning blunt fragments directly into sequencing vector cut at a blunt site (*Sma*I, *Pvu*II or *Eco*RV) should give adequate results. Alternative strategies could employ adaptors or linkers, which could give a higher efficiency of ligation into sequencing vector. Generating blunt ends requires either "filling in" (making single stranded stretches of DNA double-stranded by enzymatically copying the single strand using DNA polymerase and deoxynucleotide triphosphates) or "cutting back" using single strand specific exonucleases such as mung bean nuclease.

2.5 Sequencing Vector

Choice of sequencing vector depends on a number of factors. Often this is influenced by the needs of automated or semi-automated processes downstream of subcloning itself, in particular automated colony picking systems. Some vectors are widely available commercially in a variety of forms (including those that are ready for ligation) that can be very convenient for a large scale subcloning facility.

An example of a general purpose vector commonly used for sequencing is pUC18 (Vieira and Messing, 1982). A commercially available form of this vector, ready for ligation (having been digested with *Sma*I and treated with bacterial alkaline phosphatase), can be obtained from Qbiogene.

2.6 Template Preparation

Either single stranded or double stranded DNA vectors may be used for clone based shotgun sequencing. Both have been used in large scale genomic projects with equal success, however the advantage of double stranded DNA is that forward and reverse sequencing can be performed immediately without need for a further template preparation.

The most common and convenient single stranded DNA cloning vector is the bacteriophage M13 (Messing 1991). Double stranded DNA fragments to be sequenced are ligated into suitable M13 RFI DNA. The recombinant DNA is transformed into a suitable bacterial host (which requires an F' pilus for propagation of the phage), and phage particles containing a single stranded form of recombinant DNA are produced. The DNA needs to be removed from the phage, a method for which is described in protocol 4.

A typical plasmid vector used for propagating cloned DNA fragments is pUC18.

The most common and cheapest method for extraction of plasmid DNA is a modification of the alkaline-lysis preparation (Birnboim and Doly, 1979), a version of which is described in protocol 5 and protocol 6.

To produce successful sequence data it is essential to have high quality, quantified purified DNA. For large-scale sequence production, ensuring that the DNA you have purified contains an insert and is of a correct concentration is an important quality control step. Running an agarose gel with a standard marker will ensure that the DNA is of an adequate quality, correct concentration and size to proceed with sequencing. (Protocol 6).

2.7 Sequencing

Fluorescent automated DNA sequencing where one lane contains 4 dyes has been a great step in increasing the throughput and efficiency of reading DNA sequence, together with reducing the amount of manual intervention. There are two main fluorescent methods for sequencing. Primer chemistry uses four separate base reactions with the primer having an individual dye for each base attached to the 5' end of the sequencing primer. Four reactions are set up for each sequence, each containing one of the four dideoxynucleotides (ddATP, ddCTP, ddGTP or ddTTP), in addition to the four deoxynucleotides, a fluorescently labeled oligonucleotide primer and DNA polymerase. As the reaction progresses normal polymerisation will begin at the primer and work along the template copying the DNA until a ddNTP is incorporated, at which point the chain terminates. Although this method gives high quality sequence, it does have the limitation that specifically modified primers must be used and four separate reactions must be performed before combining into a single mix for detection. Terminator sequencing uses each of the four dideoxynucleotides individually labeled with a high-sensitivity dye and hence can be multiplexed into a single reaction (Rosenblum et al., 1997).

Historically terminator sequencing was less efficient and more prone to difficulties in purification than primer sequencing. Terminators were used primarily in the finishing processes to solve structural problems such as compressions. However advancement in dye technology has enabled terminator chemistry to give a better quality sequence with more even peak height. Two types of terminator reactions are described in protocols 7 and 8.

2.8 Sequence Assembly

Once the raw DNA sequence has been detected by electrophoresis, a base-calling program needs to be used to process the data, which is then assembled and displayed in a user friendly interface (see Chapter 12). The two interfaces that are widely used are GAP4 (Bonfield et al., 1995) and consed (Gordon et al., 1998). Once a clone has been assembled and displayed in one of these programs it is easy to determine if you have met the required objectives for the shotgun of that clone *i.e.* that the required coverage has been obtained, and that assembly of the data has formed the desired number of contigs. If the target coverage is ten-fold and only six-fold has been obtained, more reads may be sequenced to increase this coverage. The additional sequence required can be worked out by using the following equations:

$$\frac{\text{Coverage acquired}}{\text{Coverage required}} \times 100 = \text{\% coverage gained}$$

$$\frac{\text{Number of reads >2Kb}}{\text{\% coverage gained}} \times (100\text{-\%coverage gained}) = \text{Number of reads required to get desired coverage}$$

A percentage failure rate needs to taken into account to prevent over or under shotgun. This can be estimated by looking at the previous number of reads and their failure rate.

At this point a number of other important parameters of the analysis can be verified, such as whether the clone is the expected size. At the Sanger Institute a restriction enzyme digest is usually performed on a clone before it is sequenced in order to establish the size of the clone, and this information can be compared to the shotgun assembly. Deviations from the expected size can be very helpful in revealing potential contamination of the project with unwanted sequences. In addition it is important to confirm the identity of the clone, as in any large scale process mistakes will inevitably occur. This can be done through analysis of the restriction digest, but more powerfully sequence comparisons to other clones being sequenced (for instance using BLAST) will establish whether the sequence has the expected overlaps to adjacent clones in the tile path. Similarly the presence of the appropriate marker data (sequence tag sites (STS)) in the sequence is useful to confirm identity.

With large scale sequencing there is always the possibility of contamination from another clone, which is being prepared/sequenced at the same time. It is important to recognize that a clone is contaminated and remove the contaminating reads. Contamination may be recognized by:

i. Too many clone ends (vector-insert junctions) in the assembly. Shotgun of a single BAC should only contain the two vector ends. Identification of other vector sequences is also worthwhile here.

ii. The assembly of the clone may be much larger than expected from its restriction digest.

iii. Analysis of sequence overlaps with other shotgun projects or the international sequence databases may indicate large correct and incorrect overlaps, given the known position in the tile path of the clone.

Contamination of a shotgun project arises from mix up at any of the stages in the process *e.g.* the DNA preparation, ligation, transformation, clone picking, preparation or sequencing, so care must be taken at all stages. Once it has been ascertained that a clone is contaminated, the contaminating reads need to be isolated and removed to a sub-directory. This may be quite straightforward if it is clear that a defined set of sequence reads originating from specific microtitre plates do not belong to the original clone. In this case whole plates are easy to move to a sub-directory. However if the contamination is spread across the project the task of determining which reads belong where becomes more difficult. If the project has more clones ends than expected it is possible to use the overlap data to identify the correct clone ends. It can then be possible to walk through the sequence contigs using spanning pUC reads (when the pUCs have been sequenced in forward and reverse orientation) to determine which contigs are correct. The remaining contigs can then be stripped down to reads and moved to a sub-directory. Occasionally contamination cannot be removed and this must be flagged in the finishing process, so that only the correct data is submitted to the sequence databases.

Once the identity of the clone has been confirmed, contaminating reads removed and the required number of reads has been obtained, the project is ready to progress onto the next stage of the process to produce a complete finished sequence. (see Chapter 11).

3. Protocols/Techniques

Protocol 1. Large Scale DNA Preparation from Bacterial Source

Equipment and Reagents

- Glycerol stock of primary subclone
- solid and liquid selective media
- Large scale centrifuge capable of 15200 x g max.

Note: Centrifuges with lower maximum g-forces may be suitable, but the centrifugations in step 7 must be for longer than 15 minutes to achieve the same result.

- 10 mM EDTA pH 8.0
- Stocks of SDS (20% w/v) and NaOH (4 N) to make fresh lysis solution at final concentrations of 1% SDS and 0.2 N NaOH
- 3M potassium acetate (made from 100ml 7.5 M potassium acetate, 46ml glacial acetic acid and 254ml water)
- Isopropanol
- 7.5 M potassium acetate
- TE buffer at various concentrations of Tris-HCl and EDTA
- DNase-free RNase A at 10 mg/ml

Note: This may require heat inactivation of DNase activity – check the supplier's data sheet

- Phenol/chloroform 1:1 v/v; chloroform
- Ethanol (96% and 70%)

Method

1. Streak bacterial cells containing the primary DNA to be prepared from source (glycerol stock) to individual colonies on selective media. Incubate overnight at 37°C.
2. Pick a single colony into liquid culture (about 150ml) containing the selective agent. Incubate with shaking overnight at 37°C.
3. Transfer the culture to a centrifuge tube, and harvest at 5500 x g max for 5 minutes.
4. Discard supernatant (following local regulations for safe disposal of waste) and resuspend pellet in 20ml 10 mM EDTA solution. This may be achieved by vigorous pipetting of the solution.
5. Very gently add 40 ml of fresh SDS/NaOH solution, and place on ice for 10 minutes.
6. Very gently add 30ml cold 3M potassium acetate solution, incubate on ice for 15 minutes.
7. Centrifuge at 15200 x g max for 15 minutes (4°C), transfer the supernatant to a fresh tube, and repeat centrifugation.
8. Transfer supernatant to a fresh tube and add 45ml isopropanol. Centrifuge at 4500 x g max for 15 minutes.
9. Discard supernatant, and resuspend pellet in 9ml 50mM Tris-HCl 10mM EDTA (pH 7.6) transfer to a 50ml screw cap centrifuge tube (*e.g.* Falcon tube) and add 4.5ml 7.5 M potassium acetate. Incubate at -70°C for 30 minutes.
10. Thaw tube and centrifuge at 4500 x g max. Transfer supernatant to another 50ml Falcon tube and add 24ml 96% ethanol. Mix and

centrifuge at 2500 x g max. Discard supernatant. Resuspend pellet in 700μl 50mM Tris-HCl 50 mM EDTA (pH 7.6), add 10μl DNase-free RNase A (10 mg/ml) and incubate at 37°C for one hour.

11. Transfer the solution to a 1.5ml tube. Phenol/chloroform extract twice (making sure to avoid any interfacial material), chloroform extract once.

12. Precipitate DNA by adding 70μl isopropanol and centrifuging at room temperature for 5 minutes. Wash the pellet with 70% ethanol. Air dry and resuspend in 40μl TE.

Note: Depending upon what confirmation assays are required, it may be beneficial to go through a second round of precipitation and pellet washing at this stage.

Protocol 2. Large-Scale Vector Preparation

Equipment and Reagents

- 10μl high quality vector DNA (such as pUC18) at 0.5 μg/μl
- Restriction endonuclease SmaI and 10x buffer (New England Biolabs)
- 4M NaCl
- Water saturated ultra-pure phenol
- Restriction endonuclease buffer 3 (New England Biolabs)
- Calf intestinal alkaline phosphatase
- 0.5 M EGTA solution.
- 10 x TAE: 242g Tris base; 57ml glacial acetic acid; 18.6g Na_2EDTA, made to 1 litre
- Agarose (electrophoresis grade; Gibco-BRL)
- Ethidium bromide 10 mg/ml (Sigma)
- Ethanol (96% and 70%)
- TE buffer: 10 mM Tris-HCl, 1 mM EDTA pH 8.0

Method

1. To a 0.5 ml tube add 10 μl vector DNA, 20 μl 10xbuffer and 170 μl water. Remove 0.5μl to run on a subsequent gel and add 0.5μl *Sma*I. Incubate at room temperature (about 20°C) for 30 minutes.

2. Analyse a 0.5μl sample (and the undigested sample) on a 0.7% agarose gel in 1 x TBE buffer containing 0.4 μg/ml ethidium bromide. Photograph gel under a UV transilluminator. The vector should be digested, but some undigested material is aceptable.

3. To the remainder of the reaction, add 5μl 4M NaCl and 200μl buffered phenol. Mix and centrifuge at 4°C. for 30 minutes.

4. To the aqueous layer add 500μl 96% ethanol. Leave at room temperature for 30 minutes and then centrifuge at 4°C for 30 minutes. Wash the pellet with 70% ethanol, and allow to air dry.

5. Resuspend the pellet in 20μl TE. Add 20μl New England Biolabs buffer 3, 158μl water and 2μl calf intestinal alkaline phosphatase. Incubate at 37°C for one hour.

6. Add 20μl 5M EGTA solution and incubate at 65°C for 10 minutes.

7. At room temperature, add 220μl buffered phenol, mix and centrifuge. Recover the aqueous layer to a fresh tube, and add 25μl 10xTAE and 75μl Ficoll loading buffer.

8. Load the digested and dephosphorylated vector onto a large agarose gel. For the first gel it is best to use high melting point agarose. This gel can be run overnight. A gel slice containing the digested vector must be excised from the gel. This can be done by staining with ethidium bromide, but only if a long-wave UV transilluminator is available (even with long wave UV light it is recommended that multiple screens of glass and Perspex are between the gel and the light source, and UV exposure of DNA to be recovered for further processing is minimised). Alternatively, staining with Vistra Green and using a Dark Reader (Genetics Research Instrumentation) will yield satisfactory DNA.

9. Discard the gel not containing the vector, and relocate the saved slice to a position equivalent to the normal well position for the gel. Cast a fresh gel around the gel slice, but using low melting temperature gel.

10. Using fresh TAE, run the gel again (overnight).

11. Cut the vector slice from the second gel (following the same precautions as in step 8). Melt the slice at 65°C. Add 30μl 1M NaCl, and split the melted slice between 2 ml tubes (1ml per tube).

12. Add 1ml water saturated phenol to each tube, mix and stand on ice for 5 minutes. Centrifuge to separate phases and transfer the aqueous phase to a fresh tube. Repeat the extraction with water saturated phenol. Repeat the extraction using phenol chloroform (1:1).

13. Extract with isobutanol (the organic phase is the upper phase). As the isobutanol extractions are repeated, the volume of the aqueous layer will reduce. Pool the aqueous phases as the volume reduction allows. Once the volume is 330 μl, add 700μl 96% ethanol. Centrifuge and wash the pellet with 70% ethanol. Resuspend the pellet in 200μl TE, and assay the quality of the preparation by performing test ligatons.

Note: If possible, purchase or prepare enough vector DNA for the whole project during the early phase of the project. Vector DNA will be stable in small aliquots at -20°C.

Protocol 3. Construction of Libraries Using Sonicated DNA

Equipment and Reagents

- BAC, PAC, cosmid or fosmid DNA at 100ng-1ug/µl in TE
- 10 x mung bean nuclease buffer
- Ultrasonic processor with cup-horn probe and tube holder
- 10 x TBE: 108 g Tris base, 55 g boric acid, 9.3 g Na$_2$EDTA, made to 1 litre
- Agarose (electrophoresis grade; Gibco-BRL)
- Ethidium bromide 10 mg/ml (Sigma)
- Mung bean nuclease
- 1 M NaCl
- Ethanol (96% and 70%)
- Pellet paint co-precipitant
- 10 x TAE: 242g Tris base; 57ml glacial acetic acid; 18.6g Na$_2$EDTA, made to 1 litre
- Low melting point agarose
- Water saturated ultra pure phenol
- 1M Tris-HCl pH 8.0
- TE (10 mM Tris-HCl, 0.1 mM EDTA, pH8.0)
- 10x Ligase buffer
- pUC19 *Sma*I BAP 40 ng/µl (Qbiogene)
- T4 DNA ligase (5 U/µl; Roche)
- Mineral oil
- *E. coli* DH10B electroporation competent cells (Invitrogen)
- SOC medium (sterile)
- Sterile Pulser cuvettes; 0.1 cm electrode gap (Biorad)
- Pulse Controller and Gene Pulser (Biorad)

Method

1. To a 1.5ml tube add 1 to 5 µg PAC, BAC, cosmid or fosmid DNA, 6µl 10x mung bean nuclease buffer, and water to 60 µl total volume.
2. Sonicate the tube in the cup-horn probe, using chilled water in the cup to keep the tube cold. Sonicate at 12% power for 10 seconds.

Note: It will be necessary to calibrate the sonicator to obtain fragments in an appropriate size range. Factors that can be varied include power of sonication, duration of sonication and position of sample tube in the cup-horn probe. See also the troubleshooting section.

3.   Analyse a 1 µl sample on a 0.8% agarose gel in 1xTBE buffer. Use suitable markers to estimate fragment size range (for example lambda phage DNA cut with *Hin*dIII) and stain after the gel has run in an ethidium bromide bath (final concentration 10µg per ml) and photograph on a UV transilluminator. If the size range of fragments is too large, repeat steps 2 and 3.

4.   Add 0.3 µl mung bean nuclease, and incubate at 30 °C for 10 minutes.

5.   Add water to bring the volume to 200µl. Add 20 µl 1M NaCl, 1 µl pellet paint and 500 µl 96% ethanol. Precipitation at reduced temperature (for example -20°C for 2 hours) may improve recovery.

6.   Pellet the DNA by centrifugation (ideally at 4°C for 30 minutes), remove the supernatant, and wash the pellet in ethanol (96% will help the pellet to dry rapidly). Resuspend the pellet for running on a 0.8% low melting point agarose gel in 1xTAE buffer. Run with suitable markers (as for step 3).

7.   Recover slices of gel containing sonicated fragments of DNA. This can be done by staining with ethidium bromide, but only if a long-wave UV transilluminator is available. Transfer slices to 1.5ml tubes.

Note: Even with long wave UV light it is recommended that multiple screens of glass and Perspex are between the gel and the light source, and UV exposure of DNA to be recovered is minimised). Alternative methods include staining with Vistra green and using a Dark Reader (Genetics Research Instrumentation), or cutting a marker track off the gel, staining it separately and using a photograph of that as a guide to cutting the unstained gel.

8.   Melt the slice by incubating the tube at 65°C for 5 minutes. Transfer to 42°C and add 4 µl AgarACE. Incubate at 42°C for 20 minutes.

9.   Buffer water-saturated ultra-pure phenol by adding half a volume of 1M Tris-HCl pH8.0 and mixing vigorously. Allow the phases to separate, and discard the upper aqueous layer. Repeat the procedure with half a volume of TE.

10.   To the tube containing the melted and digested gel slice, add 15 µl 1M NaCl and water to 165 µl. Add 165 µl of buffered phenol and mix. Allow the phases to separate (or centrifuge to separate phases).

11.   Recover the upper aqueous phase to a 0.5ml tube. Add 1 µl pellet paint and 375 µl 96% ethanol. Mix thoroughly. Precipitation at

reduced temperature (for example -20°C for 2 hours) may improve recovery. Harvest the fragments by centrifugation (ideally at 4°C for 30 minutes), remove the supernatant and wash the pellet in 70% ethanol. Resuspend the pellet in 5 µl TE.

12. Set up ligations as follows. Prepare a master mix of pUC19 *Sma*I BAP 40 ng/µl and ligation buffer in the ratio 1:2. Aliquot 0.15 µl into 0.5ml tubes for each fragment to be ligated, and three controls.

13. To each tube add 0.7 µl fragment, except for the controls. To two of the control tubes ("A" and "B") add 0.7 µl water. To the third control ("C") add 0.7 µl control fragment (for example PhiX174 DNA digested with *Hae*III at 10ng per ml). To each tube except control B add 0.15 µl T4 DNA ligase. For ease of handling small volumes, 5 µl of mineral oil can be added to each tube before adding the T4 DNA ligase. Incubate at 16°C overnight.

14. To inactivate the ligase, incubate ligations at 65°C for 5 minutes, then at room temperature for 7 minutes, then dilute with 49 µl water.

Note: Alternative methods for ligase inactivation include digestion with proteinase K or phenol extraction.

15. Electroporate samples of diluted ligations by incubating with electrocompetent cells at 4°C and then applying 1.9kV via a 0.1cm cuvette using electroporation apparatus (*e.g.* BioRad Gene Pulser). Wash out cells using warm SOC media, and incubate at 37°C with shaking for one hour.

Note: Use of M13 vectors (for instance M13mp18) differs only in the method of transformation. Host cells must carry and F' episome (for expression of the pilus which allows propagation of the phage), and as transformants are manifest as plaques, transformed cells are added to soft agar before plating on agar plates without antibiotic.

16. Plate on TYE/Amp plates with Xgal and IPTG. Incubate at 37°C overnight.

Notes: This method can be adapted to use fine needle shearing as an alternative. 1 to 5 µg PAC, BAC cosmid or fosmid DNA would be made up to 200 µl in 1x mung bean nuclease buffer, and passed through a 30G hypodermic needle (using a 1ml syringe) at maximum hand pressure four times. This produces fragments in the range 4-12kb, and like sonication employs disposable tubes and other reagents to minimise the possibility of sample cross-contamination.

In the interests of reproducibility, a jig was made, employing a 4kg mass to operate the plunger of disposable syringes (Figure 1), but maximum hand pressure produces equally good results.

Once a project is underway, it is tempting to abandon some of the controls incorporated in the protocols. This is inadvisable since overall trends in terms of recombinant yields may be revealed by comparing control results over time.

Protocol 4. Triton Preparation for Single-Stranded M13 DNA

Equipment and Reagents

- 2XTY (1% TG1)
- 20% PEG 8000/2.5M NaCl
- Triton-TE : 0.5% Triton X-100, 10mM Tris-HCl, 1mM EDTA pH8.0
- DDW

Figure 1. A jig for needle shearing DNA. The DNA solution is placed in the syringe. The 30G needle punctures the lid of the microcentrifuge tube (which is shown away from its normal position for clarity). The 4kg weight is placed just above the syringe plunger, and allowed to fall under gravity. The process should be repeated four times. Maximum hand pressure is a suitable alternative (see text).

- Sodium Acetate pH 4.8
- 96% Ethanol
- 70% Ethanol
- Waterbath
- Centrifuge
- Vortex
- 96-well microtitre plate
- 96-deep well block/box

Method

1. Add 1.25ml of 2XTY (1% ON TG1) to each well of the 96-deep well blocks.
2. Inoculate each well with an M13 plaque, seal the block and place in incubator at 37°C, 360rpm for 5.5 hours.
3. Centrifuge block at 4000rpm for 20 minutes to pellet cells.
4. Add 200 µl of 20% PEG 8000/2.5M NaCl into a new 96-deep well block.
5. Remove 600 µl of supernatant and add to the block containing PEG, mix by pipetting.
6. Incubate at room temperature for exactly 20 minutes and then centrifuge at 4000rpm for 20 minutes at room temperature.
7. Decant the supernatant by inverting the block, leaving to drain for one minute.
8. To remove residual PEG invert block and centrifuge at 250rpm for 2 minutes.

Note: The phage/PEG pellet may be left to dry at this point.

9. Add 20 µl of Triton-TE extraction buffer to each well and seal carefully. Centrifuge briefly to collect sample at the bottom of the well.
10. Vortex for 2 minutes to resuspend phage, centrifuge briefly and repeat vortex.
11. Place the block in an 80°C water bath for exactly 10 minutes to lyse the phage, centrifuge briefly.
12. Remove the seal and add 40 µl of DDW to each well, pipette to mix. Centrifuge briefly to collect sample at the bottom of the well.
13. Transfer the contents of the block to a 96-well microtitre plate containing 10 µl of 3M NaOAc pH4.8, pipette to mix.

14. Add 160 µl of 96% ethanol to each well, pipette to mix.
15. Centrifuge at 4000rpm, 4°C for 1 hour
16. Invert the plate to remove the ethanol.
17. Add 200 µl of 70% (ice cold) ethanol. Do not mix.
18. Centrifuge at 4000rpm, 4°C for 15 minutes.

Note: It is important after this stage to remove the ethanol immediately, otherwise the centrifugation must be repeated.

19. Invert the plate to remove the ethanol.
20. Dry immediately and resuspend in 50ul of DDW.

Note: A speed vac gives the best results.

Protocol 5. Isopropanol Preparation for Plasmids

Equipment and Reagents

- 2LB medium (sterile)
- amplicillin (5g ampicillin solid, 50ml DDW – stored in refrigerator in 2ml aliquots)
- RNase A (20mg/ml) (1g RNaseA solid, 500ul 1M Tris-HCl pH7.4, 750ul 1M NaCl, 50ml DDW – store in refrigerator in 1ml aliquots)
- GTE – 20% Glucose:1M Tris-HCl pH8.0:0.1M EDTA:DDW (825ml DDW, 46ml 20% Glucose, 100ml 0.1M EDTA, 26ml 1M Tris-HCl pH8.0 – store room temperature until RNaseA added, then store in refrigerator)
- NaOH/SDS – 4N NaOH:20% SDS: DDW (900ml DDW, 50ml 4N NaOH, 50ml 20% SDS – store room temperature)
- 3M KOAc – 3.6M Potassium; 6M Acetate (14ml DDW, 14ml concentrated glacial acetic acid, 75ml 5M Potassium Acetate)
- 70% Ethanol
- 100% Isopropanol
- DDW
- 96-deep well blocks
- 96-well storage plates (Costar 3365 serocluster)
- 96-well filter plate (Costar 3504 filter plate)

Method 1 – Centrifuge Method

1. Add 1ml of ampicillin to 1L of 2LB medium

2. Inoculate E. coli into 1ml aliquots of 2LB plus ampicillin into sterile 96-deep well blocks (2ml capacity). Seal blocks, pierce to aerate and secure in an incubator. Incubate at 37°C for 18-20 hours.

Note: It is recommended to incubate a control box containing only the media to check for any potential contamination.

3. Add 3ml RNase A (20mg/ml) to 1L GTE - shake to mix.
4. Spin blocks (boxes) for 2mins at 4000rpm to pellet the cells. Decant culture supernatant to a container for proper disposal. Invert and tap blocks on paper towels to remove excess residual culture supernatant.

Note: Cell pellets may be frozen at this stage if necessary.

5. Add 140 μl Isopropanol to the wells of a 96-well storage plate
6. Add 120 μl GTE to the deep-well blocks
7. Seal boxes and resuspend the cells using a plate shaker or vortex. Complete resuspension of the cells is important so repeat step 7 until cells are resuspended.
8. Remove seal and add 120 μl NaOH/SDS, to the deep well blocks.
9. Reseal blocks and vortex briefly.
10. Remove seal and add 120 μl of KOAc, to the deep well blocks.
11. Reseal blocks and vortex briefly.
12. Place a 96-well filter plate on top of the 96-well storage plate.
13. Remove the lysate from the 96-deep well blocks (using 12-channel pipette) by lowering the pipette tips down the side of the well avoiding the precipitate until reaching the bottom. Slowly remove 140 μl of lysate and transfer into the 96-well filter plate. Try to avoid the transfer of cell debris.
14. Spin both plates on top of each other for 15minutes at 4000rpm and 4°C
15. Discard filter plate and tip off isopropanol.
16. Add 100 μl 70% ethanol to each well and spin for 5minutes at 4000rpm and 4°C.
17. Tip-off ethanol and spin 96-well storage plate upside down on a tissue for 10seconds at 250rpm.
18. Dry on bench top for maximum of 30minutes. The wells do not need to be completely dry.
19. Add 60 μl DDW to each 96-well of the storage plate to elute the plasmid.

20. Leave to stand for 15-30minutes for complete elution, plates may then be stored at -20°C, 4°C or QC checked on agarose gel.

Note: Leave to stand for 1 hour before QC testing.

Method 2 – Vacuum Method

1. Add 1ml of ampicillin to 1L of 2LB medium.
2. Inoculate *E. coli* into 1ml aliquots of 2LB plus ampicillin into sterile 96-deep well blocks (2ml capacity). Seal blocks, pierce to aerate and secure in an incubator. Incubate at 37°C for 18-20 hours.

Note: It is recommended to incubate a control box containing only the media to check for any potential contamination.

3. Add 3ml RNase A (20mg/ml) to 1L GTE – shake to mix.
4. Spin blocks (boxes) for 2mins at 4000rpm to pellet the cells. Decant culture supernatant to a container for proper disposal. Invert and tap blocks on paper towels to remove excess residual culture supernatant.

Note: Cell pellets may be frozen at this stage if necessary.

5. Add 120 μl GTE to the deep well blocks.
6. Seal blocks and resuspend the cells using a vortexer, until complete resuspension of the cells has occurred. (usually 5 minutes).
7. Remove seal and add 120 μl NaOH/SDS to the deep well blocks.
8. Re-seal boxes and vortex briefly.
9. Remove seal and add 120 μl of KOAc to the deep well blocks.
10. Reseal boxes and vortex briefly.
11. Place a 96-well storage plate in the bottom of the vacuum manifold.
12. Place the filter plate (Costar 3504 filter plate) on the top of the manifold.
13. Remove the lysate (using a 12-channel pipette) from the 96-deep well blocks by lowering the pipette tips down the side of the well avoiding the precipitate until reaching the bottom. Slowly remove 140 μl of lysate and transfer into the 96-well filter plate. (Try to avoid the transfer of cell debris.)
14. Apply the vacuum for 1.5minutes, drawing the lysate through the clearing plate into the storage plate inside the manifold.
15. Tap the filter plate top to dislodge any droplets under the plate and discard. Remove the storage plate from inside the manifold.

16. Add 140 µl Isopropanol to each well of the storage plate and leave on the bench for 15 minutes, then spin for 15 minutes at 4000rpm and 4°C.
17. Tip-off the Isopropanol and add 100 µl ethanol to the storage plate and spin for 5 minutes at 4000rpm and 4°C.
18. Tip-off ethanol and spin the storage plate upside down on a tissue for 10 seconds at 250rpm.
19. Dry for 30 minutes.
20. Re-elute the plasmid by adding 60 µl DDW to each well of the storage plate.

Note: Leave to stand for 1 hour before QC testing.

Notes: Do not overgrow the *E. coli* cells during the LB growth phase, the cells decline rapidly after reaching stationary phase and overgrowth can result in the release of chromosomal DNA and an increase in polysaccharides that co-purify with plasmids and inhibit sequencing reactions. Also overgrowth may cause the cells to start to die off, giving a decreased number of cells for DNA extraction and thus lowering the DNA yield.

All salt needs to be removed from the DNA plasmid preparation since only 10-20mM of salt can inhibit a sequencing reaction.

Before elution it is vital that all alcohol is removed from the DNA plasmid preparation as this can inhibit the sequencing reaction.

Protocol 6. Agarose Gel

Equipment and Reagents

- Agarose
- Ethidium bromide
- TBE
- pUC standard
- Loading dye – 90ml 50% glycerol, 10ml 10x TBE, 1ml 0.1M EDTA pH8.0, 0.25g Orange G.

Method

1. Add 0.8g agarose to 100ml TBE and 8 µl of ethidium bromide to make a 0.8% agarose gel.

2. Prepare a pUC standard to a concentration of 25ng/µl
3. To 4 µl of loading dye add 4 µl of DNA sample.
4. To 4 µl loading dye add 4 µl of pUC standard giving a standard concentration of 100ng.
5. Load all samples on agarose gel with the pUC standard at either end allow to run giving separate bands.
6. Visualize on an ultra-violet box.
7. The intensity of the sample band compared to the pUC standard (100ng/µl) gives an indication of the DNA concentration of the sample.

Note: If the DNA concentration is too strong (much brighter than the standard), it may be necessary to dilute in more DDW so as to prevent overloading of samples on capillary automated sequencing anylsers.

Protocol 7. ET Terminators (1/16th concentration reactions in 5 µl)

Equipment and Reagents

- 0.5 µl DYEnamic ET Dye Terminator Cycle Sequencing Kit (Amersham Pharmacia biotech)
- 1.5 µl DYEnamic ET Dye Terminator Dilution Buffer (Amersham Pharmacia biotech
- 1 µl primer M13for or pUC18rev at 3pmol concentration
- Precipitation mix: 80ml 96% Ethanol, 1.6ml 3M Sodium Acetate soln, 3.2ml 0.1mM EDTA, 17ml double deionised water
- 70% Ethanol
- 384-well sequencing plate
- thermoheat seals
- thermocycler

Method

1. To each 384-well add 3 µl DNA template, 0.5 µl DYEnamic ET terminator reagent premix, 1.5 µl DYEnamic ET terminator dilution buffer and 1 µl primer.
2. Spin plate briefly before thermocycling and seal.
3. Amplify DNA on thermocycler using the following program:
   i. 90°C 20secs
   ii. 50°C 15secs
   iii. 60°C 1minute
   iv. return to step i) for 25 cycles

4. Remove plate from thermocycler and spin briefly
5. Add 30 µl of precipitation mix to each well.

Note: An alternative clean-up method other than ethanol precipitation maybe used, spin columns or gel filtration plates.

6. Spin at 4000rpm at 4°C for 20minutes.
7. Tip off supernatant in an appropriate container for disposal and tap plate sharply on a tissue to remove all supernatant.
8. Add 30 µl of 70% Ethanol to each well.
9. Spin at 4000rpm at 4°C for 5 minutes.
10. Tip off supernatant and tap plate sharply onto tissue, to remove all supernatant
11. Allow plate to dry completely before loading onto an automated sequencing machine.

Note: Plates can be dried at room temperature, in a vaccum or in 37°C oven.

Protocol 8. Big Dye Terminators (1/16th concentration reactions in 5 µl)

Equipment and Reagents

- 0.5µl Big Dye buffer x 5 (40ml 1M Tris (Base) pH9.0, 1ml 1M $MgCl_2$, 59ml DDW)
- 0.5µl Big Dye Terminator Ready Reaction Mix (V3) (Applied Biosystems)
- 0.5µl primer M13for or pUC18rev (3pmol diluted in Big Dye Buffer)
- 0.5µl DDW
- Precipitation mix: 80ml 96% Ethanol, 1.6ml 3M Sodium Acetate soln, 3.2ml 0.1mM
  EDTA, 17ml DDW
- 70% Ethanol
- 384-well sequencing plate
- thermoheat seals
- thermocycler

Method

1. To each 384-well add 3µl DNA, 0.5µl Big Dye Terminator Ready Reaction Mix (V3), 0.5µl Big Dye Buffer x 5, 0.5µl DDW and 0.5µl primer, giving a total of 5 µl.

2. Spin plate briefly before thermocycling and seal.
3. Amplify DNA on thermocycler using the following program:
     i. 90°C 30secs
     ii. 92°C 4secs
     iii. 50°C 4secs
     iv. 60°C 1min 50sec
     v. goto step ii for another 44 cycles
     vi. 10°C forever
4. Remove plate from thermocycler and spin briefly
5. Add 30 µl of precipitation mix to each well.
6. Spin at 4000rpm at 4°C for 20minutes.
7. Tip off supernatant in an appropriate container for disposal and tap plate sharply on a tissue to remove all supernatant.
8. Add 30 µl of 70% Ethanol to each well.
9. Spin at 4000rpm @ 4°C for 5 minutes.
10. Tip off supernatant invert plate and tap sharply onto tissue, to remove all supernatant
11. Allow plate to dry completely before loading onto an automated sequencing machine.

Notes: $1/32^{nd}$ concentration reactions in 5ul also give good results.

Always use double distilled sterile water as water quality appears to be associated with T-blobs in the sequence reaction.

All salt and alcohol needs to be removed from the sequencing reaction as this may inhibit the electrophoresis of the sample. Alcohol left in a well prior to loading on a sequencing machine may give a yellow haze across the gel image.

It is a good idea to run a standard control periodically on the sequencing analysis machine to determine that the instrument is picking up dye signals correctly. Using standards of known quality such as M13 or pGEM if sequenced using the same reagents and conditions as normal can localize any problems to either template or sequencing.

4. Technology/Instrumentation/Automation

The most significant recent advance in the area of high throughput sequencing has been the use of fluorescent automated DNA sequencing analysis machines. Recently capillary based sequence analysers have replaced slab-gel machines and require no manual intervention. These give more accurate results as each capillary runs from a specific well, and so there is no problem tracking samples with their sequences.

There are three main companies that produce automated capillary sequence analysers:

Amersham Pharmacia Biotech : MegaBACE 1000
Applied Biosystems : ABI PRISM 3700, ABI PRISM 3100 and ABI PRISM 3730
Beckman Coulter : Beckman Coulter CEQ 2000XL

These systems are flexible enabling the throughput to be maximized or reduced depending on base pair (read length) requirements.

The preparation and sequencing of DNA by hand with no automated intervention, can be very time consuming and leads to manual errors, especially with the introduction of 384-well blocks(boxes)/plates. With the increase in large scale production certain elements of the protocols may also lead to repetitive strain injuries (RSI). The main areas for concern are the pipetting transfer stage in DNA preparation (Protocol 5), adding liquid to 384-well plates in sequencing (Protocol 7 and 8) and the loading of automated sequencing machines.

Most of the methods outlined in the protocol section can be semi or fully automated. Many genome centers use 'in house' automation for shotgun sequencing (see relevant websites) while others use commercial automation.

Filling blocks for growing recombinants and picking into them can be very time consuming. Automated block/box filling machines such as multi-drop (Labsystems) and Robbins Hydras are capable of supporting large genome sequencing projects. For plaque and colony picking there are a number of automated systems, one of the first being the LBNL colony picker (Uber *et al.*, 1991) that combined the Hewlett-Packard robot arm with a charged coupled device (CCD) video camera for imaging. A similar robot has been produced at the Sanger Insitute, Cambridge, UK, where a 48-pin robotic head picks colonies into blocks/boxes at a rate of 1728 colonies per hour using 24 agar source plates picking into 24 destination 96-well or 384-well blocks/boxes (Figure 2). There are many more robots on the market or ones produced 'in house', at a number of sequencing centers, these are listed below and their websites are under 'Relevant Websites'.

With the use of capillary machines it is necessary to ensure that plasimd purified DNA is of a high quality, as these machines are more sensitive to impurities. It is for this reason that an increasing number of genomic centers

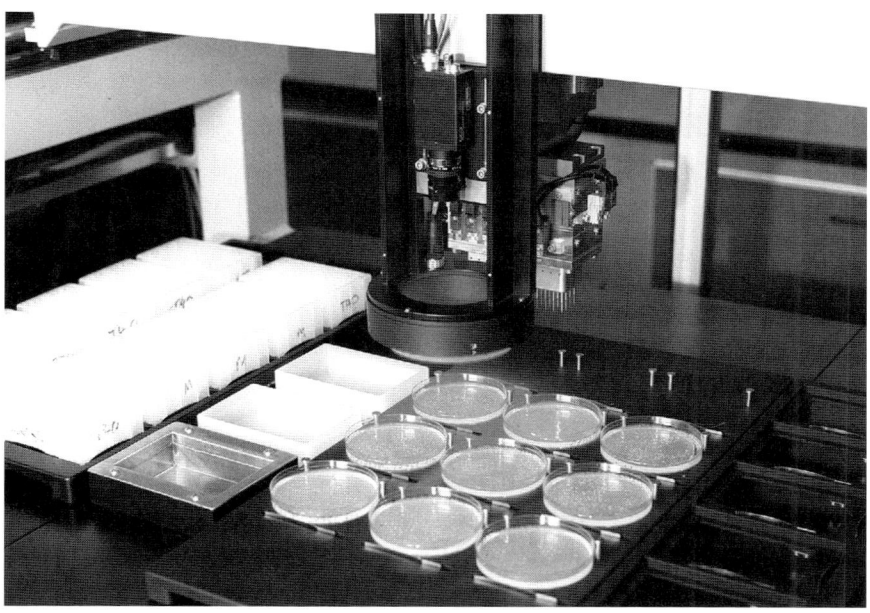

Figure 2. Picking robot designed and produced at The Sanger Insitute, Cambridge, UK. Showing the 48-pin robotic head picking recombinant colonies using a CCD camera system to guide it.

are turning to automation for plasmid template production. Many centers use a mixture of human and automated production. The alkaline-lysis DNA purification protocol outlined above can be semi-automated, with the use of multi-drop machines (Labsystems) to perform the liquid- handling stages. This method may be automated further by using a robot to perform the transfer stage of the protocol.

Other automated systems allow for automated plasmid purification using various methods. Specific commercial examples include the Hewlett Packard, Qiagen BioRobot[tm] 9600 8000 System, Beckman Biomek-2000 and Tecan Genesis robots. Other genome centres such as Baylor College of Medicine, Whitehead Institute and the Sanger Institute have 'in house' automation

Recently the use of rolling circle amplification to amplify vector DNA such as M13 or plasmid DNA from single colonies or plaques using Phi29 DNA polymerase (Templifi) has been implemented successfully by

Amersham Pharmacia biotech. This method uses random primers and Phi29 DNA polymerase to amplify circular DNA in a few hours, followed by a phosphatase treatment to inactivate unincorporated nucleotides. (Dean *et al.*, 2001).

The move from 96-well sequencing to 384-well sequencing has called for the use of automation in the liquid handling stages of the sequencing protocols, to decrease the time spent doing these stages and to increase the accuracy. Commonly used machines include:

1. Multi-drop 384 (Labsystems)
2. Robbins plate positioner (Robbins)
3. Tomtec/Tango (Tomtec)
4. Minitrak (Hewlett Packard)

Fully automated sequencing systems are also widely used in large scale sequencing production, as with plasmid purification, some are commercial and others 'in house' systems. The majority incorporate a robotic arm, pipettors and thermocyling machines to perform the standard dye primer or dye terminator chemistry. Baylor College of Medicine, MWG Biotech Inc (RoboAmp 4200), Whitehead Institute all have automated sequencing systems.

Multi-purpose automated workstations that incorporate both plasmid purification and sequencing reactions using syringe based pipetting methods for liquid handling small volumes are being improved on and developed all the time. A number of companies produce these types of workstations including The Automation Partnership, Beckman Coulter, Inc, CRS Robotics Corporation, Hamilton Company, MWG Biotech, Packard Instrument Company, and QIAGEN.

The development of automated technologies for genomic research immensely improves the throughput, quality, quantity and reduces the cost of large scale sequencing production. This in turn will increase the number of genomes that can be sequenced.

5. Troubleshooting

Below are listed a number of common problems that arise with the above techniques and some of the possible reasons and solutions to diagnose these problems.

Problem: Low yield of primary subclone DNA.

Possible causes/solutions:

i. Poor cell growth. The culture having grown overnight should be very turbid. Check concentration and type of antibiotic. Check temperature and shaking-speed of incubator. To ensure good aeration of cultures, flasks should not be more than 40% full.

ii. Phage contamination (resulting in premature lysis of cells causing slimy pellets after the first centrifugation step). Phage contamination is a very serious problem. All equipment and laboratory space must be thoroughly disinfected. Use of good aseptic technique should eliminate the possibility of phage contamination. It is possible to test for the presence of phage in imported primary clone libraries, or on the surfaces of laboratory benches and equipment. DH10B cells (from a liquid culture grown overnight) can be used to seed a soft agar overlay on LB agar plates or RODAC contact plates. Normal plates are used for testing streaks of liquid culture for the presence of phage, and RODAC plates are touched onto surfaces to assay for the presence of phage. In either case the presence of phage is indicated by plaques in the DH10B lawn following overnight incubation. The incubation should take place with a form of physical separation from other material, in order to prevent spread of phage.

iii. Detergent residue in flasks. This can also cause premature lysis. Flasks for cellular growth should be maintained separately from general laboratory glassware, and washed in such as way as to minimise the possibility of carrying over detergent when used again.

Problem: Sonication problems.

Possible causes/solutions:

i. Insufficient sonication. It is relatively easy to perform cycles of sonication and agarose gel checks until a suitable size range of fragments is obtained. However this tends to be time consuming. Careful initial calibration of your sonicator and ensuring that primary subclone DNA preparations are homogenous should overcome this problem. Some of the parameters which can be altered include the output-power of the sonicator, the length of time the DNA is exposed to the sonic output and the distance of the sample tube from the sonicator probe face. If long periods or high power are required, it is important to keep the sample cool (by for instance exchanging the water in the cup horn with freshly chilled water).

ii. Over sonication. This cannot be rectified, and the process will have to be restarted with fresh primary subclone DNA. This problem may occur if the sonicator is difficult to calibrate. A possible solution is to needle shear primary subclone samples prior to a minimal sonication.

Problem: Few recombinant colonies when transforming ligations.

Possible causes/solutions:

i. End repair failure. In this case there will be lots of recombinants on the C control plate (the positive control). Use fresh mung bean nuclease.

ii. Contamination of recovered fragments. This also will have a good positive control plate. An additional control can be performed in this instance; repeat the C control but spike the control fragment with some recovered fragment. If the recovery is at fault, the spiked ligation will have fewer recombinants that the normal C control. Carry-over of phenol is a possible cause of this problem, and increased ethanol washing steps may solve it.

iii. UV damage to recovered fragments. Again this will have a good positive control plate, but in this instance, spiking an additional control will not result in a reduced score. Alternative methods of visualizing the fragments are outlined in Protocol 3.

iv. Ligase failure. In this case the C control will not have recombinants. Use fresh ligase.

v. Transformation failure. If all plates are blank, check the antibiotic and its concentration. Attempt to transform control DNA (undigested sequencing vector) at low concentration to check transforming efficiency of the cells.

Problem: High background of non-recombinants.

Possible causes/solutions:

i. Insufficient dephosphorylation in the vector preparation. This should show up in primary vector preparation tests. A further cycle of using calf intestinal alkaline phosphatase can be employed.

ii Overloading vector preparation gels. If the preparative gel used in protocol 2 is overloaded (this depends on the size of the gel, the molecular weight of the vector and the mass of vector used), the vector band can be contaminated with uncut vector. Multiple, lower mass vector preparations can be performed and the resulting preparations pooled.

Problem: Lack of recombinant growth.

Possible causes/solutions:

i. A colony may not have been picked into a particular well, if robot picking check robot as it may need calibrating.
ii. Blocks or boxes may not have the correct seal or may not have been pricked for aeration.
iii. Incubator may not be working correctly, check temperature, shaking speed and timing.
iv. Contamination could be causing growth inhibition, clean and decontamate all robotic and other instruments used.
v. Wrong concentration of antibiotic.

Problem: Small/large recombinant colony growth or odd-looking growth.

Possible causes/solutions:

i. Size of recombinant growth could be due to the length of growth time, reduce or increase growth time. Too small growth means cells are dying off, or growth time not long enough.
ii. Contamination causing odd looking growth. All equipment and laboratory space must be thoroughly disinfected. Use of good aseptic technique should eliminate the possibility of contamination.

Problem: Growth in control block/box.

Possible causes/solutions:

i. Control block accidentally picked into, re-grow a control box using the same growth media.
ii. Contamination, all equipment and laboratory space must be thoroughly disinfected. Prepare and sequence contamination and blast against all known sequences to find sequence.

Problem: Top heavy, short read lengths.

Possible causes/solutions:

 i. Too much template DNA, elute in more water and resequence.
 ii. Extension step in cycling too short – extend this step.

Problem: Terminator blobs on traces.

Possible causes/solutions:

 i. Residual terminator mix not removed in clean up, ensure that all solutions are correct and the protocol is being followed.

Problem: Noisy/messy sequence.

Possible causes/solutions:

 i. The annealing temperature may be too low.
 ii. Template needs cleaning up, too much salt remaining.
 iii. Mixed primers, forward and reverse in the same sequencing reaction giving double peaks, resequence with correct primer.

Problem: Large DNA inserts visible on agarose gel.

Possible causes/solutions:

 i. At the lysis stage cells vortexed for too long and disrupting membrane too much.
 ii. At the lysis stage left for too long thus causing the plasmid to denature irreversibly.

Problem: Yellow tinge on sequencing gel.

Possible causes/solutions:

 i. Ethanol left behind after clean up, ensure the samples are completely dry before loading on sequencing machine.

Relevant Websites

Genome Centres

Baylor College of Medicine http://www.hgsc.bcm.tmc.edu
Joint Genome Institute http://www.jgi.doe.gov/

Sanger Institute http://www.sanger.ac.uk

Stanford University Sequencing and Technology Center http://www-shgc.stanford.edu

Washington University Genome Sequencing Center http://www. genome.wustl.edu

Whitehead Institute http://www-genome.wi.mit.edu

Companies

Amersham pharmacia http://www.amershambiosciences.com

Applied Biosystems http://home.appliedbiosystems.com

Autogen http://www.autogen.com

Beckman Coulter Inc. http://www.beckmancoulter.com

BioRobotics Ltd http://www.biorobotics.co.uk

CRS Robotics http://www.crsrobotics.com

Genetix Ltd http://www.genetix.co.uk

Genomic Solutions Inc http://www.genomicsolutions.com

Max-Planck-Institute http://www.mpimg-berlin-dahlem.mpg.de

MJ Research Inc http://www.mjr.com

MWG Biotech http://www.mwgbiotech.com

Packard Instrument Company http://www.packardinst.com

Qiagen http://www.qiagen.com

Robbins http://www.robsci.com

Tecan http://www.tecan-us.com

Tomtec http://www.tomtec.com

Sequence editing programs

Phred/phrap/consed http://www.phrap.org

Phred/phrap/gap4 http://www.mrc-lmb.cam.ac.uk/pubseq/phrap.html.

References

Anderson, S. 1981. Shotgun DNA sequencing using cloned DNase-I-generated fragments. Nucleic Acid Res. 10: 3015–3027.

Birnboim, H.C., and Doly, J. 1979. A rapid alkaline procedure for screening recombinant plasmid DNA. Nucleic Acids Res. 7: 1513–1523.

Bodnteich, A., Chissoe, S., Wang, Y-F., and Roe, B.A. 1993. Shotgun cloning as the strategy of choice to generate templates for high throughput dideoxynucleotide sequencing. In: Automated DNA Sequencing and Analysis Techniques. C. Venter, ed. Academic Press, London.

Bonfield, J.K., Smith, K.F., and Staden, R. 1995. A new DNA sequence assembly program. Nucleic Acids Res. 23: 4992–4999.

Clewell, D.B., and Helinski, D.R. 1969. Supercoiled circular DNA-protein complex in Eschericia coli: Purification and induced conversion to an open circular form. Proc. Natl. Acad. Sci. 62: 1159–1166

Deininger, P.L. 1983. Random subcloning of sonicated DNA: application to shotgun DNA sequence analysis. Anal. Biochemistry 129: 216–223

Dean, F.B., Nelson, J.R., Giesler, T.L., and Lasken, R.S. 2001. Rapid Amplification of Plasmid and Phage DNA using Phi29 DNA Polymerase and Multiply-Primed rolling circle amplification. Genome Res. 6: 1095–1099.

Ewing, B., and Green, P. 1998. Base-calling of automated sequencer traces using Phred II. Error probabilities. Genome Res. 8: 186–194.

Ewing, B., Hillier, L., Wendl, M.C., and Green, P. 1998. Basecalling of automated sequencer traces using phred I. Accuracy assessment. Genome Res. 8: 175–185.

Flint, J., Taylor, A.M., and Clegg, J.B. 1988. Structure and evolution of the horse zeta globin locus. J. Mol. Biol. 199: 427–437.

Gardener, R.C. et al. 1981. The complete nucleotide sequence of an infectious clone of cauliflower mosaic virus by M13mp7 shotgun sequencing. Nucleic Acids Res. 9: 2871–2888.

Gordon, D., Abajian,C., and Green,P. 1998. Consed: a graphical tool for sequence finishing. Genome Res. 8: 195–202.

Lander, E.S., and Waterman, M.S. 1988. Genomic mapping by fingerprinting random clones: a mathematical analysis. Genomics. 2: 231–239.

Maxim, A.M., and Gilbert, W. 1977. A new method for sequencing DNA Proc. Natl. Acad. Sci. USA. 74: 560–564.

Messing ,J. 1991. Cloning in M13 phage or how to use biology at its best. Gene. 100: 3–12.

Oefner, P.J., Hunicke-Smith, S.P., Chiang, L. Dietrich, F., Mulligan, J., and Davis, R.W., 1996. Efficient random subcloning of DNA sheared in a recirculating point-sink flow system. Nucleic Acid Res. 24: 3879–3886.

Roach, J.C., Boxser, C., Wang, K., and Hood, L. 1995. Pairwise and sequencing: A unified approach to genomic mapping and sequencing. Genomics 26: 345–353.

Rosenblum, B.B., Lee, L.G., Spurgeon, S.L., Ichan, S.H., Menchen, S.M., Heiner, C.R., and Chen, S.M. 1997. New dye-labelled teminators for improved DNA sequencing pattens. Nucleic Acids Res. 25: 4500-4505.

Sanger, F., Coulson, A.R., Hong, G.F., Hill, D.F. and Peterson, G.B. 1982. Nucleotide sequence of the bacteriophage lambda DNA. J. Mol. Biol. 162: 729–773.

Sanger, F., Nicklen, S., and Coulson, A.R. 1977. DNA sequencing with chain termination inhibitors. Proc. Natl. Acad. Sci. USA. 74: 5463–5467.

Schriefer, L.A., Gebauer, B.K., Qiu, L.Q.Q., Waterstone, R.H., and Wilson, R.K. 1990. Low pressure DNA shearing: a method for random DNA sequence analysis. Nucleic Acids Res. 18: 7455–7456.

Uber, D.C., Jaklevic, J.M., Theil, E.H., Lisharskaya, A., and McNeely, M.R. 1991. Application of robotic and image processing to automated colony picking and araying. BioTechinques 11: 642–647.

Vieira, J., and Messing, J. 1982. The pUC plasmids, an M13mp7-derived system for insertion mutagenesis and sequencing with synthetic universal primers. Gene. 19: 259–268.

From: *Genome Mapping and Sequencing*
© 2003 Horizon Scientific Press, Wymondham, UK

11

Finishing Genomic Sequence and Dealing with Problem Sequences

Adrienne R. Hunt, David L. Willey and Michael A. Quail

Abstract

Finishing is the process by which assembled shotgun DNA sequence data is manipulated and supplemented to produce a complete high quality reference sequence. This chapter highlights a number of techniques that can be used to carry out this process and their application to the finishing of both large and small genome sequencing projects. All the techniques described below are being applied, and continue to be developed, at the Wellcome Trust Sanger Institute (WTSI).

1. Introduction

Several strategies are available to perform high throughput DNA sequencing (Green, 2001). The method of generating sequence reads from random sub-clones, derived from either a whole genome, whole chromosome or from a minimal tile-path of mapped clones, remains the most popular and widely used approach. In a clone-by-clone sequencing strategy, assembled shotgun sequence is an essential pre-requisite to the finishing phase and the quantity and quality of the raw data constituting an assembly have a significant impact on the amount of additional work needed to finish the sequence and the strategy employed. Sequence reads may be derived from single-stranded (*e.g.* M13 bacteriophage) or double-stranded (*e.g.* pUC) vectors containing DNA inserts of up to 10 kilobases. The Human Genome Project (HGP) has relied on the existence of a physical map constructed primarily from overlapping fingerprints of over 300,000 Bacterial Artificial Chromosomes (BACs) derived from three libraries, combined with additional clones from seven other human DNA libraries (see Chapter 6). BACs, carrying 100-200kb inserts, are selected to provide a minimal tile path across the genome and are used as the substrate for pUC or M13 library making and the shotgun process (see chapter 'Shotgun Sequencing'). Typically, for any given BAC, shotgun data are produced to give an average sequence depth of 6-10 fold coverage in PHRED Q20 bases (Ewing *et al.*, 1998) across the clone insert. A value of Q20 refers to at least a 99% probability of the base call being correct (Ewing and Green, 1998). Assembly of this depth of sequence reads normally produces between 10 - 20 pieces, or contigs (assemblies of overlapping reads) representing at least 95% of the original clone.

Studies undertaken to evaluate the efficacy of M13 and pUC sub-cloning for shotgun sequence assembly of human BACs have shown that a combination of both M13 and pUC sub-clones gives a more complete representation of a clone than either library type alone (Chissoe *et al.*, 1997). M13 sub-clone libraries show under-representation of a clone more frequently than libraries constructed in a double-stranded vector and this is often correlated with the presence of inverted repeat sequences (where the stem repeat sequences and inter-repeat spacer sequence totals less than the average insert size of the sub-clones). These regions are often represented in pUC sub-clones, although often only one of the two strands is represented in the sequence reads. M13 has the advantage of showing a greater retention of CpG rich sequences than pUC. Although a combination of M13 and pUC sub-clones would be ideal for clones of unusual repeat and base composition, the generation of two sub-clone libraries for each large insert clone is not cost-effective for a high throughput sequencing laboratory. The WTSI therefore developed a system of high throughput shotgun sequencing based solely around pUC libraries, which in general gave higher sub-clone representation of the BACs, but retained the use of M13 libraries as a finishing technique.

Once the sub-clones have been sequenced, the processed reads are deposited in a central directory where they are assembled with the PHRAP algorithm (See 'Useful Web Sites'). During a PHRAP assembly vector-masked reads are compared against each other in order to find pairwise overlaps using the base quality information provided by PHRED. Based on this overlap information PHRAP joins the most strongly overlapping reads to make contigs while generating a consensus "quality value" based on the quality of the underlying reads.

Once a minimum 6 fold shotgun sequence coverage has been obtained and assembled the process of 'finishing' can begin. Initial assessment of the data is essential, as further work is guided by the nature of the remaining problems. Typical problems include: sequence termination due to secondary structure formation, under-representation of GC rich regions, tandemly repeated sequences and sequence gaps due to regions not being represented in sub-clones. In order to view and manipulate the data during the finishing process several viewing tools have been developed. The most widely used are GAP4 (Bonfield *et al.*, 1995) and CONSED (Gordon *et al.*, 1998). The discussion below is restricted to the use of GAP4 as the finishing tool used to manipulate a sub-clone derived shotgun assembly through the iterative process of 'finishing' genomic sequence.

2. Quality Assessment and Manipulation in GAP4

2.1 Assembled Sequence Orientation and Manipulation

Shotgun sequence for a clone is collected typically until the sequence reads assemble into twenty or fewer pieces, or contigs. The first priority is then to orientate the contigs with respect to one another. This approach is only possible if the initial shotgun is generated from DNA cloned into a double-stranded vector (such as pUC18) and subsequently sequenced using standard 'universal' priming sites on the positive strand (*e.g.* at position -20) and on the negative strand (*e.g.* at position -48). Shotgun data produced from single-stranded M13 sub-clones do not immediately lend themselves to this approach as sub-clones need to be double-stranded using PCR before a sequence read can be generated from the other end of the sub-clone (Figure 1). The first step in ordering and orientating assembled contigs in a GAP database is to note the contigs which contain reads labelled as cloning vector by PHRAP and to anchor these, for example as the clone left and right hand ends as shown in Figure 2. Observe which sequence reads fall near to the non-cloning vector end of the contig. Reads in the negative orientation will have a corresponding read-pair in the positive orientation further into the contig. Positive sequence reads at the non-vector ends should have their read-pair in a separate contig, in either the positive or negative orientation. GAP4 contains a feature 'Compliment Contig' which can be employed to achieve consistency in read-pair orientation.

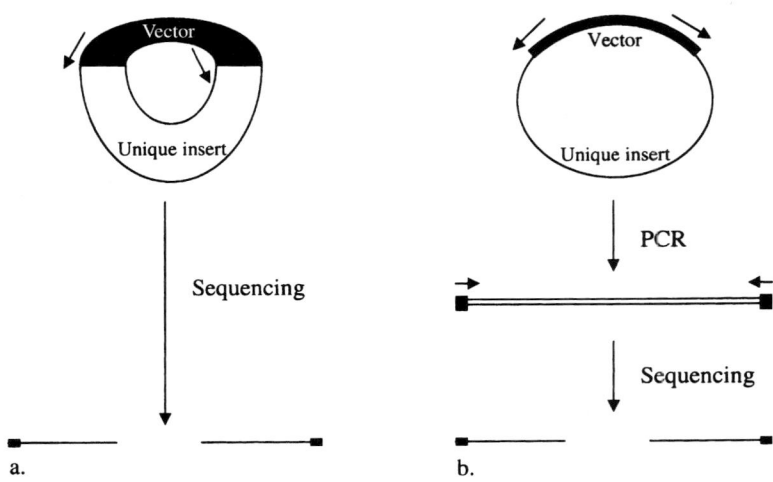

Figure 1. Generation of sequence read pairs for pUC and M13 vector inserts using universal primers. a) Double-stranded pUC vector can be used to generate sequence on both strands using forward primer 5'd(GTAAAACGACGGCCACGT)3' and reverse primer 5'd(AGCGGATAACAATTTCACACAGGA)3'. b) Single-stranded M13 can be used to generate sequence on both strands having undergone PCR to generate the second strand using forward primer 5'd(GTAAAACGACGGCCACGT)3' and reverse primer 5'd(CAGGAAACAGCTATGACC)3'. Sequence can then be generated using the same primers. Arrows indicate priming sites.

Having determined a 'link' from the first contig containing cloning vector, look in turn at the 'free-end' of the linked contig. Generally, matching up read pairs from contigs derived from clone ends will result in no more than 5 scaffolds (linked contigs).

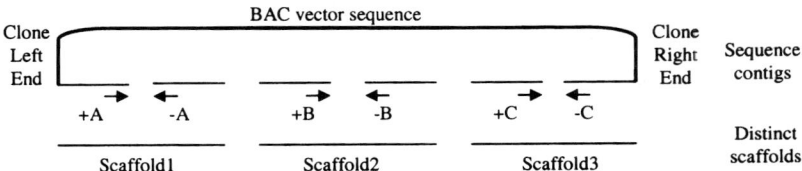

Figure 2. Representation of a BAC assembled into 6 contigs which can be ordered and orientated into 3 distinct scaffolds based on readpair information. A, B and C represent sequence reads generated from each end of a pUC sub-clone using universal pUC priming sites. One sequence read is generated from each strand of the sub-clone (+/-). Scaffold 1 and 3 can be considered anchored because they contain the BAC vector end sequence. Scaffold 2, however, could be in either orientation with respect to scaffold 1 and 3. Any contig within GAP4 can be reverse complimented in order to achieve consistency between read-pair orientation.

Two functions within GAP4 can be used to assess contig end similarity and potential joins. The function 'Find Repeats' plots matches between the given sequence along the X and Y axis. This is useful in identifying repetitive regions that may prevent contiguation, but their presence on two contig ends suggests that they might join together with further sequencing (Figure 3). 'Find Internal Joins' can be used to search for matches at contig ends based on a given percentage mismatch. This is useful as it suggests putative joins that are too weak to be made by PHRAP, when, for example, the quality of the sequence is poor, and provides a guide to contig orientation.

If the BAC of interest has a physical map position adjacent to a BAC which also has sequence data available, these can be compared and the respective orientation along the chromosome established.

2.2 Assessing Quality

The quality of a sequence assembly can be assessed on many levels. The centres participating in the Human Genome Project have drawn up a set of criteria that a sequence must meet before being described as 'finished' (See 'Useful Web Sites'). This agreement promotes consistency and reliability of finished sequence in the international community and provides a firm reference for any participating centre undertaking finishing. Compliance with the set of criteria is ensured through regular cross-checks (Felsenfeld et al., 1999.)

2.2.1 Weak Regions within Contigs

The internal quality of a contig can be assessed using a consensus quality score, which is evaluated for each nucleotide position based on four parameters over a finite window (Ewing et al., 1998). Generally a base confidence at, or above, PHRED Q30 (99.9% probability of base call being correct) will be reliable. GAP4 has a search function which, when set to consensus quality 30 will navigate to base positions which match or are lower than this value. These can then be considered and appropriate action taken to improve the quality. Weak regions can be annotated within the database from within the contig editor, which allows subsequent searching using tags. It is possible to edit any base within GAP4, but editing is generally discouraged in favour of carrying out further directed sequencing to generate high quality confirmatory sequence across a region. In cases of 'messy' sequence, where base definition is poor in part, or along the entire length, of the sequence read, repeated sequencing of the pUC sub-clones

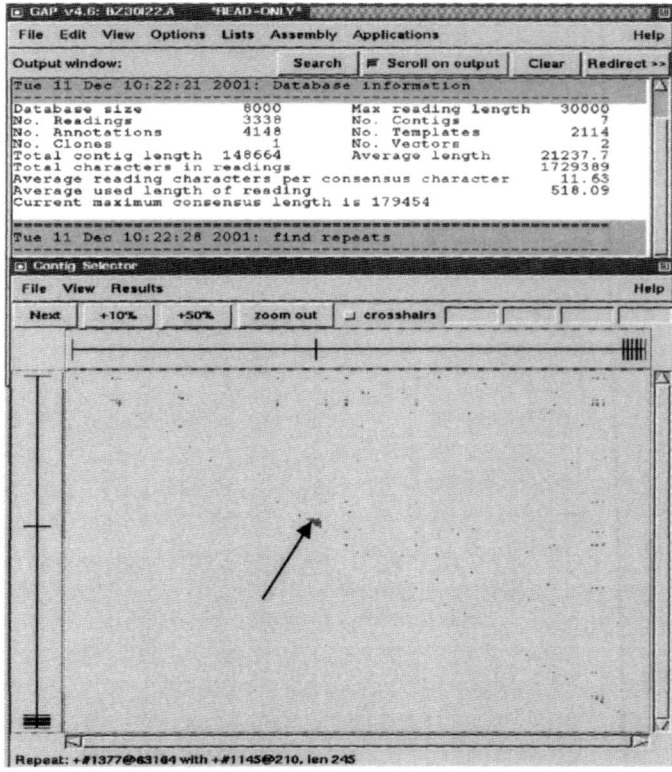

Figure 3. Image showing the two main GAP4 windows from which assembled data for zebrafish clone bZ30I22 can be manipulated. The top window shows certain statistics, for example contig number, contig depth (average reading character per consensus character) and the number of sequence reads making up the assembly. From here a variety of functions can be run. Shown in the bottom window is the results of 'Find Repeats' displayed in the Contig Selector. Here the clone can be seen to be in 2 'major' contigs (on the x-axis a horizontal line denotes a contig, vertical line denotes the start/end of a contig.) The dark specks show regions of repeat. Note the more extensive matches indicated at the contig ends (see arrow). This suggests that, while no sequence overlap is present between the two contigs the presence of similar repeat sequences suggests their possible orientation.

with universal primer may provide a cleaner read. 'Messy' sequence can be due to a number of factors which range from problems with the sequencing reactions to uneven polymerization of the slab gel used to sequence the templates. If the sub-clones over the weak area are already at the upper limit of a sequence read length (500-600 base pairs) it is advisable to re-sequence the region with a custom primer designed to anneal to the sub-clone approximately 100 base pairs away from the region to be resolved.

2.2.2 Gaps

Once contigs have been orientated using sequence read information, a strategy can be developed for BAC contiguation. Gaps fall essentially into two categories: bridged gaps, which are gaps covered by a spanning sub-clone; and unbridged gaps, which are not spanned by a sub-clone that has sequence data in the assembly. In Figure 4, gaps 1 and 2 appear to be bridged although sequence read A is only represented by one of the potential sequence reads. The pair of sequence reads may have failed to assemble due to one of the reads being of poor quality. In this case the sub-clone should be re-sequenced in order to verify that gap 1 is indeed bridged. Gap 2 is bridged by reads +B and -B and so custom primers can be designed to prime 100bp from either contig end on all available pUC bridging sub-clones, in order to maximize the chance of obtaining high quality sequence across the region. Custom primers can also be designed for gap 1 and used to generate sequence from sub-clones that are suspected to bridge the region, *e.g.* A.

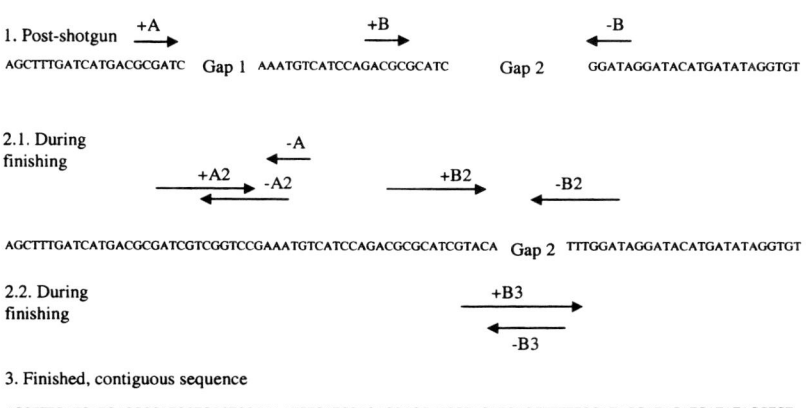

Figure 4. Illustration of iterative gap closure (2.1 and 2.2). A and B represent reads sequenced with universal pUC primers on both strands (+/-). A2 and B2 represent reads sequenced with custom primers. As a result of this sequencing Gap 1 is successfully closed but Gap 2 requires a second round of primer design and walking (B3) in order to contiguate the region.

2.2.3 Bridged Gap Problems

2.2.3.1 Gaps Resulting from Secondary Structure

In some cases a sequence gap will result from structural features within the pUC sub-clone. This occurs when inverted repeats (such as the poly A/T tails of ALU sequences) form tight hairpin loops, which the polymerase cannot sequence through. Attempts to sequence these structures using Applied Biosystems BigDye[TM] Terminator chemistry often results in premature sequence termination and so an alternative sequencing chemistry is required. The Applied Biosystems dGTP BigDye[TM] Terminator chemistry (Rosenblum *et al.*, 1997) is effective at resolving many structural problems. By replacing the dITP present in the BigDye[TM] Terminator sequencing mix with dGTP the hydrogen bonding between nucleotides in the newly synthesised DNA strand and the template are strengthened thus allowing the enzyme to continue around the hairpin and produce sequence beyond the previously terminated region. The use of dGTP BigDye[TM] Terminator chemistry does have its drawbacks, most notably the appearence of compressions in the sequence traces characterised by the visible 'stacking' of bases along the sequence reads and it also results in non-uniform base spacing (Hirao *et al.*, 1992; Sambrook *et al.*, 1989). Compressions are caused by anomalies in the migration of certain DNA fragments during electrophoresis due to the stronger intramolecular base pairing between short stretches of dyad symmetry. Compressions are frequently seen when primer chemistry is used. This chemistry uses a technique where the primers are labeled at the 5' end with fluorescent dyes that enable in-situ imaging of the dye labeled fragments on sequencing machines. The technique uses single primer linear amplification by PCR in the presence of a thermostable DNA polymerase, all four nucleotides, labeled primers and base specific ddNTPs (chain terminators.) Given the possibility of compressions it is advisable to obtain sequence from both DNA strands and to ensure that no region is represented only by uni-directional dGTP or primer sequence reads. It has been reported (Motz *et al.*, 2000) that sequencing with a nucleotide mixture containing two analogues, for example, deaza-dGTP and dITP reduces compressions. This may be due to the formation of heterogeneous DNA chains that cause either a reduction or elimination of intramolecular base pairing. The most effective mixture utilised at the WTSI contains BigDye[TM] Terminator and dGTP BigDye[TM] Terminator in a 4:1 ratio (Hall *et al.*, 2000) (Figure 5).

A further and very effective strategy for structural gap problems involves sonication of the bridging sub-clone into 300-500bp fragments followed by re-cloning into pUC or M13 to create a small insert library (SIL) (McMurray *et al.*, 1998). This provides a nested set of fragments, some of which will represent only a portion of the sequence making up the structural region and so not suffer from secondary structure and sequence termination. By sequencing approximately 96 of the 300-500bp sub-clones generated from a 1.4 - 2kb insert, good coverage of a hairpin loop is usually achieved (Figure 6a).

2.2.3.2 Gaps Resulting from GC Rich and Tandem Repeat Sequences

GC rich sequences are often repetitive and under-represented in the original pUC shotgun library. GC rich repetitive sequence or tandem repeat sequences may misassemble and are often unsuitable for unique primer

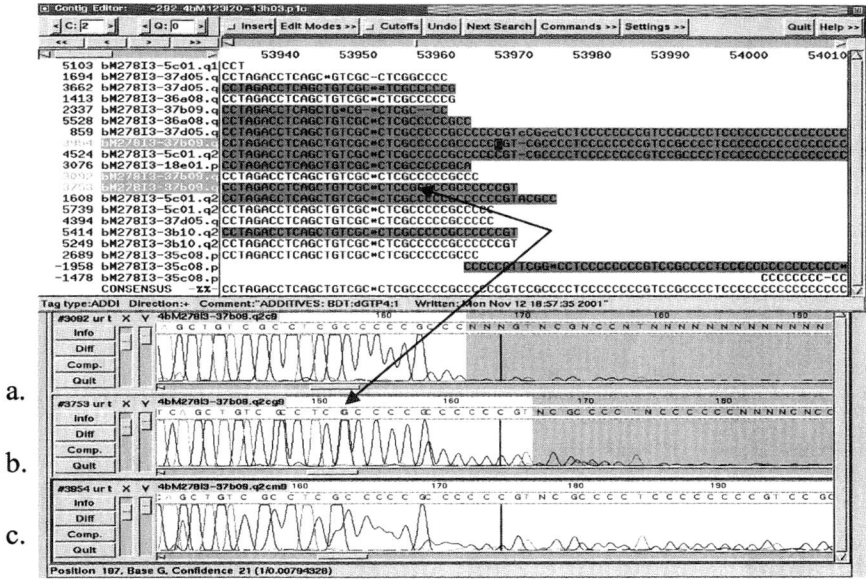

a.

b.

c.

Figure 5. Sequencing using alternative sequencing chemistries. Traces show the result of re-sequencing shotgun subclone bM278I3-37b09, derived from a mouse BAC, on the positive strand using custom primers with three different chemistry combinations: a. Applied Biosystems BigDye™ Terminator read shows sudden sequence termination. b. Applied Biosystems dGTP BigDye™ Terminator read shows termination and GC compression. c. Sequencing with a mix containing a ratio of 4:1 BigDye™ Terminator/dGTP BigDye™ Terminator. Compressions within sequence b. are notably stronger than in c. and have led to a translocation of a G for C in the called data (see arrows). The 4:1 mix does give slight visible compressions and a decrease in signal intensity after the second run of Cs. This sequence should be verified and confirmed by sequencing on the reverse strand.

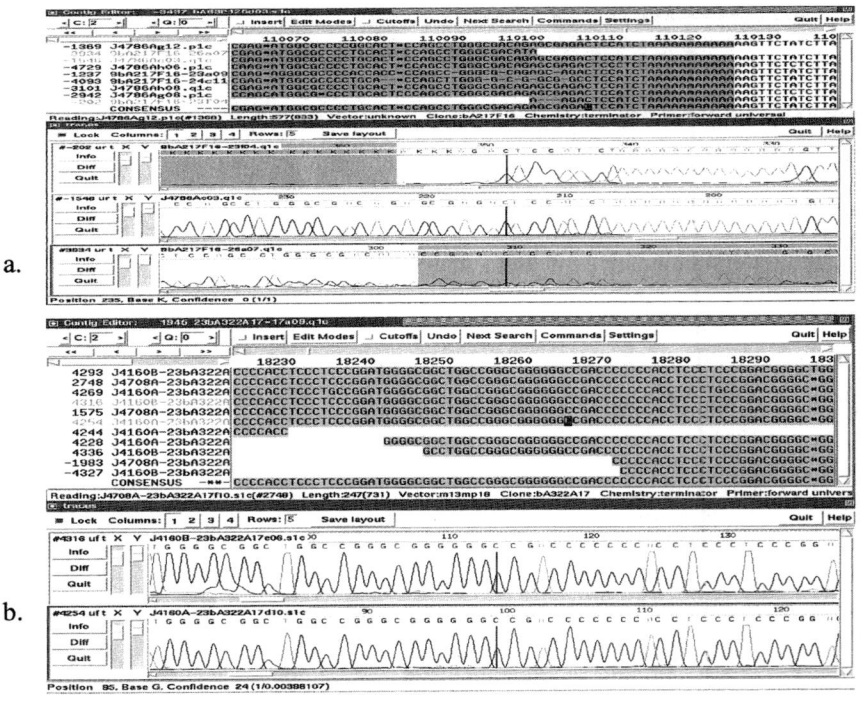

Figure 6.a. Use of a small insert pUC sub-clone library to sequence through secondary structure. Reads prefixed 9bA217F16 denote original pUC shotgun subclones of insert size range 1.4 - 2kb. The sudden sequence termination suggests the presence of secondary structure. Reads prefixed J4786 denote a pUC shotgun library of size range 300-500bp which has been derived from an original pUC sub-clone. J4786Ac03.q1c gives clear sequence through a homopolymer run of As. b. Sequence from human chromosome X after small insert M13 library production. Reads prefixed J4160 and J4708 denote M13 shotgun libraries of size range 100-300bp and 300-500bp respectively. These were derived from two different subclones generated from the original shotgun library. No sequence coverage was attained over this region during the shotgun phase but further directed sub-cloning and shotgun resolves the region with high quality data.

design. If re-sequencing and custom primer walking have not been successful there are two further strategies to be considered. Firstly, as with structural gaps, a bridging pUC sub-clone can be sonicated and sub-cloned to produce a random coverage of reads across the region. GC-rich sequences show better retention in single stranded vectors and these regions will often generate a more even distribution of sub-clones spanning a gap in M13 than in a double-stranded vector (Figure 6b).

The second strategy is creation of a transposon insertion library (TIL) (Devine *et al.*, 1997). The principle behind this strategy is that the transposon randomly inserts along the length of the target DNA producing, unique priming sites along an otherwise repetitive region. Sequencing of 96 or 192 sub-clones resulting from transposon insertion will generally produce an even representation of the region. This also has the advantage that sequencing from both sides of the insertion produces two reads that contain the duplicated flanking sequence which resulted from the insertion event. This duplication allows the two sequence reads to be joined to give a long stretch of good quality sequence over the region (for example 800bp if each read provides 400bp of good quality sequence.)

2.2.4 Unbridged Gaps

Gaps are considered 'unbridged' if there are no spanning pUC sub-clones between two contigs on which to perform the standard re-sequencing or primer walking. The initial strategy used for such a region involves either direct primer walking on BAC DNA or generation of a PCR product from the BAC for subsequent sequencing. If these approaches are unsuccessful, the oligonucleotide primers designed for PCR can be used to generate radioactively-labelled primers (P^{32}) and used to screen a pUC library to identify a clone or clones not represented in the initial shotgun. This is achieved by using replica copies of the pUC colonies on a nylon membrane. Replica copies are used for screening against sets of primers which may be, for example, at the contig ends of an un-cloned gap. Two or more replicas can be directly compared and any pUC sub-clone containing both sequences can be identified by the presence of a hybridisation signal at the same position on both replica filters. Once the 'positive' sub-clones have been identified they can be 'picked off' the original master filter, grown up, sequenced and subjected to a SIL/TIL if required. This is a very powerful approach for identifying bridging sub-clones.

The final approach that is used to produce DNA covering unbridged gaps involves the isolation of a restriction fragment from the BAC DNA. Using a tool in GAP4 called 'Restriction Enzyme Map' a 'virtual' restriction enzyme digest of the assembled consensus sequence can be generated and by reference to a restriction digest of the original clone, an estimate of gap size made. The virtual digests can then be used to select an enzyme which will generate a unique fragment containing the DNA to be bridged. Following enzyme digestion of the clone, the appropriate fragment of

DNA can be isolated from an agarose gel and used to produce a small insert library for subsequent sequencing.

3. Sequence Finishing Reactions- Sample Loading, Processing and Evaluation

At the WTSI, finishing sequence samples are sequenced either on ABI PRISM 377 slab gel sequencers or on Applied Biosystems 3700 or 3100 capillary sequencers (PE). Other sequencing machines are available, for example the MegaBACE 1000 (Amersham International).

After samples have been loaded, processed and deposited into the project directory there are a number of ways in which they can be incorporated into the database. Running PHRAP2GAP (see 'Useful Web Sites') will generate a new GAP4 database, which will contain both shotgun and finishing reads assembled by PHRAP. The advantage of this technique is that previously unassembled shotgun reads may be incorporated into the main contigs, for example over regions generated by primer walking. The disadvantage is that any tags created in the original database will not be present in this new version, however, the two databases can be compared. A second method is to use the 'Normal Shotgun Assembly' option within the current GAP4 assembly. This program assembles either a list or a file of sequence read names under definable match criteria (A list can be created within GAP4 but only exists for the duration of that session. A file can be created external to GAP4, within the project directory and remains for future sessions). This assembly technique has the advantage of building on the existing database and so tags will be unaffected and searchable.

Once all reads have been assembled it is advisable to re-run 'Find Repeat' and 'Find Internal Joins' as PHRAP can occasionally miss joins. For example, if the data are weak at the start of a primer walk they may be assembled as an independent contig rather than joined to the 'parent' contig. When all finishing reads have been accounted for the clone should be re-assessed for gaps and weak regions. Remaining problem regions can be tackled using an alternative sequencing chemistry, further primer walking, BAC PCR, SILs or TILs as judged to be most appropriate for the problem type (Figure 7).

PROBLEM TYPE

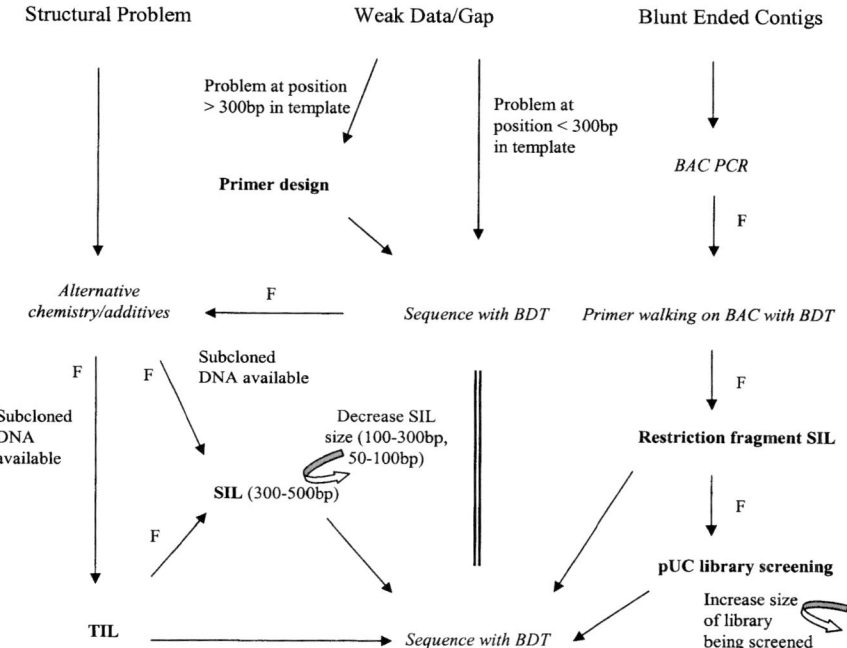

Figure 7. A flow diagram of the sequencing finishing process. Italics indicate a sequencing/PCR step, bold type indicates a process and F indicates a failed attempt. BDT refers to BigDye™ Terminator chemistry. SIL and TIL indicate the production of a small insert and transposon library respectively.

4. Protocols

Protocol 4.1 BigDye™ Terminator Sequencing and Precipitation

Equipment and Reagents

- PTC-225 96 well DNA ENGINE TETRAD THERMAL CYCLER (MJ Research)
- HYBAID OMNISEAL TD MATS (Hybaid Ltd)
- A centrifuge capable of spinning 96 well microtitre plates
- 60 ng/μl source DNA subcloned into pUC18 (prepared according to Chapter 11 'Shotgun Sequencing', Protocol 5).
- 10x Reaction Buffer: 400 mM Tris-HCl, pH 9.0, 7 mM MgCl$_2$.

- BigDye™ Terminator Cycle sequencing Ready Reaction Version 2, (PE Applied Biosystems) Stored at -20°C, light sensitive.
- Primer (one of the following dependent on requirements):
- Universal pUC18 Sequencing Primer (-18) (17MER) 5'd(GTAAAACGACGGCCACGT)3'. Store at -20°C.
- Universal pUC18 Reverse Sequencing Primer (-48) (24MER) 5'd(AGCGGATAACAATTTCACACAGGA)3'. Store at -20°C.
- Custom primer 12 pmol/μl in ddH$_2$0. Store at -20°C once resuspended in water.
- 60% isopropanol
- 80% isopropanol

Method

4.1.1 Sequencing

1. Place 1 μl (12 pmol/μl in ddH$_2$0) of each primer required into separate wells of 96 well sequencing plate.
2. Add 2 μl (60 ng/μl) of each DNA sample required.
3. Add 1 μl of 10x Reaction Buffer to all reaction wells.
4. Add 2 μl of BigDye Terminator reaction mix to all reaction wells.
5. Add 4 μl of ddH$_2$O to all reaction wells.
6. Place Hybaid lid on top of the plate and press down to seal all the wells
7. Pulse centrifuge the samples at 300 rpm until the samples are collected at the bottom of the plate.
8. Place the reaction plate onto a thermocycler programmed for the following conditions: 95°C for 15 sec, 50°C for 5 sec and 60°C for 2 min for a total of 35 cycles.

Note: For difficult templates as previously described alterations to the above protocol may be used:

i) dGTP BigDye™ Terminator version 2 (PE Applied Biosystems) protocol modification: Replace the BigDye™ Terminator reaction mix with 2 μl of dGTP BigDye™ Terminator reaction mix.

ii) SequenceR$_X$ Enchancers A, E, F (3x 100 μl GibcoBRL Life Technologies) protocol modification: Add 2 μl of the chosen enhancer last after the ddH$_2$0. Also decrease the volume of ddH$_2$0 by 2 μl accordingly.

For blunt-ended gaps, BAC walks protocol modification: Add 2 μl BAC DNA (60 ng/μl), stored at -20°C.

4.1.2 Isopropanol Precipitation for Terminator Sequencing

1. Add 60 μl of 80% isopropanol (room temperature) to each reaction well after cycle sequencing.
2. Leave to stand at room temperature for 10 min.
3. Centrifuge at 4000 rpm at room temperature for 30 to 45 min.
4. Discard supernatant and drain on to tissue paper.
5. Add 100 μl of 60% isopropanol (room temperature) to each reaction well.
6. Centrifuge at 4000 rpm at room temperature for 10 min.
7. Discard supernatant and drain on to tissue paper.
8. Place the inverted plate onto tissue paper and spin in the centrifuge for 1 min at 1000 rpm.
9. Store dried plate at -20°C until loading.

Protocol 4.2 BigDye^{TM} Primer Sequencing and Precipitation

Equipment and Reagents

- PTC-225 96 well DNA ENGINE TETRAD THERMAL CYCLER (MJ Research)
- HYBAID OMNISEAL TD MATS (Hybaid Ltd)
- A centrifuge capable of spinning 96 well microtitre plates
- 60 ng/μl source DNA subcloned into pUC18 (prepared according to Chapter 11 'Shotgun Sequencing' Protocol 5.)
- BigDye^{TM} Primer (PE Applied Biosystems)
- 100% Ethanol, store at -20°C in a spark proof freezer

Method

4.2.1 Sequencing

1. Place 2 μl of each pUC template DNA to be sequenced in 4 wells.
2. Add 4 μl of a single base BigDye primer A, T, G and C mix to each well (one row of 12 wells for each base).
3. Place a Hybaid lid on top of the plate and press down to seal all the wells.

4. Pulse centrifuge the samples at 300 rpm until the samples are collected at the bottom of the plate.
5. Place the reaction plate on to the thermocycler programmed for the following conditions: 95°C for 30 sec, 55°C for 30 sec and 70°C for 1 min for a total of 15 cycles then 95°C for 30 sec, 70°C for 1 min, for a total of 15 cycles.

4.2.2 Precipitation for BigDye™ Primer Sequencing

1. Pool the four reactions for each template into one well.
2. Add 50 µl 100% Ethanol (chilled) to each reaction well.
3. Stand at room temperature for 10 min.
4. Centrifuge for 20 min at 4000 rpm.
5. Discard supernatant onto tissue paper.
6. Leave plate to dry at room temperature for 10 to 15 min.
7. Store the plate in dry conditions at room temperature until loading.

Note: Amersham Biosciences Thermosequenase BigDye™ primer chemistry requires a specifc run module on the sequencing machine and there is a different run module for BigDye™ terminators.

Protocol 4.3 PCR of Cloned DNA Using Ready-to-go™ PCR Bead

Equipment and Reagents

- PTC-225 96 well DNA ENGINE TETRAD THERMAL CYCLER (MJ Research)
- HYBAID OMNISEAL TD MATS (Hybaid Ltd)
- A centrifuge capable of spinning 96 well microtitre dishes
- BAC DNA (60 ng/µl), stored at -20°C
- Custom primer 12 pmol/µl in ddH$_2$0 (Sigma Genosys Ltd) Store at -20°C once resuspended in water.
- Ready-to-go™ PCR beads, 96 reactions (Amersham Biosciences)

Method

1. Aliquot 0.5 µl of BAC DNA (60 ng/µl) into a 96 well sequencing plate.

2. Add 1 μl (12 pmol) of each of the two primers (suspended in H_2O) to each well.

3. Add one PCR ball to each reaction by tapping the ball gently into the well (store beads dry at room temperature).

4. Add 25 μl ddH_2O to each well and allow ball to dissolve before mixing with a pipette.

5. Place a Hybaid lid on top of the plate and press down to seal all the wells.

6. Pulse centrifuge the samples at 300 rpm until the samples are collected at the bottom of the plate.

7. Place the reactions plate onto a thermocycler programmed for the following conditions: Hot start at 95°C for 5 minutes then cycle at 95°C for 30 seconds, 50°C for 30 seconds, 72°C for 1 minute for a total 35 cycles.

Note: Sequencing of PCR product without clean up step can be carried out as in Protocol 1 using 0.5 μl of the product.

Protocol 4.4 Transposon Library Production

Equipment and Reagents

- 10 x TBE: 900mM Tris- borate, 20mM EDTA, pH 8.3
- Agarose (high gelling temperature; Gibco-BRL)
- *E.coli* DH10B electrocompetent cells (Gibco-BRL)
- Sterile Pulser cuvettes; 0.1 cm electrode gap (Bio-Rad)
- Pulse Controller and Gene Pulser (Bio-Rad)
- EZ::TN <KAN-2> Insertion Kit (Epicentre Technologies)

Method

1. Analyse 1 μl of pUC template on a 0.8% HGT agarose gel in 1 x TBE buffer containing 0.4 μg/ml ethidium bromide. Photograph gel on a UV transilluminator. Estimate DNA concentration by reference to DNA size markers.

2. To a 0.5 ml tube add X μl target DNA (0.2 μg), 1 μl EZ::TN 10 x reaction buffer, 0.5 μl EZ::TN<KAN-2> transposon, 0.5μl EZ::TN transposase, sterile water to 10 μl. Incubate for 2 h at 37°C. Stop the reaction by adding 1 μl EZ::TN 10 x Stop Solution. Mix and heat at 70°C for 10 minutes.

3. Transform 1 µl of the transposon reaction into *E.coli* DH10B cells as described for small insert library transformation. Plate 1 µl and 25 µl of the transformed cells on TYE plates containing 50 µg/ml kanamycin.

Protocol 4.5 Small Insert Library Production

Equipment and Reagents

- pUC18 - *Sma*I (40 ng/µl; Q.BIOgene)
- 10 x NEB4 buffer (New England Biolabs)
- *Eco*RI (20U/µl; New England Biolabs)
- *Hin*dIII (20U/µl; New England Biolabs)
- *Sac*I (20U/µl; New England Biolabs)
- *Xba*I (20U/µl; New England Biolabs)
- Mung bean nuclease (150,000 U/ml; Amersham Biosciences)
- 10 x Mung bean buffer: 0.5 M NaCl, 0.3 M NaOAc (pH 5), 10 mM ZnCl$_2$, 50% glycerol
- T4 DNA ligase (5U/µl; Roche)
- 10 x ligase buffer (Roche)
- Pellet paint (Novagen)
- AgarACE (0.29U/µl; Promega)
- 10 x TBE: 900mM Tris- borate, 20mM EDTA, pH 8.3
- 10 x TAE: 400mM Tris- acetate, 10mM EDTA, pH 8.2
- Agarose (high gelling temperature; Gibco-BRL)
- Agarose (low melting point; Gibco-BRL)
- T$_{0.1}$E buffer: 10 mM Tris-HCl, 0.1 mM EDTA, pH 8.0
- SOC medium: 2% (w/v) bactotryptone, 0.5% (w/v) yeast extract, 0.05% (w/v) NaCl, 2.5mM KCl, 10mM MgCl$_2$, 20mM glucose, pH 7.0
- TYE plates: 1% (w/v) bactotryptone, 0.5% (w/v) yeast extract, 0.8% NaCl, 1.5% bactoagar
- Phenol, buffered with T0.1E
- 1M NaCl
- Ethanol (96% and 70%)
- Ethidium bromide (10 mg/ml; Sigma)
- Vistra Green (Amersham Biosciences)
- Ficoll loading dye: 0.1% (w/v) bromophenol blue, 10% (w/v) Ficoll 400, 1x TBE
- *E.coli* DH10B electrocompetent cells (Gibco-BRL)

- *E.coli* TG1 electrocompetent cells
- Pulser cuvettes; 0.1 cm electrode gap (Bio-Rad)
- Pulse Controller and Gene Pulser (Bio-Rad)
- Proteinase K (15.6 mg/ml, 30units/mg; Roche)
- Dark Reader (Genetics Research Instruments)
- Sonicator (Heat Systems Inc. Farmingdale, NY.)

Method

4.5.1 Analytical Restriction Digest

1. To a 0.5 ml tube add 1 µl of pUC subclone DNA, 2 µl x2 buffer (*Eco*RI buffer for *Hin*dIII/*Eco*RI digest or buffer 4 + BSA for *Sac*I/*Xba*I digest), 0.6 µl of double distilled water, 0.2 µl of each restriction enzyme, either *Hin*dIII + *Eco*RI or *Sac*I + *Xba*I. Incubate at 37°C for at least 1 hr.
2. Analyse the sample on a 0.8% HGT agarose gel in 1 x TBE containing 0.4 µg/ml ethidium bromide. Photograph gel on a UV transilluminator. Identify the restriction digest which releases the intact insert DNA from the pUC subclone.

4.5.2 Preparative Restriction Digest

1. To a 0.5 ml tube add 5 µl of pUC subclone DNA, 6 µl x2 buffer (either *Eco*RI buffer or buffer 4 + BSA), 0.5 µl of each of the two appropriate restriction enzymes. Incubate at 37°C for 2 hr. Restriction digests can be stored at -20°C at this stage.
2. Restriction fragments may be isolated from BAC/PAC DNA clones. To a 0.5 ml tube add 5 µl BAC/PAC DNA (60 ng/µl), 2 µl x10 buffer, 1 µl of the appropriate restriction enzyme, 12 µl water. Incubate for 2 hr at the appropriate temperature for the chosen restriction enzyme.
3. Add 2 µl of 10 x TAE and 4µl of Ficoll loading dye to each sample. Load the samples onto a 0.8% low melting point agarose gel with a blank lane between each sample. Electrophorese in 1x TAE at 20V for 2 hr.
4. Stain with Vistra Green for 20 min. Visualise the DNA on a Dark Reader and cut out the gel slices corresponding to the human DNA inserts or the BAC/PAC restriction fragment with a sharp, ethanol-washed, scalpel blade. The gel slices may be stored at -20°C at this stage.

4.5.3 DNA Extraction and Sonication

1. Incubate the gel slices at 65°C for 5 min to melt them. Transfer the molten gel slices to a 42°C water bath and equilibrate for 5 min.
2. Add 4 µl AgarACE (Promega) to each molten gel slice and leave at 42°C for 20 min to digest. Place the samples on ice and, if required, add the appropriate amount of distilled water to bring the volume to 54 µl. Add 6 µl 10 x mung bean buffer to each sample.
3. Sonicate each sample at maximum power using a cup-horn probe for 2 min in 2 x 1 min bursts at 4°C.
4. Analyse a 5 µl sample on a 0.8% HGT agarose gel in 1 x TBE buffer containing 0.4 µg/ml ethidium bromide. Photograph the gel under a UV transilluminator to determine the extent of sonication.
5. A smear of DNA fragments should be visible with a size range from the size of the DNA insert down to approx 50bp. Repeat the sonication if there is still a significant amount of intact insert remaining.

4.5.4 End Repair of Sheared DNA Fragments

1. To the sonicated DNA add 0.3 µl (45 units) of mung bean nuclease and incubate at 30°C for 10 min. Place the samples on ice, then add $T_{0.1}E$ to bring the volume of the sample up to 200 µl.
2. Add 20 µl 1M NaCl , 2 µl pellet paint and 550 µl of 96% ethanol to each sample. Precipitate the DNA at -20°C overnight or at -70°C for 30 min.
3. Pellet the DNA in a microfuge at 13000 rpm for 30 min at 4°C. Wash the pellet with 500 µl of 70% ethanol and re-spin for 15 min. Dry the DNA pellet in a desiccator for 15 min.
4. Dissolve the sonicated end-repaired DNA in 6.25 µl $T_{0.1}E$, 0.75 µl 10 x TAE and 2 µl Ficoll loading dye.
5. Load the samples onto a 0.8% LMP agarose gel in 1 x TAE, leaving a blank lane between each sample. Electophorese at 20V for 2 hr. Visualise the DNA by staining with Vistra Green for 20 min and viewing on a Dark Reader. By reference to the 100bp marker, cut out gel slices corresponding to 100-300bp, 300-500bp and 500-800bp.

4.5.5 AgarAce Digestion and Phenol Extraction

1. Melt the appropriate gel slice at 65°C for 5 min, then equilibrate at 42°C for 5 min.

2. Add 4 µl of AgarAce and continue the incubation at 42°C for 20 min. Transfer the samples to ice and add $T_{0.1}E$ to 200 µl. Phenol extract the DNA with 200 µl of neutralised phenol on ice. Equilibrate the samples for 5 min on ice and separate the phases by centrifugation at 13000 rpm for 2 min.

3. Remove the upper aqueous phase carefully. Add 20 µl 1M NaCl, 2 µl pellet paint and 550 µl 96% ethanol. Precipitate the DNA at -20°C overnight or -70°C for 30 min.

4. Pellet the DNA in a microfuge at 13000 rpm for 30 min at 4°C. Wash the pellet with 500 µl of 70% ethanol and re-spin for 15 min. Dry the DNA pellet in a desiccator for 15 min.

4.5.6 Ligations

1. Dissolve the DNA pellet in 4 µl $T_{0.1}E$. For ligations into pUC, in a 0.5 ml tube, ligate 2 µl of sonicated, gel-purified DNA to 0.25 µl pUC18/*Sma*I in a 4 µl reaction containing 0.4 µl of 10 x ligase buffer, 0.3 µl T4 DNA ligase and 1.05 µl water. Incubate the ligations at 16°C for 16 hr.

2. Add 1 µl proteinase K (0.47 units) and 46 µl $T_{0.1}E$ and incubate at 50°C for 1 hr.

3. For ligations into M13, in a 0.5 ml tube, ligate 2 µl of sonicated, gel-purified DNA to 0.2 µl M13mp18/*Sma*I in a 10 µl reaction containing 1 µl of 10 x ligase buffer, 0.5 µl T4 DNA ligase, 3 µl 50% Ficoll and 3.3 µl water. Incubate the ligations at 16°C for 16 hr.

4. For M13 ligations only, add 40 µl $T_{0.1}E$ and phenol extract with 50 µl neutralised phenol. Add 5 µl 1M NaCl, 1 µl pellet paint and 125 µl 96% ethanol. Precipitate the DNA at -70°C for 30 min. Pellet the DNA in a microcentrifuge at 13000 rpm for 30 min at 4°C. Wash the pellet with 100 µl 70% ethanol and dry the DNA pellet in a desiccator for 15 min. Dissolve the pellet in 20 µl $T_{0.1}E$.

4.5.7 Transformations

1. For transformations of pUC ligations, electroporate 40 µl *E.coli* DH10B electrocompetent cells with 0.5 µl of the ligation (1900 V; 200 ohms; 25 µF). Immediately, add 0.5 ml of SOC medium to the cells and mix gently.

2. Transfer the cells to a clean tube and incubate at 37°C for 1 hr with

shaking at 300 rpm. Plate out 25 µl and 100 µl of the transformed cell mix onto 2 x TYE plates containing 50 µg/ml ampicillin, X-gal and IPTG. Incubate at 37°C overnight.

3. For transformation of M13 ligations, electroporate 40 µl *E.coli* TG1 electrocompetent cells with 0.5 µl of the ligation mix (1700 V; 200 ohms; 25 µF). Immediately, add 0.5 ml warm (45°C) SOC medium to the cells and mix gently. Transfer the cells to a clean tube. Transfer separately 50 µl and 450 µl of the transformed cell mix to 3 ml of molten (45°C) 0.7% agar containing X-gal and IPTG. Mix gently and pour onto a warmed (37°C) TYE plate. Incubate at 37°C overnight.

Protocol 4.6 Oligonucleotide Screening of pUC Shotgun Libraries

Equipment and Reagents

- TYE plates (8 x 12 cm) containing 50 µg/ml ampicillin
- Glycerol (20%) plates (8 x 12 cm)
- Pulser cuvettes; 0.1 cm electrode gap (Bio-Rad)
- Pulse Controller and Gene Pulser (Bio-Rad)
- Nytran N nylon transfer membranes (Schleicher & Schuell)
- Autoclaved velvet
- 3MM chromatography paper (Whatman)
- pUC ligation
- Custom primer 12 pmol/µl in ddH$_2$0
- *E.coli* DH10B electrocompetent cells (Gibco-BRL)
- SOC medium: 2% (w/v) bactotryptone, 0.5% (w/v) yeast extract, 0.05% (w/v) NaCl, 2.5mM KCl, 10mM MgCl$_2$, 20mM glucose, pH 7.0
- 10% SDS
- 20 x SSPE: 3M NaCl, 0.2M NaH$_2$PO$_4$, 20mM EDTA, pH 7.4
- Denaturing solution: 0.5M NaOH, 1.5M NaCl
- Neutralising solution: 0.5M Tris-HCl pH 7.4, 1.5M NaCl
- Master mix for labelling primers: 1.5 µl 10x PNK buffer, 0.5µl 0.1 DTT, 1.5 µl [γ-^{32}P] ATP, 0.2 µl PNKinase (Roche, 10U/µl), 11.5 µl dd H$_2$O
- 20 x SCP: 2M NaCl, 0.6M Na$_2$HPO$_4$, 20mM EDTA, pH 6.2
- Hybridisation mix: 6 x SCP, 3% (v/v) Sarkosyl NL 30, 10% (w/v) dextran sulphate, 50 mg/ml denatured, sheared salmon sperm DNA
- Post-hybridisation wash solution: 0.5 x SCP, 0.5% SDS

Method

4.6.1 Transformations

1. Electroporate 60 µl *E.coli* DH10B electrocompetent cells with 3 µl of the ligation (1900 V; 200 ohms; 25 µF). Immediately, add 0.5ml SOC, mix gently and transfer the cells to a clean tube.
2. Incubate at 37°C for 1hr with shaking at 300 rpm.
3. For each transformation, plate out 10 µl of transformed cells onto TYE plates containing 50 µg/ml ampicillin, X-gal and IPTG.
4. Incubate overnight at 37°C (15 hr). Store the remainder of the transformed cell suspension at 4°C.

4.6.2 Master Plate Production

1. Plate out the appropriate volume of transformed cells onto TYE/Amp plates to give approximately 1500 colonies on each master plate. Incubate at 37°C for 15 hr
2. Place the corresponding master filter face down onto the master plate holding the filter by the edges using tweezers.
3. Leave until the entire filter is wet and then very carefully pull the filter away from the plate.
4. Lay the filter face up on a TYE/Amp plate and either store at -70°C or use immediately to make the replicas.
5. To store at -70°C, place the filter, colony side up, onto glycerol plates. After 2 hr at 4°C the filters should be placed in an empty 8 x 12 cm culture plate and stored at -70°C.

4.6.3 Preparation of Replica Filters

1. If the master filters have been stored at -70°C, then place them on TYE/Amp plates and incubate at 20°C for 2 hr to allow the glycerol to diffuse out of the colonies.
2. Place the master filter colony side up on chromatography paper which is itself on a piece of glass.
3. Lay the replica filter face down on a TYE/Amp plate to wet it. Remove the replica filter and line up the two filters (replica filter face down) and then carefully place the replica filter face down onto the master filter.

4. Cover with autoclaved velvet, lay another piece of glass over the velvet and press firmly together.

5. Using tweezers, carefully pull the filters apart and lay the replica face up on the TYE/Amp plate used to wet it. Replace the master filter colony side up on a TYE/Amp plate.

6. Incubate all plates at 37°C. Colonies on master filters require 1-2 hr to recover. Colonies on replica filters require 4 - 6 hr to recover.

7. When colonies on masters have recovered, store at -70°C, following glycerol impregnation.

8. When colonies on replicas have grown, store at 4°C (store for no longer than 16 hr).

4.6.4 Treating Filters

1. Disruption and denaturation of colonies on the replica plates are carried out on 3MM paper in trays.

2. Place the filters, colony side up, for 5 min on each of the following solutions in turn: 10% (w/v) SDS; 0.5M NaOH/1.5M NaCl;0.5M Tris-HCl pH 7.4, 1.5M NaCl twice; 2 x SSPE.

3. Following the last treatment, lay the filters on chromatography paper and leave to air dry for 2 hr. Once dry, pile up the filters separating them by tissue paper and wrap in tin foil.

4. Place filters at 120°C for 1 hr to fix the DNA to the filter.

4.6.5 Hybridisation and Autoradiography

1. Transfer 0.75 µl of the appropriate primer into a 0.5 ml tube, dry at 120°C for 5 min followed by dessication for 5 min. Add 15.2 µl of labelling Master mix to each primer.

2. Incubate at 37°C for 45 min.

3. Labelled primers can either be stored at -20°C or used immediately for hybridisation.

4. Add labelled primer to the appropriate volume of hybridisation mix (10 ml for 1 filter, 13 ml for 2 filters, 16 ml for 3 filters). Gently pour the primer/hybridisation mix into a clean 8 x 12 cm culture plate then place the filter, DNA side up, onto the solution. Allow the filter to wet thoroughly and then carefully turn the filter over.

5. Cover with a precut plastic sheet, remove all air bubbles and incubate at 42°C for at least 1 hr.

6. After hybridisation, wash the filters at room temperature in 0.5 x SCP, 0.5% SDS for 5 min. Repeat twice with fresh wash solution.

7. Dry filters on 3MM chromatography paper at room temperature for 2hr. Subject filters to autoradiography for 2 days at room temperature.

4.6.6 Identification and Picking of Positive Colonies

1. Picking of oligo-hybridising colonies is carried out in an ice bucket containing dry ice to keep the glycerol-impregnated colonies frozen. Identify colonies on the master plate which hybridised with both oligonucleotide probes.
2. Overlay the replica autoradiographs on a light box to identify colonies which hybridise with two or more oligonucleotide probes.
3. Locate the correct colony on the master filter, touch with a cocktail stick and streak onto a TYE / Amp plate. Incubate at 37°C overnight to produce single colonies.

Protocol 4.7 Plasmid Preparation (for Small Insert Libraries, Transposon Libraries and Screens)

Equipment and Reagents

- Centrifuge capable of spinning 96 well Corning costar assay box
- Centrifuge capable of spinning 96 well microtitre plates
- 90°C oven
- Corning Costar assay box (Non-treated) (Corning Incorporated)
- Polypropylene untreated 96 well round bottom plate (Corning Incorporated)
- Filter plates (Corning Incorporated)
- Falcon lid (Becton Dickinson Labware)
- Mylar plate sealer (ThermoLabsystems)
- Scotch plate sealer (3M)
- Circle Grow (B10 101) containing ampicillin (Sigma) (100 µg/ml)
- GTE solution: 1.3ml 1M Tris pH 7.4, 5ml 0.1M EDTA, 2.3ml 20% Glucose, 42ml ddH$_2$0
- RNase A: 200mg RNase A (solid, Sigma), 100 µl 1M Tris pH 7.4, 150 µl 1M NaCl, 10 ml ddH$_2$0, store at -20 °C
- NaOH/SDS solution: 2.5ml 20% SDS, 2.5ml 4M NaOH, 45ml ddH$_2$0
- KOAc: 147.21g Potassium acetate, 7.5ml Glacial Acetic Acid, ddH$_2$0 to give final volume 500ml, store at -20 °C
- 70% Ethanol (BDH Laboratory Supplies), store at -20°C in a spark proof freezer

- Isopropanol (BDH Laboratory Supplies)

Method

1. Fill the wells of a Corning Assay box with 1ml of Circle Grow containing ampicillin (100 µg/ml).
2. Pick colonies into the media, seal the box with mylar plate sealer and pierce each well with a needle to allow aeration during growth. Place the box in a shaker at 37°C at 320 rpm for 22 hr.
3. Remove box from shaker and spin at 4000 rpm for 5 min to pellet cells and discard supernatant. Leave to drain on a tissue for 1 minute.
4. Add to each well 250 µl of GTE solution and vortex the cells for 2 min to resuspend.
5. Spin box at 4000 rpm for 5 min to pellet cells. Discard supernatant and leave to drain on a tissue for 1 min.
6. Add to each well 250 µl of GTE solution and vortex the cells for 2 min to resuspend.
7. Add to each well of a polypropylene plate 4 µl of RNase A (20mg/ml). Transfer to this 60 µl of the resuspended cells.
8. Add to each well 60 µl of NaOH/SDS solution and 60 µl of 3M potassium acetate (stored at 4°C). Seal the plate with a 3M scotch plate sealer, mix by inversion 10 times and leave for 10 minutes.
9. Spin plate briefly up to 1000 rpm. Remove the plate sealer and place in a preheated 90°C oven for exactly 30 min and then cool plate on ice for a minimum of 5 min.
10. Transfer the full volume to a filter plate, secure with tape on top of a polypropylene plate which contains 150 µl of 100% isopropanol. Place a Falcon lid on top and spin at 4000 rpm, 4°C for 2 hr.
11. Discard the supernatant and add 200 µl of 70% ethanol (stored at 4°C). Spin at 4000rpm, 4°C for 15 min.
12. Discard the supernatant and place in a 90°C oven for 10-15 min. Resuspend in 60 µl of ddH$_2$0.

Protocol 4.8 M13 Triton Preparation (for Small Insert Libraries and Screens)

Equipment and Reagents

- Centrifuge capable of spinning 96 well Corning costar assay box
- Centrifuge capable of spinning 96 well microtitre plates
- Vortex

- 90°C oven
- Water bath
- Corning costar assay box (Non-treated) (Corning Incorporated)
- Polypropylene untreated 96 well round bottom microtitre plate (Corning Incorporated)
- Mylar plate sealer (ThermoLabsystems)
- Scotch plate sealer (3M)
- *E.coli* TG1
- 2x TY: 1.6% (w/v) Bactotryptone, 1% (w/v) Yeast extract, 0.5% (w/v) NaCl
- 20% PEG 8000/2.5M NaCl
- Triton-TE extraction buffer: 0.5% Triton-X (Sigma); 10mM Tris-HCl and 1mM EDTA, pH 8.0)
- M13 Precipitation Mix: 10 µl 3M Sodium acetate and 160 µl ethanol

Method

1. Innoculate 2x TY with TG1 (1ml of TG1 per 100 ml 2x TY).
2. Pick plaques into a Corning Assay box containing 1.25 ml of the 2x TY/TG1 mix. Seal the box with a mylar plate sealer and pierce each well with a needle to allow aeration during growth. Place in a shaker at 37°C, 360 rpm for 5.5 hr.
3. Centrifuge the box at 4000rpm for 20 minutes at 20°C.
4. Add 200 µl of 20% PEG 8000/2.5M NaCl into a new Corning assay box. Transfer into this 600 µl of supernatant and mix by pipetting.
5. Incubate at room temperature for exactly 20 min and then spin at 4000 rpm at room temperature for exactly 20 min.
6. Decant the supernatant by inverting the box, leave to drain on a piece of towel for 1 min and then centrifuge inverted boxes at 250 rpm for 2 min to remove the residual PEG.
7. Add 20 µl of Triton-TE extraction buffer to each well and cover with 3M silver tape. Spin briefly.
8. Vortex for 2 min and then spin briefly at 1000 rpm. Place the box in an 80°C water bath for exactly 10 min and then spin briefly at 1000 rpm.
9. Remove silver foil and add 40 µl of ddH$_2$0 to each well, mix by pipetting.
10. Add 170 µl M13 Precipitation Mix to an untreated Corning microtitre plate. Transfer to this the contents of the box and mix by pipetting.

11. Precipitate at -70°C for 20 min or -20°C overnight.
12. Spin at 4000 rpm, 4°C for 1 hr and then invert plate to empty.
13. Add 200 µl of 70% ethanol (stored in freezer) spin at 4000rpm, 4°C for 1 hr and then invert plate to empty. Dry in 90°C oven for 10-15 min.
14. Re suspend in 60 µl of ddH$_2$0.

5. Trouble-Shooting

A key part of the finishing process is identifying problems that are not resolved through recourse to the usual finishing strategies. Some of the problems that can arise and solutions are discussed below.

5.1 Mis-Assemblies

Sequence mis-assemblies are caused most frequently by the presence of extensive regions of highly conserved repetitive sequence that are assembled together by PHRAP, rather than being arranged in tandem array. Misassembled sequence can be identified by the presence of high quality base discrepancies and a greater than average depth of sub-clones assembled over a region. The main focus during assessment of such a region should be sequence read pair information and a sequence comparison of the clone against itself. By using the sequence comparison to establish the outer limits of the repeat and then selecting pUC sub-clones that have repeat sequence at one end and unique sequence at the other, a scaffold of sub-clones across the repetitive regions can be constructed. Accurate assembly relies heavily on the quality of the shotgun data and the number of base pair discrepancies between repeat sequences. Additional sequence can be generated from the sub-clones by custom primer walking if the repeat is represented only once in any one 1.4 - 2kb sub-clone. Alternatively, small insert or transposon libraries constructed from an anchored sub-clone can give high quality sequence of the entire region onto which further repeat sequence reads can be assembled. Most repeats can be correctly assembled in this way. However, in cases where nearly identical repeats in tandem extend over regions that exceed the length of the sub-clones used for assembly, it may be necessary to screen a larger insert pUC library, for example 4-6kb or 10kb, to identify pUC sub-clones that read from and into the repeat. The sub-clone(s) which covers the repeat most completely should be identified by sequencing all positive pUCs on both strands and by sizing the pUC insert. The largest sub-clone whose sequence read falls nearest to the unique-repeat junction should be used to create a TIL. This method will give firmly anchored repeat data on which to assemble the original shotgun data (Figure 8).

Restriction digest data of the original clone should also be utilised to assist with and confirm the assembly of repetitive regions. Care needs to be taken with enzyme selection to take into account the frequency of restriction sites within the repeat. Too many restriction sites will result in multiple bands of the same size, the true number of which will be hard to resolve. Too few cut sites will result in large bands and increased inaccuracy of sizing when compared back to the assembled consensus digest data.

Restriction digests are not only useful for repetitive clones but also for general confirmation of BAC assembly. At the WTSI all assembled clones

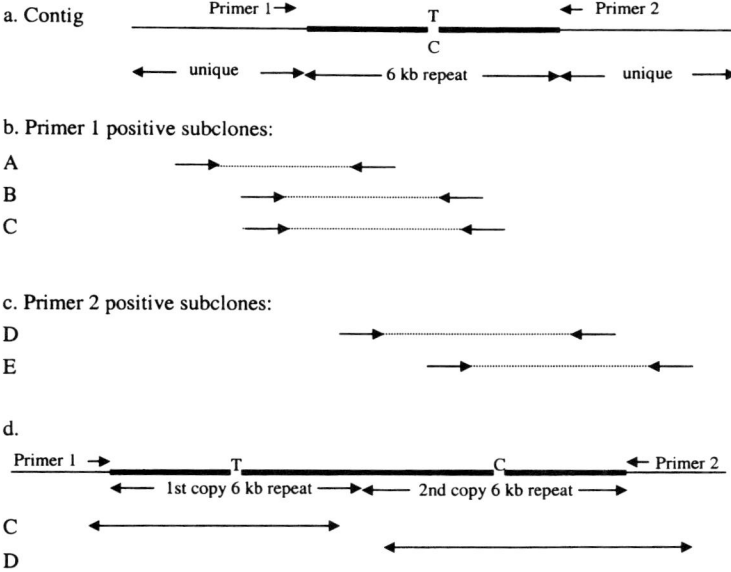

Figure 8. Strategy for resolving large scale misassemblies. a) The thin line represents unique sequence and the thick line represent a 6kb repeat region which has a greater than average coverage of sub-clones and a T/C base pair discrepancy suggesting misassembly. Screening of a 10kb pUC library can be performed using primers selected at the unique/repeat junction (1 and 2). b) and c) Positive sub-clones can be sequenced in order to determine their position along the clone. An arrow represents the result of sequencing on the forward and reverse strand. The dotted line represents the inferred region covered by the whole sub-clone insert. In the above example sub-clones C and D give the most extensive coverage of the repeat, covering the T/C base pair discrepancy while still being anchored in unique. d) C and D can be used for transposon library making in order to attain good quality sequence across the repeat. This sequence can then be used as a scaffold on which to assemble the original shotgun reads.

are compared back to a 'real' digest performed on the clone DNA before it was prepared for sequencing (Figure 9). For further details on the program, Confirm, which is used for digest confirmation and other WTSI finishing software refer to 'Useful Web Sites'.

5.2 Primer Design

Additional sequence reads for extension or confirmation of sequence contigs are frequently generated by primer walking from shotgun sub-clones. There are a number of primer design programs available (Hyndman *et al.*, 1996; Rychlik *et al.*, 1989). GAP4 has the advantage of a built in primer selection programme called OSP (Hillier *et al.*, 1991), which has been extensively used during the Human Genome Project and has proved to be very reliable. The primer should be selected approximately 100bp away from the problem region/contig end and used to sequence at least four sub-clones, which are known to span the region. This avoids walking on an insert and back into pUC and also allows for an occasional sequence failure.

If primers are being designed independently there are a number of factors to consider. Firstly, primer selection within known repeats, such as *Alu* or LINE within human sequence, should be avoided. Also the primer itself

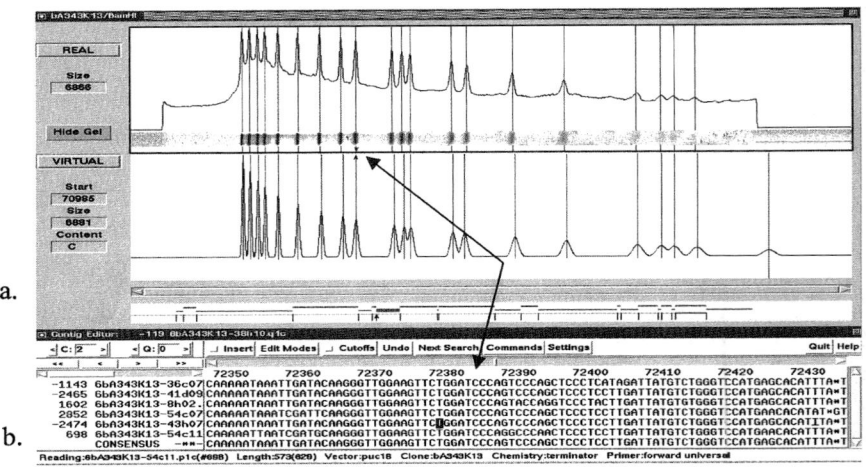

a.

b.

Figure 9. Confirmation of sequence assembly by restriction enzyme analysis. a. A view from 'Confirm' in which a 'virtual' BamHI digest has been simulated from the finished consensus of clone bA343K13. This is then matched against the 'real' digest extracted from the program Image. This is a powerful tool for clone verification as it can be used to refer back to the clone assembly- for example the highlighted 6.8kb band refers to the cut site as seen in GAP4 (b.) See arrows.

should not contain repetitive sequence or homopolymeric runs of more than 4-5 nucleotides. It is advisable to avoid the chance of secondary binding by screening the selected primer against the BAC sequence, in particular if the primers are being used for PCR or walking on the BAC DNA. Ideally the primer should be between 18 and 25 bases in length with a GC content of 40%-50%, a T_m of around 50°C ($T_m = 2(A+T) + 4(G+C)$) and with a 1-2 nucleotide 3' GC clamp. Although adherence to the above criteria usually produces efficient primers, a recent study has shown that the physico-chemical characteristic of the primer may not be as important as template and primer purity and quality in producing good sequence (Buck et al., 1999).

5.3 Sequencing Chemistry Modifications

5.3.1 DMSO

Dimethyl sulphoxide, a member of the group of organic sulphoxides, is well documented for having beneficial effects when used in conjunction with PCR and sequencing reactions (Pomp et al., 1991; Winship, 1989). Incorporation of DMSO into a standard Taq polymerase-mediated PCR reaction at a final concentration of 5% (vol/vol) has been shown to permit correct amplification of complex secondary structures where altering extension time, annealing temperature and/or Mg^{2+} concentration in the reaction buffer resulted only in a shorter than anticipated product (Shen et al., 1992). DMSO can also facilitate sequencing of GC rich templates (Sheidl et al., 1995) giving up to a 60% increase in readlength in comparison to an identical reaction without DMSO. While the exact method of action of DMSO is uncertain, it has been suggested that its hydroscopic nature results in a dehydration effect on the double-stranded DNA helix allowing the denatured template to be acted upon by DNA polymerase by eliminating the intramolecular secondary structures. It has also been suggested that DMSO may affect the melting temperature (T_m) of the primers, the thermal activity profile of the polymerase or the degree of product strand separation achieved at a particular denaturing temperature (Gelfand, 1989). DMSO can prove to be an inexpensive solution to apply to non-optimal standard PCR and sequencing reactions.

5.3.2 Enchancer Additives

SequenceR$_X$ Enhancers have been developed by GibcoBRL Life Technologies to improve sequence quality over regions that may not sequence readily. The most commonly used are enhancers A, E and F. A can be applied to a range of problems such as GC rich regions, direct repeats, CT repeats. E is best for direct repeats and short CT rich repeats while F should be applied to regions of high GC content. The chosen additive is used during Protocol 4.1 by reducing the volume of ddH_2O in step 5 to $2\mu l$ and then adding $2\mu l$ of additive to the required reactions. SequenceR$_X$ Enhancer A produces more consistent results over hairpin structures when it is combined with dGTP BigDye^{TM} Terminator rather than with BigDye^{TM} Terminator.

6. Automation

A number of advances have increased the efficiency of finishing. These are often specific to a particular sequencing centre, for example at the WTSI, an Oracle database is used to track DNA sample locations, finishing reactions, primer orders and to generate sample loading sheets for DNA sequencers. The nature of such automation and tracking does not lend itself to a smaller laboratory. However, automation of reaction selection is more accessible with the use of Autofinish (Gordon *et al.*, 2001.) This program has been used extensively at the Genome Sequencing Center, St Louis (See Chapter 12). The WTSI is using an alternative selection program that is currently under development with the Staden group. The output of this program will be directly supplied to a finishing robot to provide the basis of an automated pre-finishing service. The robot has the capacity to store 3360 384-well DNA plates and will re-array selected templates into 96 well format ready for re-sequencing. Reaction selection is currently limited to primer walking, relying on the program OSP for primer design. Reactions are performed with a 4:1 BigDye^{TM} Terminator and dGTP BigDye^{TM} Terminator mix.

7. Gap Closure and Finishing in Small Genome Projects

In addition to the high profile large genome projects such as the HGP, there are in progress, and now complete, a great number of smaller genome projects (http://www.sanger.ac.uk/Projects/ and http://www.tigr.org/tigr-scripts/CMR2/CMRHomePage.spl). These are typically concerned with sequencing the genomes of bacteria and lower eukaryotes. Both the extreme diversity and often the lack of prior research into these organisms can create difficulties for the finisher. Base content can vary from organism to organism, *e.g.* from an extreme of 72% G/C in *Streptomyces coelicolor* to 80% A/T in *Plasmodium falciparum*, and can vary widely within an organism, *e.g.* centromeric and intergenic regions of *P. falciparum* that approach 100% A/T and the PGRS elements of *M. tuberculosis* that are >85% G/C. These extremes of base composition cause several problems: 1) the cloning host *E.coli* has an inherent bias against sequence of extreme composition resulting in such sequences often being under-represented in any shotgun sequencing project. 2) Elements of extreme composition, especially those that contain repeat regions, often cause the polymerase used in the sequencing reaction to stall or skip (*e.g.* in homopolymeric tracts). 3) It is often difficult to choose unique stable primers for oligo walking, especially in extreme A/T rich regions. The lack of prior research on an organism also has a profound effect on the finisher's task as complete or even partial physical maps are not always available. This has the consequence that instead of having to finish the sequence within a defined 100-200kb BAC insert, the finisher is faced with piecing together and closing whole genome or whole chromosome shotguns of >1Mb and often >4Mb in size. The assemblies are often complicated by families of uncharacterised repeats and cross-contamination with DNA from other chromosomes, or host DNA, if looking at an intracellular parasite.

The techniques used to overcome these problems can be split into practical and strategic/*in-silico* approaches. The latter can be used to reduce the complexity of the finishing task by giving an indication of contig order. For example in whole genome shotgun projects one might generate end-sequences from a larger insert library (*e.g.* in a BAC, cosmid/fosmid, lambda or low copy number plasmid vector) and use those end sequences to scaffold the shotgun by linking contigs that contain sequence matches to the opposing ends of such a clone. In practice it is common to make use of all the possible information that is available. For example one could exploit the probable syntenic positioning by comparison to an organism for which the sequence is known. In this way a contig order might be postulated and thereafter attempts to close gaps by PCR can be attempted. Genetic maps containing ordered STS markers and known genes can also be used to assist contig ordering. An extreme and more accurate version of this is the use of the HAPPY map, which displays an accurately ordered set of markers at a typical resolution of 1 marker every 10kb across an entire genome or chromosome. Such a map is currently being applied to aid finishing of chromosome 6 of the social amoeba *Dictyostelium discoideum* at the Sanger Institute (http://www.sanger.ac.uk/Projects/D_discoideum/). The *D. discoideum* genome is being sequenced by an international consortium with different laboratories finishing individual whole or part-chromosomes after sequencing from libraries prepared from pulsed-field separated chromosomal DNA. Due to co-migration during separation these libraries are between 10 and 50% contaminated with DNA from other chromosomes and so contig ordering is further complicated by the presence of contigs from those other chromosomes. Two strategies are being employed at The Sanger Institute and Baylor College of Medicine to reduce this complexity: 1) A HAPPY map of this chromosome has been published (Konfortov *et al.*, 2000), which incorporates many of the sequences generated within the chromosome 6 sequencing project as markers. 2) Although the published YAC map of this organism is inaccurate, YAC clones are available that cover regions of the chromosome. By sample sequencing YAC clones, contigs from the chromosome shotgun that show matches to these reads can be assumed to be from the same region of the chromosome and can be 'binned' together.

The majority of sequencing problems in these projects are solved on a practical basis by standard finishing techniques such as oligo walking, PCR (standard, long-range and combinatorial) and using different sequencing chemistries or additives. However there remains a significant fraction of problems that require a different solution. As in the human genome project the main tool for tackling such problems is the SIL (McMurray *et al.*, 1998). This technique is particularly useful for G/C rich regions, repetitive regions and those containing elements predicted to have potential secondary structure. It is not, however, useful for A/T rich sequences. For cloned A/T rich and repetitive G/C rich regions where the extreme base composition makes primer design for oligo walking impossible, transposon insertion libraries (TILs) are invaluable (Devine *et al.*, 1997). Uncloned A/T rich sequences can be sequenced by sub-cloning into M13 (Quail, 2002). Here a set of nested deletions of A/T rich sequences are created by digestion with Mung Bean nuclease, prior to cloning in M13 and re-sequencing. Finally, walking using the Genome Walker kit from Clontech (http://www.clontech.com/products/cat/HTML/1112.shtml) has allowed recovery of sequences that extend contigs. Using the prescribed procedure genomic DNA from the sequenced organisms is cut with enzymes that generate blunt-ends and specific adapters are ligated to these termini to form genome walker libraries. PCR of these using one primer directed out from a contig end and another specific adapter sequence thereby allows amplification of a product extending from the contig end to the nearest recognition sequence for the particular enzyme used.

8. Conclusion and Discussion

Finishing is an evolving, iterative process of refining assembled shotgun data in order to produce contiguous sequence with a maximum error rate of 1 in

10000 base pairs. The techniques described were born from early sequencing projects (Bowman *et al., 1997*; The *C. elegans* Sequencing Consortium 1998) and have more recently been applied to the finishing of human chromosomes 22 (Dunham *et al.*, 1999) and 20 (Deloukas *et al.*, 2001) and in a modified version to human chromosome 21 (Hattori *et al.*, 2000), as well as over 30 genomes of human pathogens and other model organisms.

Throughout the course of the Human Genome Project there have been increasing improvements to assembly algorithms which, along with an increase in compute power, can now generate a 6-7x assembly of a 3 Gigabase genome allowing early release of genomic sequence covering 95% of a genome. Scientists awaiting the sequence of genes value this as an early product of a genome sequencing programme. If large insert sequence read pairs and a BAC map are available, such an assembly provides sequence contigs with partial order and orientation but with gaps.

A few genomes have now gone on to be finished and comparisons have been made between the 'draft' and complete sequence. For example, a recent study has compared six draft genome assemblies over a 5.6Mb region of chromosome 4 (Semple *et al.*, 2002). Data were assessed from two consecutive releases by the International Human Genome Consortium taken from the browser mounted at University of California, Santa Cruz, two consecutive releases from the National Center for Biotechnology Information, the release from Celera Genomics and a hybrid assembly made by the authors using the International Human Genome Consortium and Celera Genomics assemblies. The authors found that all assemblies contained some degree of mis-assembly as measured by deviation from both their framework marker order and the order observed in the completed BAC sequences. Significantly the differences between assemblies resulted in differences in genome annotation. On completion of the sequence of chromosome 20, an analysis of the discrepancies between the International Human Genome Consortium draft and finished assemblies identified several sequence positions that had shifted more than 10 megabases between versions (Hattori and Taylor, 2001). Such discrepancies have significant potential implications for accurate gene identification and for studies of chromosome organization and long range regulation.

Comparison has previously been made between the predicted transcript sequences annotated in the Celera Genomics and the International Human Genome Consortium draft assemblies. It was found that more than a third of the genes identified in one assembly were not found in the other (Hogenesch *et al.*, 2001). While differences in the draft sequence assemblies may have influenced the gene predictions, it would not be accurate to assume finished sequence will provide definitive conclusions (Goodman, 2002), however reliable sequence is an essential 'foundation' on which further research can be confidently built.

It has been argued that sequence finishing is equally, if not more important for microorganisms on account of their gene density and the occurrence of extremes of base composition which, due to cloning bias, may well be absent or under-represented in a shotgun assembly. This problem has recently been described during the *Mycobacterium tuberculosis* sequencing project for a region encoding a family of large repetitive genes termed PE-PGRS (polymorphic GC-rich repetitive sequences (PGRS). With more than 60 members in the gene family and the potential to contain several kilobases of repetitive sequence of >80% GC, the initial shotgun data were found to contain very large gaps encompassing nearly all members of the gene family. As the author states, while the function of these genes is not understood, they are known to be both antigenic and variable and as such may play an important role in the interaction between *M. tuberculosis* and its host (Parkhill, 2000). It is only by considerable manual effort that such regions can be fully elucidated and this process is required if future research is not to be limited.

As more sequence becomes available the potential for comparative work increases. A number of comparisons have already been performed between complete and incomplete genomes with particular respect to the presence or absence of genes or groups of genes. While such studies provide a good deal of information there should also be the consideration of whether the genes found are capable of being translated into functional proteins. A recent paper states that *Salmonella typhi (S. typhi)* and *Salmonella enterica serovar Typhimurium (S. typhimurium)* have >200 and 45 predicted pseudogenes respectively (Parkhill, 2002). Furthermore the pseudogenes in *S. typhi* are over represented in certain functional classes, for example pathogenicity island- and cell surface-associated genes, leading the author to suggest these inactivated genes may be involved in virulence and host range. This is supported further by the fact that several genes that have been shown to be important for these phenotypes in *S. typhimurium* appear to be inactive in *S. typhi* suggesting the functionality of these genes accounts for the difference in virulence and host range between the two strains. While our knowledge is incomplete and with such possibilities to investigate it is clear that the aim of genomic sequencing should be the production of the highest quality sequence to provide a reference against which comparisons can be made.

As finishing experience has been gained aspects of automation, as with the shotgun process, have been introduced. For example using an automated oligo design programme in conjunction with a robot capable of re-arraying and performing the reactions. However, it remains clear that both problematic assemblies and the final quality assessment still need to be made by well trained individuals.

The cost of sequence finishing has always constituted at least 50% of the cost of production of finished sequence. This remains true even though the cost of shotgun has fallen over the past two years by roughly 60%. This continuing decrease in cost has the effect of making finished sequence a more viable end point of a sequencing programme.

9. Useful Web Sites

Wellcome Trust Sanger Institute- with links to other major genome sequencing centers and genomic resources:

http://www.sanger.ac.uk

International Human Genome Sequencing Consortium agreed finishing criteria:

http://genome.wustl.edu/Overview/glbstand.php

Software

For information on the PHRED basecaller, PHRAP, the database assembly engine, Consed, the graphical tool for editing PHRAP assemblies and Autofinish, the automatic reaction selection program, http://www.phrap. org/

For information on GAP, the graphical tool for editing PHRAP assemblies, (Staden Group) http://www.mrc-lmb.cam.ac.uk/pubseq/phrap.html

Acknowledgements

The authors are funded by The Wellcome Trust.

References

Bonfield, J. K., Smith, K. F., and Staden, R. A. 1995. A new DNA sequence assembly program. Nucleic Acids Res. 23: 4994–4999.

Bowman, S., Churcher, C., Badcock, K., Brown, D., Chillingworth, T., Connor, R., Dedman, K., Devlin, K., Gentles, S., Hamlin, N., Hunt, S., Jagels, K., Lye, G., Moule, S., Odell, C., Pearson, D., Rajandream, M., Rice, P., Skelton, J., Walsh, S., Whitehead, S. and Barrell, B. 1997. The nucleotide sequence of Saccharomyces cerevisiae chromosome XIII. Nature. 387 (6632 Suppl):90–93.

Buck, G.A., Fox, J.W., Gunthorpe, M., Hager, K.M., Naeve, C.W., Pon, R.T., Adams, P.S., and Rush, J. 1999. Research Report: Design strategies and performance of custom DNA sequencing primers. BioTechniques. 27: 528–536.

Chissoe, S.L., Marra, M.A., Hillier, L., Brinkman, R., Wilson, R.K., and Waterston, R.H. 1997. Representation of cloned genomic sequences in two sequencing vectors: correlation of DNA sequence and subclone distribution. Nucleic Acids Res. 25 (15): 2960–2966.

Deloukas, P., Matthews, L.H., Ashurst, J., Burton, J., Gilbert, J.G., Jones, M., Stavrides, G., Almeida, J.P., Babbage, A.K., Bagguley, C.L., Bailey, J., Barlow, K.F., Bates, K.N., Beard, L.M., Beare, D.M., Beasley, O.P., Bird, C.P., Blakey, S.E., Bridgeman, A.M., Brown, A.J., Buck, D., Burrill, W., Butler, A.P., Carder, C., Carter, N.P., Chapman, J.C., Clamp, M., Clark, G., Clark, L.N., Clark, S.Y., Clee, C.M., Clegg, S., Cobley, V.E., Collier, R.E., Connor, R., Corby, N.R., Coulson, A., Coville, G.J., Deadman, R., Dhami, P., Dunn, M., Ellington, A.G., Frankland, J.A., Fraser, A., French, L., Garner, P., Grafham, D.V., Griffiths, C., Griffiths, M.N., Gwilliam, R., Hall, R.E., Hammond, S., Harley, J.L., Heath, P.D., Ho, S., Holden, J.L., Howden, P.J., Huckle, E., Hunt, A.R., Hunt, S.E., Jekosch, K., Johnson, C.M., Johnson, D., Kay, M.P., Kimberley, A.M., King, A., Knights, A., Laird, G.K., Lawlor, S., Lehvaslaiho, M.H., Leversha, M., Lloyd, C., Lloyd, D.M., Lovell, J.D., Marsh, V.L., Martin, S.L., McConnachie, L.J., McLay, K., McMurray, A.A., Milne, S., Mistry, D., Moore, M.J., Mullikin, J.C., Nickerson, T., Oliver, K., Parker, A., Patel, R., Pearce, T.A., Peck, A.I., Phillimore, B.J., Prathalingam, S.R., Plumb, R.W., Ramsay, H., Rice, C.M., Ross, M.T., Scott, C.E., Sehra, H.K., Shownkeen, R., Sims, S., Skuce, C.D., Smith, M.L., Soderlund, C., Steward, C.A., Sulston, J.E., Swann, M., Sycamore, N., Taylor, R., Tee, L., Thomas, D.W., Thorpe, A., Tracey, A., Tromans, A.C., Vaudin, M., Wall, M., Wallis, J.M., Whitehead, S.L., Whittaker, P., Willey, D.L., Williams, L., Williams, S.A., Wilming, L., Wray, P.W., Hubbard, T., Durbin, R.M., Bentley, D.R., Beck, S., and Rogers, J. 2001. The DNA sequence and comparative analysis of human chromosome 20. Nature. 414: 865–71.

Devine, S.E., Chissoe, S.L., Eby, Y., Wilson, R.K., and Boeke, J.D. 1997. A Transposon-based strategy for sequencing repetitive DNA in

Eukaryotic Genomes. Genome Res. 7: 551–563.

Dunham, I., Hunt, A.R., Collins, J.E., Bruskiewich, R., Beare, D.M., Clamp, M., Smink, L.J., Ainscough, R., Almeida, J.P., Babbage, A., Bagguley, C., Bailey, J., Barlow, K., Bates, K.N., Beasley, O., Bird, C.P., Blakey, S., Bridgeman, A.M., Buck, D., Burgess, J., Burrill, W.D., O'Brien, K.P., and *et al.* 1999. The DNA sequence of human chromosome 22. Nature. 402: 489–95.

Ewing, B., and Green, P. 1998. Base-calling of automated sequencer traces using phred. II. Error probablities. Genome Res. 8. (3): 186–194.

Ewing, B., Hillier, L., Wendl M.C., and Green, P. 1998. Base-Calling of automated sequencer traces using phred. I. Accuracy assessment. Genome Res. 8 (3): 175–185.

Felsenfeld, A., Peterson, J., Schloss, J., and Guyer, M. 1999. Assessing the quality of the DNA sequence from the Human Genome Project. Genome Res. 9: 1–4.

Gelfand, D.H. 1989. PCR Technology: Principles and Applications for DNA Amplification (Erlich, H.A., ed.), pp17–22, Stockton Press.

Goodman, N. 2002. Column: The IT guy. Genome Technol. 18: 54–58.

Gordon, D., Abajian, C., and Green, P. 1998. Consed:a graphical tool for sequence finishing. Genome Res. 8:175–185.

Gordon, D., Desmarais, C., and Green, P. 2001. Automated finishing with autofinish. Genome Res. 11 (4): 614–625.

Green, E. D. 2001. Strategies for the systematic sequencing of complex genomes. Nature Review Genet. 2: 573–583.

Hall, L.S., Thomas, E., Lilley, K., and Grills, G. 2000. Advances in DNA sequencing. J. Biomol. Tech. 11(4): 230 (abstract).

Hattori, M., Fujiyama, A., Taylor, T.D., Watanabe, H., Yada, T., Park, H.S., Toyoda, A., Ishii, K., Totoki, Y., Choi, D.K., Soeda, E., Ohki, M., Takagi, T., Sakaki, Y., Taudien, S., Blechschmidt, K., Polley, A., Menzel, U., Delabar, J., Kumpf, K., Lehmann, R., Patterson, D., Reichwald, K., Rump, A., Schillhabel, M., Schudy, A., Zimmermann, W., Rosenthal, A., Kudoh, J., Schibuya, K., Kawasaki, K., Asakawa, S., Shintani, A., Sasaki, T., Nagamine, K., Mitsuyama, S., Antonarakis, S.E., Minoshima, S., Shimizu, N., Nordsiek, G., Hornischer, K., Brant, P., Scharfe, M., Schon, O., Desario, A., Reichelt, J., Kauer, G., Blocker, H., Ramser, J., Beck, A., Klages, S., Hennig, S., Riesselmann, L., Dagand, E., Haaf, T., Wehrmeyer, S., Borzym, K., Gardiner, K., Nizetic, D., Francis, F., Lehrach, H., Reinhardt, R., and Yaspo, M.L. 2000. The DNA sequence of human chromosome 21. Nature. 405: 311–9.

Hattori, M. and Taylor, T.D. 2001. Part three in the book of genes. Nature. 414: 854–855.

Hillier, L., and Green, P. 1991. OSP: An oligo selection program. PCR

Methods Appl. 1: 124–128.

Hirao, I., Nishimura, Y., Tagawa, Y., Watanabe, K., and Miura, K. 1992. Extraordinarily stable minin-hairpins: electrophoretical and thermal properties of the various sequence variants of d(GCGAAAGC) and their effect on DNA sequencing. Nucleic Acids Res. 20 (15): 3891–3896.

Hogenesch, J.B., Ching, K.A., Batalov, S., Su A.I., Walker, J.R., Zhou, Y., Kay, S.A., Schultz, P.G. and Cooke, M.P. 2001. A comparison of the Celera and Ensembl predicted gene sets reveals little overlap in novel genes. Cell. 106:413–415.

Hyndman, D.A., Cooper, S., Pruzinshky, D., Coad. and Mitsuhashi, M. 1996. Software to determine optimal oligonucleotide sequences based on hybridisation simulation data. BioTechniques. 20: 1090–1097.

Konfortov, B.A., Cohen, H.M., Bankier, A.T., and Dear, P.H. 2000.A high-resolution HAPPY map of *Dictyostelium discoideum* chromosome 6. Genome Res. 11: 1737–42.

McMurray, A.A., Sulston, J.E., and Quail, M.A. 1998. Short-insert libraries as a method of problem solving in genome sequencing. Genome Res. 8 (5): 562–566.

Motz, M., Paabo, S., and Kilger., C. 2000. Short technical reports. Improved cycle sequencing of GC-rich templates by a combination of nucleotide analogs. Biotechniques. 29: 268–270

Parkhill, J. 2000. In defense of complete genomes. Nat. Biotechnol. 18: 493–494.

Parkhill, J. 2002. The importance of complete genome sequences. Trends Microbiol. 10: 219–220.

Pomp, D., and Medrano, J.F. 1991. Organic solvents as facilitators of polymerase chain reaction. BioTechniques. 10 (1): 58–59.

Quail, M.A. 2002. M13 cloning of mung bean nuclease digested PCR fragments as a means of gap closure within A/T-rich genome sequencing projects. DNA Seq. 12: 355–359.

Rosenblum, B.B., Lee, L.G., Spurgeon, S.L., Khan, S.H., Menchen, S.M., Heiner, C.R., and Chen, S.M. 1997. New dye-labeled terminators for improved DNA sequencing patterns. Nucleic Acids Res. 25: 4500–4504.

Rychlik, W. and Rhoads, R.E. 1989. A computer program for choosing optimal oligonucleotides for filter hybridisation sequencing and in vitro amplification of DNA. Nucleic Acids Res. 17: 8543–8551.

Sambrook, J., Fritsch, E.F., and Maniatis, T. 1989. Molecular Cloning: A Laboratory Manual. Cold Spring Harbour Laboratory Press, Cold Spring Harbour, New York.

Semple, C.A.M., Morris, S.W., Porteous, D.J., and Evans, K.L. 2002. Computational comparison of human genomic sequence assemblies for a region of chromosome 4. Genome Res. 12: 424–429.

Sheidl, T.M., Miura, Y., Yee, H.A., and Tamaoki, T. 1995. Automated fluorescent dye-terminator sequencing of G+C-rich tracts with the aid of dimethyl sulfoxide. BioTechniques. 19: 691–694.

Shen, W.-H., and Hohn, B. 1992. DMSO improves PCR amplification of DNA with complex secondary structure. Trends Genet. 8: 227.

The *C. elegans* Sequencing Consortium. 1998. Genome sequence of the nematode *C. elegans*: A platform for investigating biology. Science. 282: 2012–2018.

Winship, P.R. 1989. An improved method for directly sequencing PCR amplified material using dimethyl sulphoxide. Nucleic Acids Res. 17 (3): 1266.

From: *Genome Mapping and Sequencing*
© 2003 Horizon Scientific Press, Wymondham, UK

12

Software for Genomic Sequencing

Michael C. Wendl, Asif T. Chinwalla,
and LaDeana W. Hillier

Abstract

Modern high-throughput DNA sequencing laboratories increasingly rely on software to automate the processing, analyzing, storing, and retrieving of sequence data. We outline the issues and tasks faced in such an environment and describe software solutions that have been developed to address them. Major applications used in the human sequencing project are emphasized.

1. Introduction

DNA molecules are the fundamental hereditary material of most organisms and contain the complement of instructions for building and maintaining life. Knowledge of DNA sequences provides perhaps the most promising avenue to date for delving into many of the longstanding mysteries in biology, genetics, and medicine. The potential applications of complete sequences, along with technology advances which enable their acquisition, have motivated continuing expansion of international sequencing projects. Perhaps the most visible efforts have been those working towards deciphering our own human sequence, which is now known in draft form (International Human Genome Sequencing Consortium, 2001; Venter *et al.*,

2001). However, there are many more projects planned or underway for various scientifically, medically, and commercially relevant organisms. Scientists, as well as the general public, are eagerly anticipating advances these data will lead to. In fact, this fundamental knowledge has been likened to the biological analog of chemistry's periodic table (Lander, 2000).

Every sequencing venture relies heavily on computer software to manage the millions of repetitive operations involved in sequencing an organism. We shall attempt to outline these software tools and the environments in which they are used to automate lab procedures. Some of this discussion will be qualitative, as there are many small-scale generic tasks that have given rise to home-grown software solutions. Sample sheet generation is one obvious example; There are perhaps as many different sample sheet scripts as there are sequencing labs. Other computational tasks are more formidable and these have led to specific software implementations that are widely used. We shall describe some of these programs in more detail.

1.1 The Need for Constant Software Advances

The modern era of DNA sequencing began with the implementation of the chain-terminating "dideoxy" sequencing reaction (Sanger et al., 1977) and the random shotgun strategy (Sanger et al., 1980; Anderson, 1981; Deininger, 1983). Sequences on the order of 10^4 nucleotides were typical of what was attempted in this era (Fiers et al., 1978; Reddy et al., 1978; Sanger et al., 1978; Sanger et al., 1982). Even at this comparatively early stage, investigators recognized the need for software to automate the processing and they applied rudimentary tools where possible(Staden, 1977). The introduction of fluorescent dye labels to the sequencing reaction several years later (Strauss et al., 1986) led to the development of the first automated sequencing machine by the Applied Biosystems Corporation. A follow-on to this machine, the model 373 (ABI, 1994) was successfully used by many laboratories for sequencing strands on the order of 10^6 nucleotides (Wilson et al., 1994; Fleischmann et al., 1995). Automated sequencers, coupled with growing deployment of robotic systems for sample preparation, placed increasing demands upon computer software to fully exploit higher throughput capacities (Krauter et al., 1995). By the late 1990's, full-length sequences of many larger genomes were becoming available. Examples include the yeast genome S. cerevisiae at 12 megabases (Krauter et al., 1995), the roundworm C. elegans at 100 megabases (CESC, 1998), and the fruit fly D. melanogaster at 180 megabases (Adams et al., 2000). All of these projects relied on substantially higher levels of software automation than their predecessors.

The human genome, at roughly 3 gigabases, represents probably the most formidable project yet undertaken. While its impact will be unprecedented, particularly in high profile areas such as clinical medicine (Green and Waterston, 1991), the pinnacle for scientists will be the availability of *many* "base perfect" plant and animal sequences. Thus, much of the driving force for automation, including at the software level, is the trend toward sequencing more DNA per unit time coupled with the ability to handle larger more complex genomes. Mathematical analyses (Lander and Waterman, 1988; Wendl *et al.*, 2001b) indicate that sequencing progress is approximately proportional to e^{-LC}, where C is a constant based upon genome size and number of reactions and L is the length of DNA that can be read by a sequencing machine from a single reaction. At a 500 base read length, initial estimates determined that the human genome would require on the order of 10^8 individual reactions (Venter *et al.*, 1998). Unfortunately, there is an upper limit on the read length imposed by the machine's ability to resolve base-pair differences in length for long chains. Many laboratories now consistently obtain 600 base read lengths, but it is unclear how much more is to be expected. Consequently, higher throughputs have been handled almost exclusively by increasing the number of reactions carried out per unit time.

Because of these circumstances, the most successful labs have evolved into factory-style operations and are now capable of carrying out and processing several million reactions per month. There is a fairly standard blueprint. Slow and error-prone manual steps are replaced as much as possible by automatic ones carried out by robotic hardware. Reaction products are fed to groups of automated capillary sequencing machines, primarily "third-generation" ABI Prism 3700 and Megabace 1000 instruments. Raw data are then analyzed and, if necessary, assembled into contiguous elements and finished by closing gaps. All of these operations must be supported by reliable and robust software.

1.2 Overview of Shotgun Sequencing

To get a better idea of the software requirements and what role software automation plays, let us consider the basic methodology of shotgun sequencing using the dideoxy sequencing reaction. We briefly describe the procedure for large-insert clones, *e.g.* Bacterial Artificial Chromsome (BAC) clones (Shizuya *et al.*, 1992), although the process is similar for the whole genome shotgun method (Weber and Myers, 1997). For our purposes, the entire process can be divided into the following major steps, which we will refer to by number in succeeding sections.

1. Bacterial cells containing a clone of interest are streaked onto an agar plate and incubated, after which a single colony is picked and further grown in media. The cells are then lysed, permitting the clone DNA to be isolated. If the DNA yield is sufficient and the clone's identity confirms via a fingerprint comparison, clone copies are broken up into smaller fragments using ultrasound. This step is known as *sonication*. In the ideal case, the set of fragments is uniformly distributed over the original source clone. Mung bean nuclease is applied to obtain blunt ends.

2. Fragments are recovered by phenol extraction and ethanol is used to precipitate the DNA. After centrifugation, the DNA pellet is re-suspended in buffer and fractionated on an agarose gel to isolate sizes in the appropriate range to be handled by the sequencing machines, usually a few kilobases. The selected DNA is again phenol extracted and re-suspended in buffer and the concentration is determined.

3. The random fragments are ligated into an appropriate vector. Each vector-fragment combination is now referred to as a "subclone".

4. Subclones are inserted into *E. coli* competent cells (so-called because they are especially adept at this job) and grown up on plates. As cells multiply and neighboring cells are infected, a colony (plaque) forms. Individual colonies are picked into trays, usually of 96-well format.

5. Another set of transfer and purification steps yields sequenceable DNA.

6. Subclone DNA is sequenced with the dideoxy chain-terminating reaction. Essentially, strands of complementary DNA are synthesized from the subclone template using a primer to initiate the reaction, a polymerase for incorporating deoxy residues, and dideoxy residues to terminate the reaction. Reaction products consist of a set of dye-labeled sub-fragments, where all possible positions in the sequence are represented by the set of terminating bases. The dye label resides either on the primer (dye-primer reaction), or it may be embedded in the terminating base (dye-terminator reaction). With both reaction types, four different dye colors are used and their correspondence to each of the four bases is known in advance.

7. Reaction products for each subclone are analyzed with an automated sequencing machine. Sub-fragments migrate through a capillary tube containing a polyacrylamide gel under an applied voltage, their individual speeds being inversely proportional to their respective lengths. Consequently, they separate or *fractionate* according to length. The order of bases in the original subclone can then be

inferred from the order of arrivals at a detector, which reads dye labels as they are excited by laser emission. The output is a "trace" of the four-color digital signal.

8. Raw trace data from the machines are processed into an assembly-ready form. Minimally, this involves basecalling, screening out vector and other contaminated sequences, filtering low quality sequences, logging base quality statistics, and storing the results on a permanent file system (usually *Unix*). According to commonly used plain-text formats, the result is called either an "exp file" (Bonfield and Staden, 1996) or a "phd file" (Gordon *et al.*, 1998).

9. For large-insert clones, such as BACs, reads are assembled to infer the consensus sequence of the original clone. Additional reactions are usually required to close remaining gaps. Once the clone is contiguous, it is passed along for analysis. For whole genome shotgun reads, assembly may be delayed until the full shotgun is complete.

1.3 Software Languages

We touch briefly on the topic of procedural programming languages. It is probably fair to say that most, if not all of the mainstream languages are represented in the vast compendium of sequence-related software. Except for specific applications that are in common usage, language choice remains very much linked to local laboratory conventions. However, when pressed to identify the closest things to standards, we probably would have to cite *C* (Kernighan and Ritchie, 1988), and perhaps *C++*, as being the most commonly used languages for compute-intensive applications and *Perl* (Wall *et al.*, 2000) as the language for "everything else". This is certainly the case for the major public centers participating in human sequencing and will likely continue to be for quite some time. Honorable mentions go to *Python* (Lutz, 2001) and the various shell programming languages. The latter encompasses most of the earliest legacy software, some of which is still being used in sequencing labs.

2. Laboratory Information Systems

Before discussing specific pieces of software used in sequencing, let us touch upon the data management systems that underlie many of these tools. In genomics, the term "database" is often associated with various annotation and protein databases, for example SWISS-PROT (Bairoch and Apweiler, 2000). But in regards to the sequencing pipeline, "database" connotes laboratory information management systems used to schedule and queue processes and track, analyze, and report data. As pointed out by Searls (2000), the history of database usage in genomics largely reflects the evolution of database technology itself. The earliest laboratory management systems were simply spreadsheets and text files, usually with specific formats which enabled some level of parsing. Throughput increases eventually overwhelmed such systems. Most laboratories and centers now employ dedicated information management systems.

There are two main categories of structured database: the object model, which implements data in the form of objects and specifies associated methods of operation, and the relational model, which implements data atomically and specifies relationships between relevant atoms. Object-oriented databases are known for their capability in handling complex data structures and are therefore often chosen for molecular biology analysis, for example the *ACeDB* package (Stein and Thierry-Mieg, 1998; Walsh *et al.*, 1998). Yet the lack of widely-accepted standards, especially in query language, coupled with the fact that lab data are abundant, but not too complex, has prompted widespread use of the relational architecture for laboratory tracking, although some laboratories and centers use *ACeDB* for a subset of laboratory tracking tasks. This is likely to remain the case for some time yet. Of the many robust commercial and freeware implementations of the relational model, we have found *Oracle* (Loney and Koch, 2000), *Sybase* (Worden, 2000), and *MySql* (Dubois, 2000) to be particularly popular in the sequencing community. All adhere to the SQL query language standard (Date and Darwen, 1997).

These systems merely provide a basis for laboratory database design. An actual database instance is implemented according to an instruction set known as a *schema*, which describes data entities and their relationships. It is at this point where uniformity among sequencing labs ends. Design methodology remains an art rather than a science. Moreover, schema requirements vary by laboratory, being somewhat proportional to the level of automation. For example, a small lab may have only a few sequencing machines and may rely on manual performance of most of the lab work. A few simple tracking applications based upon a rudimentary schema is typical of this situation. Conversely, large labs having a highly automated environment will require a much more ambitious schema. In addition to extensive tracking and analysis capabilities, provisions for scheduling and queueing would be built into such schema. These capabilities are routinely supported by the integration of barcodes (Venter *et al.*, 2001).

One concept that is especially useful for large centers is *workflow* (Goodman *et al.*, 1998). Here, a data management system is extended with two additional features. First is the ability to manage time-ordered chains of events, sometimes referred to as workflow pipelines. The pipeline concept involves identifying materials, defining the steps which operate upon them, and specifying the states in which materials and steps are allowed to exist. This design lends itself to process-oriented environments such as sequencing labs in that it facilitates scheduling, queuing, and analysis of history. The second requirement is the ability to readily reconfigure or expand pipelines as evolving lab protocols and technologies dictate. Goodman *et al.* (1998) implemented what appears to be the first such system used in the sequencing community and other laboratories, including our own, utilize variations of this theme. Further details are discussed below for specific applications that interface to lab databases.

There are a number of ways to handle interfacing, but all fall broadly into two categories: proprietary and non-proprietary. For example, *Oracle* database programmers can use the Developer forms toolkit (Koletzke and Dorsey, 1999) with PL/SQL (Feuerstein and Pribyl, 1997), both of which are proprietary, to write graphical database-specific applications and PL/SQL for standalone scripts to be called by other laboratory applications. The primary advantage of proprietary tools is that they offer significant integration with the database platform. Some labs opt for this choice, but many prefer the more general non-proprietary tools. This is especially true if the expense of a commercial database license compels the selection of an open-source platform or if maximal flexibility in the applications is a design requirement. Here again, *Perl* is one of the more popular options. Its *DBI.pm* extension (Descartes and Bunce, 2000) provides drivers for almost all the relational database platforms currently available. The extension provides an application programming interface (API) through which SQL commands are passed to execute database transactions. Other popular scripting languages, notably *Python* (Lutz, 2001), handle transactions in a similar manner. Individual programmers who favor this approach cite the ease with which database access can be integrated directly into their scripts. Non-proprietary tools also facilitate the free exchange of software between labs participating in large distributed sequencing projects. Such software exchange has been a major factor in all publically sponsored projects to date.

3. Sample Preparation Software

In terms of workflow, Steps 1 through 6 above can be thought of as the sample preparation pipeline. Numerically, they represent the majority of a sequencing operation, yet these steps are not as software intensive as those in succeeding phases of processing. Software at the clone level (Step 1) focuses predominantly on simple tracking: for example managing lists of clones for growing, streaking, digesting, etc. In most environments, tracking applications take the form of graphical user interfaces tied to the laboratory database. Some can be quite intricate, allowing the user to prioritize lists, customize views of the data, and query histories. Facilities for handling clones that must be regrown, redigested, or those which do not confirm their mapping fingerprint are also incorporated. Succeeding steps primarily involve fluid transfers and are performed either manually or semi-automatically (using table-top pipetting robots), or in the case of larger labs, by fully automatic prepping systems (*e.g.* The *PlateTrak* system from Packard Biosciences). Here, tracking is accomplished primarily by barcode interfaces to the laboratory database. Barcode tags are affixed directly to tubes and plates and are read by hand scanners or barcode readers mounted on robotic hardware. Many lab databases are designed to track data at very high resolution, *e.g.* lot numbers of various reagents used in the pipeline, etc. This allows for very sophisticated analysis and troubleshooting.

Additional software packages of a more generic nature support sample preparation too. For example, before loading samples onto the sequencing machines (Step 7), a script generates names for the yet to be created trace files according to local naming conventions. Various other scripts manage filesystem directories and generate automated reports regarding preparation results.

4. Trace Processing Software

Before the advent of modern capillary sequencing machines, certain aspects of Step 7 above would have been included here. Some of these tasks, such as lane tracking, have been completely eliminated. Others, such as signal deconvolution and dye mobility-shift correction are now ably performed by the machine itself and can be treated as "black boxes". Therefore, we assume that trace processing formally begins when trace files arrive on a lab's central filesystem. This brings us to Step 8. From here on, sequencing operations are increasingly dependent upon robust software.

Implementations of trace processing pipelines run largely along the same lines, at least for the larger sequencing laboratories. Little to no human intervention is involved. Rather, there is a primary piece of code that is responsible for all high-level tasks and it calls or loads modules for specific processing tasks (Lawrence *et al.*, 1994; Bonfield and Staden, 1996; Smith *et al.*, 1997; Giddings *et al.*, 1998; Wendl *et al.*, 1998). For example, the main code typically manages input validation for a set of traces, disk checking, file output, batch database interactions, and reporting. Often, it is set up to run automatically, *e.g.* under the auspices of a Unix *daemon* process. As data processing requests arrive, sets of traces are requisitioned and enter the pipeline. Frequently these will be sets of 96 corresponding to the 96-well plate capacity of the ABI Prism 3700 and Megabace 1000 instruments. Specific modules (discussed below) are called sequentially on each trace file. In some cases, including our own, the pipeline is configured to process many small batches in parallel.

The main code can also be executed outside of the automatic pipeline. There is a constant need to process some traces manually, for example those resulting from in-house protocol change experiments. The code handles these via a graphical interface, which allows users to customize their processing session. For example, a user may wish to omit certain modules or view traces individually for manual quality clipping (Gleeson and Hillier, 1991).

Finally, we include under this heading various other in-house scripts that serve in a support role. Most of these are concerned with the file system. One class handles directory structures: creating directories and subdirectories for BAC projects or whole genome shotgun reads and shuttling these among various physical disks as needed. The latter requirement arises directly from shear volume of data. For example, traces are commonly stored in the Standard Chromatogram Format (SCF) (Dear and Staden, 1992), a machine-independent binary file architecture. Average read lengths of 500 to 600 bases translate into trace file sizes of 40 to 50 kilobytes (The companion ascii file containing the processed data is on the order of 10 % of the trace file size). Therefore, if a sequencing center processes 50,000 reactions per day, the daily disk consumption by trace files alone would be roughly 2.5 gigabytes. File compression helps to some degree, however, because disk loads are dynamic, such "load balancing" software is inevitably needed. Most labs also have some level of permanent off-line archiving capability and other scripts which check the integrity of the filesystem's directories, symbolic links, etc.

Let us now examine some of the modular components in greater detail along with some representative pieces of software. Not every laboratory implements all of these steps and some have additional ones not listed here. However, this list is representative of most of the larger centers.

4.1 File Conversion

Several commercial sequencers are in use by the community, and each utilizes its own format for trace files. As mentioned above, the standard

SCF format (Dear and Staden, 1992) is preferred, not only for the sake of uniformity, but also because its file size is comparatively small. The first step is thus conversion from machine-specific format to SCF format. A variety of programs will handle this task, notably *Phred* (Ewing *et al.*, 1998) and *MakeSCF* from the Staden Package (Bonfield and Staden, 1996).

4.2 Base Calling

Data generated by the sequencing machines from analyzing reaction products are not actually strings of the four bases. Rather, data for each trace are in the form of a four channel digital signal, where each channel represents the dye excitation corresponding to one of the nucleotide bases. These signals are indicators of the four dye concentrations as a function of time and therefore encode the order of the bases. The next step is then to analyze these signals to determine the actual sequence they represent. This is known as *base calling*.

Ideally, a trace would exhibit a composite signal having evenly spaced non-overlapping peaks, where each peak would represent one group of identical dye-labeled fragments as they arrive at the detector. Inferring the actual sequence would then be trivial. Unfortunately, actual traces deviate from ideal behavior in several ways, each of which correlates to specific parts of the trace. The beginning of a trace is marked by anomalous migration of short fragments and unreacted dye-containing molecules, which result in irregular, noisy peaks. This part must be removed. The middle section of the trace tends to be the most well behaved, although *compressions* can cause base calling problems when they arise. A compression occurs when a portion of the sample doubles back upon itself by binding with complimentary sequence upstream or downstream. A hairpin loop forms, causing the sample to migrate faster than expected by its true sequence. Its peak is then artificially shifted to the left in the trace and may overlay a genuine peak in its correct position. Late in the trace diffusion effects start to smear the signals. Moreover, this region is associated with the longest fragments whose number arriving at the detector is small compared to the shorter fragments. Signal strength to noise ratio thus deteriorates. Ewing *et al.,* (1998) describe additional anomalies. The objective is then to faithfully translate the composite signal into the correct sequence of bases despite these obstacles.

Considerable attempts have been made toward understanding the generalized physical interactions underlying DNA fractionation (Babskii *et al.*, 1989; Levene and Zimm, 1989; Deutsch, 1990; Luckey and Smith, 1993), but a universal theory has yet to be developed. Consequently, software implementations are based more on empirics rather than first principles. A number of different algorithmic approaches have been developed (Giddings *et al.*, 1993; Golden *et al.*, 1995; Berno, 1996; Ewing *et al.*, 1998), however, owing to its demonstrated accuracy, *Phred* (Ewing *et al.*, 1998) has evolved as perhaps the most widely-used application for calling bases. The algorithm consists of four sequential phases. The first phase predicts local peak locations by identifying the most prominent Fourier mode using a Fast Fourier Transform (FFT) of the data. Actual peaks are identified in the second phase by determining where signals are concave (according to a simple finite difference approximation of the second derivative) and by applying empirical rules for the total area under the signal. A dynamic programming algorithm is used in the third phase to match predicted peaks with observed peaks. The fourth phase captures any remaining peaks that have not yet been called as bases by comparing them to a set of empirical rules.

Another aspect of the overall problem is the ability to assign quality (confidence) values for the correctness of each base call. These values are important in a number of respects. For the assembly problem (discussed below), they allow full read lengths to be used and provide a basis for statistical criteria to guide finishing. They provide a basis for statistical quality analysis of the raw data as well. Unfortunately, there are many ways to calculate such a quantity. At least two properties are required. A method must be predictive, that is, it must be able to perform calculations on the data without knowing the true sequence. Secondly, calculated error probabilities must correspond closely to error rates found by comparing to the correct sequence. Lawrence and Solovyev (1994) carried out an extensive study based upon linear discriminant analysis of a large group of parameters, such as fluorescent intensity, peak height, and peak spacing. Most parameters were examined locally, that is, at a given base position and perhaps the two surrounding positions. They found that parameters with the best resolving power varied by the three types of errors studied: substitution, deletion, and insertion. Ewing and Green (1998) instead employ only four parameters, which are evaluated over a finite window rather than locally. This algorithm calculates quality quite well according to the above criteria and it has been incorporated into the base calling package (Ewing *et al.*, 1998).

4.3 Template and Signal Information

There may be one or more modules that extract specific information from various sources for incorporation into the results file of each trace.

Template information of the subclone source and signal information of the raw data are frequently collected for this purpose. The former is queried from the laboratory information database (and usually cached by the main processing program), while the latter can be parsed directly from the trace file.

4.4 Filtering and Tagging

Another responsibility of the main processor is to coordinate modules that filter or tag unwanted data. These data may or may not be related to the specific sequence content. For example, poor trace quality is not related to sequence *per se*. It could be caused simply by low signal-to-noise ratio or irregular peak spacing. However, the resulting sequence must be tagged as being of poor quality so that it can be removed before assembly if desired. Many trace processing pipelines actually bundle this particular filter in with the base calling module, for example, by using *Phred* (Ewing *et al.*, 1998) output to evaluate the overall quality of the trace. Others implement this step as a standalone module, for example using the *Trace_Clip* utility in the Staden Package (Staden, 1996). This task is straightforward since it is strictly related to the data itself; *i.e.* it is independent of the sequence that is represented.

The problem is somewhat more involved for anomalies related to sequence content. One must first know the specific sequences that are to be screened out, were they to be found in any raw data. Because of the inherent imperfections of base calling discussed above, these sequences will not necessarily appear in the data as base-perfect strings. Therefore, a method of determining sufficient similarity to declare a match between two sequences is required. This leads to the so-called sequence alignment problem, which arises repeatedly in genomics. In this case, an algorithm takes two strings, representing two nucleotide sequences, and attempts to align them optimally given a specific system for scoring substitution, insertion, and deletion errors. An excellent overview of the subject is given by Waterman (2000). There are a number of generalized sequence alignment programs, notably *Fasta* (Pearson and Lipman, 1988), *Blast* (Altschul *et al.*, 1990; Altschul *et al.*, 1997), and *Cross_Match* (Ewing *et al.*, 1998), which are based on the well-known Smith-Waterman algorithm (Smith and Waterman, 1981). Filtering modules are frequently just "wrapper" scripts that set up the problem, call one of these sequence alignment programs to do the actual comparison, then post-process and report results. We mention several such scenarios.

Perhaps the most common filtering function relates to vector artifacts, which can occur both at the clone level ("cloning vector" artifact) and at the subclone level ("sequencing vector" artifact). For example, manufacture of a sequenceable library requires sonicating BAC clones as discussed in Step 1 above. Assuming typical lengths for source DNA and BAC cloning vector of 200 and 10 kilobases, respectively, one would expect on average about 5 % of the inserts to be BAC vector DNA in actuality. At the subclone level, phage or plasmid vectors sometimes do not take up a DNA insert (or perhaps only a particularly short insert), in which case the vector itself is re-sequenced. All methods used in DNA vector filtering are variations upon two themes: the wrapper script approach mentioned above, or the use of packages tailored specifically at vector screening, such as the *Vector_Clip* utility in the Staden Package (Staden, 1996). Even here the difference is largely semantic because all approaches to this problem must, at their lowest level, implement a comparison between candidate base-called sequence and a known string of vector sequence. However, programs such as *Vector_Clip* exploit coordinates of the so-called cloning site where the vector and DNA insert are joined. Any matches to vector will trigger a specific tag to be inserted into the results file for that particular trace. Moreover, trace and results files containing vector may be separated on the filesystem from files representing genuine data. These steps ensure that vector sequences are not included in the assembly phase of sequencing discussed below.

A related filtering task is the removal of any *E. coli* DNA resulting from library contamination. Here it is also common to implement a wrapper around *Fasta*, *Blast*, or *Cross_Match* to compare candidate sequence to an *E. coli* sequence library or to use a dedicated program, such as the *Screen_Seq* utility in the Staden Package (Staden, 1996). Matches are once again tagged and segregated.

Another task is tagging various repeated elements that may be found in the data. Although they are not considered contamination in the sense of vector or *E. coli* sequence, these repeats cause significant difficulties in assembly through false positive associations of unrelated sequences (Myers, 1999). As an example, draft sequencing of the human genome indicates about 45 % transposon-derived repeat sequence as well as approximately 3 % microsatellite repeats and large segmental duplications (International Human Genome Sequencing Consortium, 2001). Researchers had substantial difficulties in obtaining an accurate assembly using shotgun data alone and had to employ long range mapping data to obtain closure (Venter *et al.*, 2001). This is yet another example of a sequence comparison problem and the above software solutions are once again applicable. A number of specific packages for this problem have also been developed, notably *Hmmer* (Eddy, 1998), which is based upon Hidden Markov Models, *Seg* (Wootton and Federhen, 1993), which searches low complexity A-T rich sequence, and *RepeatMasker* (Bedell *et al.*, 2000), which can handle complicated nested repeats. Graphical representations are also useful in examining these repeated structures (Parsons, 1995).

4.5 Statistical Analysis

There is essentially no manual review of results up to this point, so having some means of statistical quality assessment is critical. A convenient resource which many pipelines employ is the set of base quality values discussed above. The convention associated with *Phred* defines log-transformed error probabilities as $q = -10\log_{10}p$, where p is the estimated error probability of a base call. No units are associated with either p or q. Because these quality values have already been calculated in the base calling module, they are readily available for compilation. This involves simply counting the number of bases which exceed a threshold quality value, usually $q = 20$ representing a 1 % chance of error, and storing the result in the lab database. Supplementary analyses of quality 30 and 40 are also sometimes performed. Storage is usually determined by a balance between level of normalization in the database versus transaction speed with which users demand for their analyses. At one extreme is maximal normalization whereby analyses are stored individually on a per-read basis, however, this design is also the slowest. At our own center, we have found denormalization to the level of 96 well trays to be an optimal compromise. In particular, we not only store compilations on a per-read basis, but we have implemented an auxiliary table in our database to store compilations on a per-tray basis. Technicians find that roughly 90 % of the analyses they are interested in can be constructed from this table. Figure 1 shows a representative analysis for 3 recent individual months of our data. The large peak on the left represents very short, essentially "failed" reads, while the peak to the right represents good reads having long stretches of high-fidelity base calls. Such data are optimal for the assembly process described below.

The real leverage of a denormalized table is obtained by attaching as much common information as possible to the batches of 96 reads, either directly or through relationships in the database. For instance, one can record tray and sequencing machine numbers, clone, library, and vector names, sequencing group, dates, employee names, reagents, etc. This enables fast yet powerful queries to track a diverse set of performance metrics according to data quality. The notable exception to this design is statistical analysis of individual sequencing machine capillaries, which must be derived from read-specific compilations.

Many labs also implement an analogous system based upon pass/fail statistics. These consist in most cases of simple tallies of the read outcomes, *e.g.* number passed, number that were sequencing vector, number that were poor data quality, etc. Histogram-based analysis similar to Figure 1 can then be performed.

5. Assembly and Finishing Software

Here, we focus on Step 9. The power of shotgun sequencing lies in the fact that overdetermined information (redundant sequencing) from many small individual experiments enables one to infer a much longer original sequence from overlaps. This is called the *assembly* problem. Of course, the basic assumption is that the set of fragments covers the entire source sequence, *i.e.* there are no gaps. Lander-Waterman theory (Lander and Waterman, 1988) predicts that the number of gaps in a sequencing project is roughly proportional to Ne^{-NC} where C is a constant based upon clone and genome length and N is the number of clones processed. As N becomes large, the exponential term dominates and the number of gaps

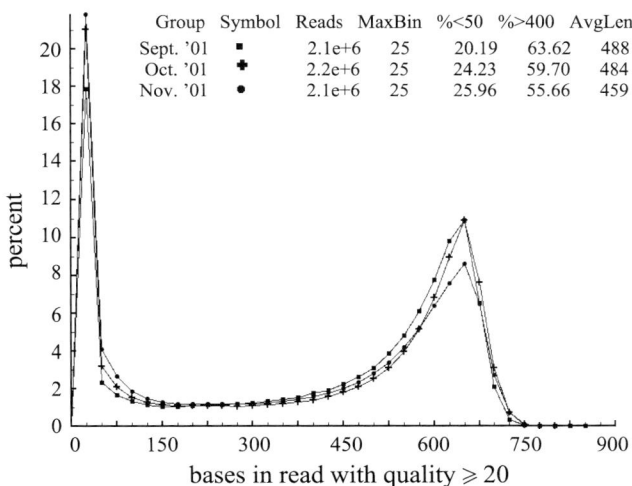

Group	Symbol	Reads	MaxBin	%<50	%>400	AvgLen
Sept. '01	∎	2.1e+6	25	20.19	63.62	488
Oct. '01	✦	2.2e+6	25	24.23	59.70	484
Nov. '01	●	2.1e+6	25	25.96	55.66	459

Figure 1. Statistical base quality curves for three recent months of data (excluding vector reads) from the Genome Sequencing Center. The abscissa displays number of bases per read having quality 20 or better, while the ordinate shows percentage of reads in the sample.

tends asymptotically to zero. Of course, it is also intuitive that, given enough clones, one would expect that all gaps should eventually be closed by pure chance.

This argument fails if the distribution of subclones in a library is not uniform. In particular, any region against which a bias exists will tend to remain uncovered. This can be caused by a number of factors: sonication of a clone may be biased, insertion of fragments into vectors may be biased, or certain vector-insert combinations may prevent proper cloning. Another factor in this equation is that there are practical limits on N in terms of cost. Therefore, the fact that a certain percentage of sequencing reactions will simply fail as shown in Figure 1 will also promote the existence of gaps. Closing these remaining gaps is known as the *finishing* problem.

Although this whole section is covered in Step 9, most labs implement three rather distinct phases of this process: automatic assembly, an automated "pre-finish" phase of gap closing, and a final finishing process which typically involves manual direction and input on the part of a skilled user. Let us now examine these three phases in more detail along with some representative pieces of software.

5.1 Assembly Software

At its foundation, assembly is essentially a very large multiple sequence alignment problem. Assuming N available subclone sequences, one has to make on the order of $N(N-1)/2$ sequence alignment calculations. This figure is derived as follows: for each of the N sequences there are $N-1$ possible comparisons (since sequences are not compared to themselves), however, only ½ of these are independent since a comparison is commutative. That is, aligning A→B is identical to B→A. It is approximate because sequences are divided according to read direction whereby forward and reverse read comparisons are typically exclusive. Given sufficiently high N according to Lander-Waterman theory (Lander and Waterman, 1988), *e.g.* 7X to 10X coverage, minimal subclone bias, and reasonable data quality, one would expect a "good" assembly containing a limited number of gaps. This is typically considered the completion of the shotgun phase of a project. Some labs elect to run assembly software on a regular basis during the shotgun phase as well in order to monitor progress.

Of course, with the presence of real-world anomalies, *e.g.* repeats and a mixture of data quality, the problem is not so simple. In fact, development of assembly software has been one of the more active areas of informatics research for several decades (for review see *e.g.* (Wendl *et al.*, 2001a) and work is ongoing. Earlier generations of software tools required considerable user interaction to compensate for various deficiencies, including the inability to effectively handle non-uniformities in data quality. As such, assembly was a highly labor intensive job. Current tools are more sophisticated and are tailored in considerable degree to the type of sequencing being performed.

At one end of the spectrum is large-insert BAC clone sequencing of the type used in the human genome project (International Human Genome Sequencing Consortium, 2001). This has evolved from older methods such as cosmid sequencing. Here, one typically starts with highly redundant coverage of a 100 to 200 kb BAC clone. The advantage of BAC clones from an assembly standpoint is that long-range misassembly issues are eliminated and short-range ones are reduced because sequence information is local. There are a number of well-known software implementations for this problem, *e.g. Cap3* (Huang and Madan, 1999), *Phrap* (Gordon *et al.*, 1998), and *Gap4* (Bonfield *et al.*, 1995; Bonfield *et al.*, 1998). All of these operate along similar lines. Overlaps between reads are computed, after which reads are joined to form contigs according to their overlap scores. Read-based constraints for forward and reverse directions are applied if available. A multiple sequence alignment is then constructed from which a consensus is derived for each base position.

The primary goal of the software developer has been to automate the assembly process while at the same time maintaining error levels at or below an expert user employing an interactive system. An important advance in this regard has been the use of base quality values discussed above. The *Cap3*, *Phrap*, and *Gap4* systems all exploit log-transformed values as defined by Ewing and Green (1998) for determining overlaps, computing multiple alignments, and deriving consensus sequence. The availability of quality for consensus sequence is particularly useful for the pre-finishing and finishing phases as discussed below.

A related but not identical assembly problem is realized using the whole genome shotgun approach. This procedure is of a more recent origin than map-based large-insert clone sequencing and has likewise been applied to human sequencing (Venter *et al.*, 2001). In qualitative terms, the problem is simply a scaled-up version of BAC sequence assembly if the organism has a sparse repeat structure. Conversely, a repeat-rich genome presents both short and long-range risks for misassembly, complicating the task substantially. Large-scale repeats are particularly problematic because single reads having unique stretches outside the repeat cannot span that region. Early bacterial projects were successfully assembled with software that utilized basic Smith-Waterman alignments (Smith and Waterman, 1981) and forward-reverse constraints (*e.g.* Fleischmann *et al.*, 1995; Bult *et al.*, 1996). More recently Myers *et al.*, (2000) have developed a system that was applied to data from the *Drosophila* genome project (Adams *et al.*, 2000). The underlying difference from packages mentioned above is that here the assembler makes use of long-range linking information from the end sequences of large-insert 10 kb and 100 kb clones to construct ordered contig *scaffolds*. This is followed by progressive attempts to resolve repeats. Although results were comparable to regions of the *Drosophila* genome that had previously been finished, a variation of this algorithm did not perform as well on human data (Venter *et al.*, 2001). This has led to hybrid approaches using both whole genome shotgun and clone-based data (Huson *et al.*, 2001).

Lastly, a strategy has recently been devised that may potentially resolve most of the difficulties posed by repeats during assembly (Pevzner *et al.*, 2001). All the classical "overlap-layout-consensus" approaches discussed above are formally equal to a Hamiltonian path problem, where every read corresponds to a vertex in an overlap graph. Vertices representing overlapping reads are connected by edges. Thus, assembly requires finding a path in the graph visiting each vertex exactly once. This is an NP-complete problem. The new approach discretizes individual reads such that a graph can be constructed whereby each repeat corresponds to an edge rather than a collection of vertices. The problem simplifies to finding a path that visits every edge once, *i.e.* an Eulerian path problem, which is solvable in linear time. Thus far, the software has been successfully tested on a number of bacterial whole genome shotgun assembly problems (Pevzner *et al.*, 2001). Independent tests indicate that it is still too memory intensive and dependent upon deep coverage for production usage. Furthermore, it does not yet have a facility to view multiple alignments of an assembly. This algorithm will likely make a significant impact if these issues can be resolved.

5.2 Pre-Finishing Software

After completion of the shotgun phase of a project, an assembly will rarely consist of entirely contiguous sequence. Rather, it usually is in the form of several segments separated by gaps of unknown sequence. Moreover, errors in the raw data lead to low quality regions where the consensus sequence is uncertain. The remaining job is to finish the project such that it is a single contiguous segment of high quality consensus sequence. Before the advent of widespread software automation, this was largely a manual task. Now, an initial round of automatic finishing, or pre-finishing, is usually undertaken. In some cases closure can actually be obtained at this stage by software alone.

Both the pre-finishing and finishing phases revolve intimately around error probabilities of the consensus sequence. These probabilities enable an objective statistical approach toward closure. Specifically, the standard equation of expected value yields the expected number of errors in a segment. When this value drops below a previously defined threshold, the project is by definition finished. Therefore, the most efficient method is to determine regions that make the highest contribution to the overall error and to target these regions with additional data and/or editing. Many labs use the *Autofinish* tool (Gordon *et al.*, 2001) for automatically choosing finishing reads. The program's output is usually integrated with the laboratory database and robotic hardware such that the selected reads are merged directly into the regular pipeline. This arrangement saves numerous man-hours of labor, dramatically accelerating the finishing process.

Essentially, *Autofinish* first evaluates the error rate of all bases in the region to be finished. It then constructs simulated reads as candidates to resolve problem areas. Estimates are made of their length and position-specific quality values based upon statistics gathered from existing data. Next, subclone templates that are acceptable for obtaining new reads are identified. This is somewhat of a process of elimination: the software rejects any template having a chimeric read, insert length shorter than 300 bases, from a library known to be bad, etc. The program also finds the location of each template's insert with respect to contigs in the assembly by examining universal primer reads derived from that template and determining their vector-insert junctions. The next step is identifying which contigs to attempt to finish. This is based on heuristic rules associated with contig length and depth. For every candidate contig, the program then derives additional ordering and orientation information and then suggests reads that would be expected to make the greatest contribution to the problem based on available templates and the virtual read analysis.

Figure 2 shows results we have obtained at the Genome Sequencing Center for 904 clones using *Autofinish*. The number of gaps remaining after assembling all shotgun data was significant: over one third had 3 or more gaps and no clone was gap free. The software significantly reduced the most problematic clones, *i.e.* those with 6 or more gaps, and completely finished an astounding 400 clones! In this case, the total number of gaps that must be addressed manually has been reduced by about 60 %.

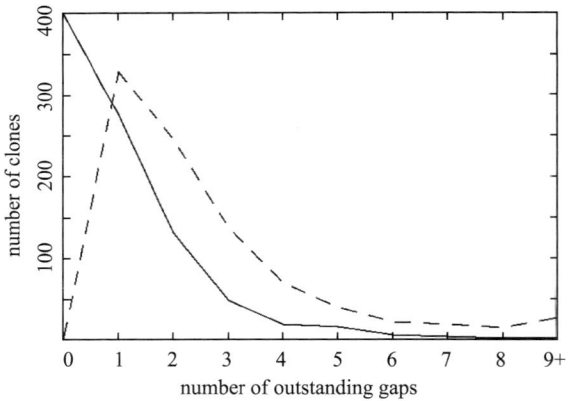

number of outstanding gaps

Figure 2. Gap closure results using *Autofinish* (Gordon *et al.*, 2001) on 904 large-insert clones at the Genome Sequencing Center. Dashed and solid curves indicate gap distributions before and after application of *Autofinish*, respectively.

5.3 Finishing Software

Even after several iterations of automated pre-finishing, a project frequently remains in several pieces. Moreover, misassemblies may still be present. For example, finished sequences of human chromosomes 20 through 22 (Dunham *et al.*, 1999; Hattori *et al.*, 2000; Deloukas *et al.*, 2001) revealed significant anomalies in the draft assemblies (Hattori and Taylor, 2001). These problems are the most difficult to resolve and skilled personnel are expected to complete the project manually. At present, the whole genome shotgun approach is not considered adequate for large repeat-intensive genomes unless other data and anchoring information are available. Therefore, we shall concentrate on the finishing process as applied to large-insert clones.

A finisher's main tool is a graphical program customized for viewing assembly and trace data, facilitating detection of problem areas, assisting in read selection, and editing errors. As with most of the other components of the pipeline discussed thus far, a number of programs have been developed for finishing. For large scale efforts, *Consed* (Gordon *et al.*, 1998) and *Gap4* (Bonfield *et al.*, 1995; Bonfield *et al.*, 1998) have evolved as the standard applications. Like assembly and pre-finishing applications, these programs utilize base quality values to guide the finishing effort according to an objective statistical measure of expected error. According to the "Bermuda Rules" established early in the human sequencing project (International Human Genome Sequencing Consortium, 2001), the accepted standard is a maximum of 1 error per 10,000 bases and no gaps. Various tags may be attached to individual reads or portions of an assembly: *e.g.* to identify chimeras, vector, compressions, contamination, repeats, etc. These tags are automatically appended at various stages, including base calling and assembly, and manually by the user as a part of assembly editing.

In general, the finishing process proceeds as follows: inspection of the assembly to assess anomalies and the need for additional data, obtaining required data via additional reads, and final editing of assembly and consensus sequence. For large or difficult projects, several iterations of this cycle can be required. To run the software, a finisher imports all the trace files and read files (experiment files for *Gap4* or phd files for *Consed*) and an existing assembly. Contigs are graphically displayed in a viewer, which the user can use to navigate to various regions of interest. The displays are rich with intuitive visual cues to emphasize various kinds of information for the user. For instance *Consed* has several color display modes. In one, bases in the assembly are overlayed with varying colors and shading to indicate base quality, discrepancies, and tags. In another, colors symbolize information relevant to the underlying *Phrap* assembly. In a third, colors show edit-related information. Underlying trace data can be called up to the display by clicking specific reads. Supplemental data can be ordered by specifying additional experiments. Options also exist for overriding the assembly in terms of forcibly breaking or joining contigs.

In addition to the statistical aspect of finishing, heuristic standards were derived by the human genome sequencing consortium defining the terms under which sequence can be considered "finished" (see http://genome. wustl.edu/Overview/glbstand.php for the definitive document maintained for the International Consortium by the Washington University Genome Sequencing Center). Briefly, all regions must be double-stranded, sequenced with an alternate chemistry, or covered by high quality data, *i.e.* $q \geq 30$. The assembly must also be confirmed by restriction digest. The rules also describe a set of annotation tags to be applied to regions which cannot be resolved.

The main finishing applications are supported by an array of utility programs. The set usually includes scripts for reporting contig numbers and lengths, "stealing" data from overlap regions of contiguous clones, ordering custom oligos for finishing reads, and comparing simulated *in silico* digests to actual restriction digests. Other programs assist in final validation of the finished sequence, for example by comparing assemblies from the various assemblers discussed above. Special graphical displays are used to facilitate interpretation of these comparisons since string lengths are so long (Parsons, 1995).

6. Post-Sequencing Software

The sequencing operation formally ceases once a project is deemed finished. It then moves onto several post-processing phases which involve release, review, annotation, and submission of the sequence and creation of genome-wide consensus.

6.1 Data Release, Review, Annotation, and Submission

A hallmark of public sequencing efforts established in the early days of the *C. elegans* project (Sulston *et al.*, 1992; C. elegans Sequencing Consortium, 1998) has been the immediate release of all sequence data via both the World Wide Web (WWW) and through the public databases, *e.g.* GenBank, the European Molecular Biology Laboratory (EMBL) database, and the DNA Databank of Japan (DDBJ). All assembled consensus sequences for large-insert clones and all individual reads for whole genome shotgun projects are published nightly on ftp sites at individual sequencing centers. As assembled sequences are generated for whole genome shotgun projects, those sequences are released immediately as well. Related mapping data used to place clones and sequences within their genomic context are also made available via the internet. Moreover, preliminary analyses are offered in many cases, including results of similarity searches of individual reads. For example, initial identification of each read was furnished for the Washington University Salmonella project (McClelland *et al.*, 2000; McClelland *et al.*, 2001). Other analyses provided are dependent on data type and organism.

Submission to the public databases involves various phases, including release of unfinished assembled sequence (and whole genome shotgun reads), and for finished sequence, a quality review phase followed by an annotation phase and finally the data submission phase. Unfinished sequences, both assembled clone consensus sequences and individual shotgun reads which are to be submitted, are again subjected to vector and contamination screens as discussed above before being submitted. Sequence, mapping and quality information are then formatted appropriately and uploaded to the appropriate data repositories. For example, GenBank utilizes the ISO standard Abstract Syntax Notation One (ASN.1) data format. See http://www.ncbi.nlm.nih.gov/Sitemap/Summary/asn1.html for documentation.

To prepare "finished" data for submission, a wide range of tools and methods for sequence analysis and annotation have been developed. As with sequencing pipelines, increasing throughput has driven the ongoing automation of these tools. There has been some debate over the usefulness of automatically generated annotation in the public databases (Wheelan and Boguski, 1998), resulting in varying degrees of annotation on database entries. Some laboratories provide no annotation, others provide only automated annotation, while other labs provide manually reviewed gene predictions.

Finishers typically forward their clone and its relevant information to the analysis pipeline via a home-grown script. This information would include coordinates of clone endpoints and a specific length of overlapping sequences of neighboring clones. Also, specific comments the finisher wishes to accompany the GenBank entry would be included. A set of standard analyses are then performed (Wendl *et al.*, 2001a): identification of repetitive elements and other significant sequence features, homology (similarity) searching, and gene prediction. As an example, *C. elegans* clones are treated with the following protocol. Consensus sequence is compared via *Blast* (Altschul *et al.*, 1990; Korf and Gish, 2000) to various *C. elegans* and *C. briggsae* DNA and cDNA databases. A related program, *Blastx* (Altschul *et al.*, 1997), is used to search non-redundant protein databases. Visualization tools are used to assist in interpreting the results (Sonnhammer and Durbin, 1994). *TrnaScan-SE* (Fichant and Burks, 1991; Lowe and Eddy, 1997) is used to search for tRNAs. Repeats, especially those from known *C. elegans* repeat families, are located using the *Tandem* and *Inverted* tools in the *ACeDB* package and *Scan* and *Hmmfs* (Eddy, 1998). Finally, *Genefinder* identifies candidate genes using systematic statistical criteria to characterize and display probable exons (P. Green, unpublished. Source code is available from Phil Green at the University of Washington Genome Center). Additional gene finding software is also used, in particular the *Genscan* (Burge and Karlin, 1997) program. Analysis pipelines are similar for other publically funded projects, notably those for yeast (Johnston *et al.*, 1997) and *Arabidopsis* (Tabata *et al.*, 2000).

The protocol is somewhat modified for bacterial projects. Different prediction tools are used, specifically *Glimmer* (Salzberg *et al.*, 1999) and *Genemark* (Lukashin and Borodovsky, 1998). Predicted proteins are searched against the *Pfam* protein family database (Bateman *et al.*, 2000) and the *COG* database (Tatusov *et al.*, 2001) to find putative orthologues in other completed genomes. Protein localization prediction software such as *Psort* (Nakai and Horton, 1999) is also applied. For refinement, the annotation can then be compared with predictions for other bacteria. For example, salmonella (McClelland *et al.*, 2001) was compared with the *S. typhi* annotation (Parkhill *et al.*, 2001). All differences in CDS predictions and start predictions were reassessed on the basis of choice of start codon, length, conservation in *Enterobacteriaceae* and presence of identifiable motifs.

In bacterial and yeast genome sequences, computational analysis and annotation have led to near complete catalogs of known or potential genes. Because base composition of coding and noncoding DNA sequence is easily distinguished and transcriptional start points match well-conserved motifs, gene structures are readily identified. There are few, if any, introns and a much lower proportion of noncoding sequence in these genomes than those of more complex organisms. In contrast, predictions made by these *ab initio* methods are appreciably less reliable when applied to human sequence. Analysis must instead rely more on the results of homology searches against other DNA and protein sequences. Therefore, while the basic steps remain the same for the human and mouse genomes, there are some variations. One difference is the early identification of repetitive elements using the *RepeatMasker* and *MaskerAid* applications (Bedell *et al.*, 2000). Searches against human and mouse mRNAs and human and mouse EST sequences are also performed (Adams *et al.*, 1992; Hillier *et al.*, 1996; Marra *et al.*, 1999). While significant matches would not be expected to uncover the structure of entire genes, they often indicate regions that contain coding exons. In addition to *Genefinder* and *Genscan* mentioned above, predictions are derived from other programs as well. These include *Fgenesh+* (Salamov and Solovyev, 2000), *Slam* (Pachter *et al.*, 2002), and *Twinscan* (Korf *et al.*, 2001), which takes advantage of similarity information from related organisms during gene prediction. Of course, annotation provided with individually submitted clones is necessarily of a local nature.

Annotation of all genomic sequence at Washington University is carried out using implementations of the *ACeDB* database environment (Stein and Thierry-Mieg, 1998; Walsh *et al.*, 1998), into which are loaded all the results of gene predictions and sequence searches. Sequences are linked to the appropriate position on the map. A fully integrated display of all data and analysis information is then available. Each sequence is manually reviewed and disparate predictions are reconciled. Once the best gene structure is determined, corresponding protein predictions are compared to protein databases and families (Bateman *et al.*, 2000). Results are used to describe the proteins encoded by the predicted genes. For final submission of these annotated finished sequences, *ACeDB* is used to translate all information into formats required by the public databases.

6.2 Generating Genome-Wide Consensus Sequence

Generation of genome-wide consensus sequence for large genomic projects relies on integration of all relevant mapping and sequence data. Even for whole genome assembly projects, it is typically the integration of available mapping data that allows sequence contigs to be anchored to the map and to each other. For example, read pair data (reads from opposite ends of a single clone) can be used to orient and position contigs relative to one another. Moreover, markers can be found in contigs using sequence similarity searching, enabling them to be anchored to existing maps. For genomes sequenced on a clone by clone basis where an underlying map of those clones is available, it is integration of the map, relevant mapping data and sequence data that allows generation of consensus genomic sequences.

In the *C. elegans* and *S. cerevisiae* projects, physical maps had already been built before sequencing commenced (Coulson *et al.*, 1986; Olson *et al.*, 1986). Clones for sequencing were then selected from these maps so that, as each clone was verified and then sequenced, consensus sequence could be immediately integrated into the overall chromosomal sequence by virtue of the map. In the *D. melanogaster* sequencing project (Adams *et al.*, 2000), *S. typhimurium* project (McClelland *et al.*, 2001), and for the current mouse effort, there is a combination of whole genome shotgun data, large-insert clone data, and various types of mapping information. This necessitates more careful integration to construct consensus sequence. Likewise, no complete map existed at the start of the human sequencing effort (International Human Genome Sequencing Consortium, 2001; Venter *et al.*, 2001). In fact, there were multiple maps with disparate data types. Even with the generation of the fingerprint map (International Human Genome Sequencing Consortium, 2001), many clones having sources outside of the map were inadvertently sequenced. In terms of data integration, assembly of the human draft consensus sequence (International Human Genome Sequencing Consortium, 2001) was a significantly more ambitious exercise than any that had taken place previously. The first task was filtering data to eliminate sequences where data from one BAC clone was substantially contaminated with data from another BAC clone and to eliminate non-human contamination and other artifacts. Second, sequenced clones were associated with specific clones on the physical map to produce a "layout" using primarily two methods. In the first method, the accuracy of restriction fragment sizes was exploited. Sequence data were used to calculate in silico restriction fragments which could then be compared to BAC fingerprints from the physical map. In the second method, sequenced clones were searched for matches to BAC ends. Multiple matches were required as a safeguard against errors known to exist in the BAC end database and against repeated sequences. The third step in assembly was associating fingerprint clone contigs with chromosomal locations via sequence similarity to Sequence Tagged Sites (STSs). These sites were derived from radiation hybrid maps (Hudson *et al.*, 1995; Stewart *et al.*, 1997; Deloukas *et al.*, 1998; Olivier *et al.*, 2001), genetics maps (Dib *et al.*, 1996; Broman *et al.*, 1998) and FISH (Bentley *et al.*, 2001; Cheung *et al.*, 2001; International Human Genome Mapping Consortium, 2001). The process was performed iteratively by comparing order and orientation of the STSs in the fingerprint clone contigs and the various STS-based maps to identify and refine discrepancies.

Next, the layout of sequenced clones along the physical map was used as a template for merging sequence data into chromosomal consensus. The program *GigAssembler* (Kent and Haussler, 2001) performs this process. It considers nearby sequenced clones, detects overlaps between the initial sequence contigs in these clones, merges the overlapping sequences and attempts to order and orient the sequence contigs. The program then selects the highest quality path through the sequence data. Finally, it creates sequence-contig scaffolds by ordering and orienting the sequence contigs using available additional information. Such information includes sequence data from paired-end plasmid and BAC reads, known messenger RNAs and ESTs, as well as any additional linking information provided by the centers. Fingerprint clone contigs, therefore, may contain several sequenced-clone contigs, where spanning clones remain to be sequenced. Laying out these sequence-contig scaffolds in order along the chromosomes based on the layout suggested by the physical map results in an integrated draft sequence of the human genome.

The approach to developing a consensus for an entire genome is clearly dependent on the sequencing method and on integration of existing mapping data, which is in turn dependent on the types of mapping data that are available. The difficulty of the assembly process is in part dependent on the amount of repetitive DNA in the organism, but is highly dependent as well on quality of the sequence data and that of the underlying maps.

Conclusion

Work is ongoing in most of the areas we have discussed. Databases, assembly, and sequence analysis are among the most active topics at the moment. The overall focus continues to be extending both the capacity for automation and the accuracy and robustness of the implementations.

Future progress will also be driven by advances in algorithms, hardware, and developments in software languages. While the level of manual intervention required from expert human scientists and technicians continues to fall, we are probably a ways from seeing their participation completely eliminated. And, barring a series of quantum leaps, we certainly will not see anything akin to "The ACME Pocket Sequencer" in the near future. For the moment, software developments have helped resolve enough bottlenecks to keep pace with our capacity for generating sequence.

Acknowledgements

The authors wish to thank Dr. R. H. Waterston, Director of the Genome Sequencing Center, for continued guidance, Dr. J. Wallis for insightful discussions, and Catrina Strowmat for compiling autofinishing data from the Genome Sequencing Center. This work was supported by a grant from the National Human Genome Research Institute (HG02042).

References

ABI 1994. 373 DNA Sequencing System: User's Manual. Applied Biosystems, 903204a Edition.

Adams, M.D., Celniker, S.E., Holt, R.A., Evans, C.A., Gocayne, J.D., Amanatides, P.G., Scherer, S.E., Li, P.W., Hoskins, R.A., Galle, R.F., George, R.A., Lewis, S.E., Richards, S., Ashburner, M., Henderson, S.N., Sutton, G.G., Wortman, J.R., Yandell, M.D., Zhang, Q., Chen, L.X., Brandon, R.C., Rogers, Y.H., Blazej, R.G., Champe, M., Pfeiffer, B.D., Wan, K.H., Doyle, C., Baxter, E.G., Helt, G., Nelson, C.R., Gabor Miklos, G.L., Abril, J.F., Agbayani, A., An, H.J., Andrews-Pfannkoch, C., Baldwin, D., Ballew, R.M., Basu, A., Baxendale, J., Bayraktaroglu, L., Beasley, E.M., Beeson, K.Y., Benos, P.V., Berman, B.P., Bhandari, D., Bolshakov, S., Borkova, D., Botchan, M.R., Bouck, J., Brokstein, P., Brottier, P., Burtis, K.C., Busam, D.A., Butler, H., Cadieu, E., Center, A., Chandra, I., Cherry, J.M., Cawley, S., Dahlke, C., Davenport, L.B., Davies, P., de Pablos, B., Delcher, A., Deng, Z., Mays, A.D., Dew, I., Dietz, S.M., Dodson, K., Doup, L.E., Downes, M., Dugan-Rocha, S., Dunkov, B.C., Dunn, P., Durbin, K.J., Evangelista, C.C., Ferraz, C., Ferriera, S., Fleischmann, W., Fosler, C., Gabrielian, A.E., Garg, N.S., Gelbart, W.M., Glasser, K., Glodek, A., Gong, F., Gorrell, J.H., Gu, Z., Guan, P., Harris, M., Harris, N.L., Harvey, D., Heiman, T.J., Hernandez, J.R., Houck, J., Hostin, D., Houston, K.A., Howland, T.J., Wei, M.H., Ibegwam, C., Jalali, M., Kalush, F., Karpen, G.H., Ke, Z., Kennison,

J.A., Ketchum, K.A., Kimmel, B.E., Kodira, C.D., Kraft, C., Kravitz, S., Kulp, D., Lai, Z., Lasko, P., Lei, Y., Levitsky, A.A., Li, J., Li, Z., Liang, Y., Lin, X., Liu, X., Mattei, B., McIntosh, T.C., McLeod, M.P., McPherson, D., Merkulov, G., Milshina, N.V., Mobarry, C., Morris, J., Moshrefi, A., Mount, S.M., Moy, M., Murphy, B., Murphy, L., Muzny, D.M., Nelson, D.L., Nelson, D.R., Nelson, K.A., Nixon, K., Nusskern, D.R., Pacleb, J.M., Palazzolo, M., Pittman, G.S., Pan, S., Pollard, J., Puri, V., Reese, M.G., Reinert, K., Remington, K., Saunders, R.D., Scheeler, F., Shen, H., Shue, B.C., Siden-Kiamos, I., Simpson, M., Skupski, M.P., Smith, T., Spier, E., Spradling, A.C., Stapleton, M., Strong, R., Sun, E., Svirskas, R., Tector, C., Turner, R., Venter, E., Wang, A.H., Wang, X., Wang, Z.Y., Wassarman, D.A., Weinstock, G.M., Weissenbach, J., Williams, S.M., Woodage, T., Worley, K.C., Wu, D., Yang, S., Yao, Q.A., Ye, J., Yeh, R.F., Zaveri, J.S., Zhan, M., Zhang, G., Zhao, Q., Zheng, L., Zheng, X.H., Zhong, F.N., Zhong, W., Zhou, X., Zhu, S., Zhu, X., Smith, H.O., Gibbs, R.A., Myers, E.W., Rubin, G.M., and Venter, J.C. 2000. The genome sequence of Drosophila melanogaster. Science. 287: 2185–95.

Adams, M.D., Dubnick, M., Kerlavage, A.R., Moreno, R., Kelley, J.M., Utterback, T.R., Nagle, J.W., Fields, C., and Venter, J.C. 1992. Sequence identification of 2,375 human brain genes. Nature. 355: 632–4.

Altschul, S.F., Gish, W., Miller, W., Myers, E.W., and Lipman, D.J. 1990. Basic local alignment search tool. J Mol Biol. 215: 403–10.

Altschul, S.F., Madden, T.L., Schaffer, A.A., Zhang, J., Zhang, Z., Miller, W., and Lipman, D.J. 1997. Gapped BLAST and PSI-BLAST: a new generation of protein database search programs. Nucleic Acids Res. 25: 3389–402.

Anderson, S. 1981. Shotgun DNA sequencing using cloned DNase I-generated fragments. Nucleic Acids Res. 9: 3015–27.

Babskii, V.G., Zhukov, M.Y., and Yudovich, V.I. 1989. Mathematical Theory of Electrophoresis. Consultants Bureau, New York, NY.

Bairoch, A., and Apweiler, R. 2000. The SWISS-PROT protein sequence database and its supplement TrEMBL in 2000. Nucleic Acids Res. 28: 45–8.

Bateman, A., Birney, E., Durbin, R., Eddy, S.R., Howe, K.L., and Sonnhammer, E.L. 2000. The Pfam protein families database. Nucleic Acids Res. 28: 263–6.

Bedell, J.A., Korf, I., and Gish, W. 2000. MaskerAid: a performance enhancement to RepeatMasker. Bioinformatics. 16: 1040–1.

Bentley, D.R., Deloukas, P., Dunham, A., French, L., Gregory, S.G., Humphray, S.J., Mungall, A.J., Ross, M.T., Carter, N.P., Dunham, I., Scott, C.E., Ashcroft, K.J., Atkinson, A.L., Aubin, K., Beare, D.M.,

Bethel, G., Brady, N., Brook, J.C., Burford, D.C., Burrill, W.D., Burrows, C., Butler, A.P., Carder, C., Catanese, J.J., Clee, C.M., Clegg, S.M., Cobley, V., Coffey, A.J., Cole, C.G., Collins, J.E., Conquer, J.S., Cooper, R.A., Culley, K.M., Dawson, E., Dearden, F.L., Durbin, R.M., de Jong, P.J., Dhami, P.D., Earthrowl, M.E., Edwards, C.A., Evans, R.S., Gillson, C.J., Ghori, J., Green, L., Gwilliam, R., Halls, K.S., Hammond, S., Harper, G.L., Heathcott, R.W., Holden, J.L., Holloway, E., Hopkins, B.L., Howard, P.J., Howell, G.R., Huckle, E.J., Hughes, J., Hunt, P.J., Hunt, S.E., Izmajlowicz, M., Jones, C.A., Joseph, S.S., Laird, G., Langford, C.F., Lehvaslaiho, M.H., Leversha, M.A., McCann, O.T., McDonald, L.M., McDowall, J., Maslen, G.L., Mistry, D., Moschonas, N.K., Neocleous, V., Pearson, D.M., Phillips, K.J., Porter, K.M., Prathalingam, S.R., Ramsey, Y.H., Ranby, S.A., Rice, C.M., Rogers, J., Rogers, L.J., Sarafidou, T., Scott, D.J., Sharp, G.J., Shaw-Smith, C.J., Smink, L.J., Soderlund, C., Sotheran, E.C., Steingruber, H.E., Sulston, J.E., Taylor, A., Taylor, R.G., Thorpe, A.A., Tinsley, E., Warry, G.L., Whittaker, A., Whittaker, P., Williams, S.H., Wilmer, T.E., Wooster, R., and Wright, C.L. 2001. The physical maps for sequencing human chromosomes 1, 6, 9, 10, 13, 20 and X. Nature. 409: 942–3.

Berno, A.J. 1996. A graph theoretic approach to the analysis of DNA sequencing data. Genome Res. 6: 80–91.

Bonfield, J.K., Rada, C., and Staden, R. 1998. Automated detection of point mutations using fluorescent sequence trace subtraction. Nucleic Acids Res. 26: 3404–9.

Bonfield, J.K., Smith, K., and Staden, R. 1995. A new DNA sequence assembly program. Nucleic Acids Res. 23: 4992–9.

Bonfield, J.K., and Staden, R. 1996. Experiment files and their application during large-scale sequencing projects. DNA Seq. 6: 109–17.

Broman, K.W., Murray, J.C., Sheffield, V.C., White, R.L., and Weber, J.L. 1998. Comprehensive human genetic maps: individual and sex-specific variation in recombination. Am J Hum Genet. 63: 861–9.

Bult, C.J., White, O., Olsen, G.J., Zhou, L., Fleischmann, R.D., Sutton, G.G., Blake, J.A., FitzGerald, L.M., Clayton, R.A., Gocayne, J.D., Kerlavage, A.R., Dougherty, B.A., Tomb, J.F., Adams, M.D., Reich, C.I., Overbeek, R., Kirkness, E.F., Weinstock, K.G., Merrick, J.M., Glodek, A., Scott, J.L., Geoghagen, N.S., and Venter, J.C. 1996. Complete genome sequence of the methanogenic archaeon, Methanococcus jannaschii. Science. 273: 1058–73.

Burge, C., and Karlin, S. 1997. Prediction of complete gene structures in human genomic DNA. J Mol Biol. 268: 78–94.

C. elegans Sequencing Consortium 1998. Genome sequence of the nematode C. elegans: a platform for investigating biology. The C. elegans

Sequencing Consortium [published errata appear in Science 1999 Jan 1;283(5398):35 and 1999 Mar 26;283(5410):2103]. Science. 282: 2012–8.

Cheung, V.G., Nowak, N., Jang, W., Kirsch, I.R., Zhao, S., Chen, X.N., Furey, T.S., Kim, U.J., Kuo, W.L., Olivier, M., Conroy, J., Kasprzyk, A., Massa, H., Yonescu, R., Sait, S., Thoreen, C., Snijders, A., Lemyre, E., Bailey, J.A., Bruzel, A., Burrill, W.D., Clegg, S.M., Collins, S., Dhami, P., Friedman, C., Han, C.S., Herrick, S., Lee, J., Ligon, A.H., Lowry, S., Morley, M., Narasimhan, S., Osoegawa, K., Peng, Z., Plajzer-Frick, I., Quade, B.J., Scott, D., Sirotkin, K., Thorpe, A.A., Gray, J.W., Hudson, J., Pinkel, D., Ried, T., Rowen, L., Shen-Ong, G.L., Strausberg, R.L., Birney, E., Callen, D.F., Cheng, J.F., Cox, D.R., Doggett, N.A., Carter, N.P., Eichler, E.E., Haussler, D., Korenberg, J.R., Morton, C.C., Albertson, D., Schuler, G., de Jong, P.J., and Trask, B.J. 2001. Integration of cytogenetic landmarks into the draft sequence of the human genome. Nature. 409: 953–8.

Coulson, A., Sulston, J., Brenner, S., and Karn, J. 1986. Towards a physical map of the genome of the nematode *Caenorhabditis elegans*. Proc Natl Acad Sci U S A. 83: 7821–7825.

Date, C.J., and Darwen, H. 1997. A Guide to the SQL Standard. Fourth Edition. Addison-Wesley, Boston, MA.

Dear, S., and Staden, R. 1992. A standard file format for data from DNA sequencing instruments. DNA Seq. 3: 107–10.

Deininger, P.L. 1983. Random subcloning of sonicated DNA: application to shotgun DNA sequence analysis. Anal Biochem. 129: 216–23.

Deloukas, P., Matthews, L.H., Ashurst, J., Burton, J., Gilbert, J.G., Jones, M., Stavrides, G., Almeida, J.P., Babbage, A.K., Bagguley, C.L., Bailey, J., Barlow, K.F., Bates, K.N., Beard, L.M., Beare, D.M., Beasley, O.P., Bird, C.P., Blakey, S.E., Bridgeman, A.M., Brown, A.J., Buck, D., Burrill, W., Butler, A.P., Carder, C., Carter, N.P., Chapman, J.C., Clamp, M., Clark, G., Clark, L.N., Clark, S.Y., Clee, C.M., Clegg, S., Cobley, V.E., Collier, R.E., Connor, R., Corby, N.R., Coulson, A., Coville, G.J., Deadman, R., Dhami, P., Dunn, M., Ellington, A.G., Frankland, J.A., Fraser, A., French, L., Garner, P., Grafham, D.V., Griffiths, C., Griffiths, M.N., Gwilliam, R., Hall, R.E., Hammond, S., Harley, J.L., Heath, P.D., Ho, S., Holden, J.L., Howden, P.J., Huckle, E., Hunt, A.R., Hunt, S.E., Jekosch, K., Johnson, C.M., Johnson, D., Kay, M.P., Kimberley, A.M., King, A., Knights, A., Laird, G.K., Lawlor, S., Lehvaslaiho, M.H., Leversha, M., Lloyd, C., Lloyd, D.M., Lovell, J.D., Marsh, V.L., Martin, S.L., McConnachie, L.J., McLay, K., McMurray, A.A., Milne, S., Mistry, D., Moore, M.J., Mullikin, J.C., Nickerson, T., Oliver, K., Parker, A., Patel, R., Pearce, T.A., Peck, A.I., Phillimore, B.J., Prathalingam, S.R., Plumb, R.W., Ramsay, H., Rice, C.M., Ross, M.T., Scott, C.E., Sehra,

H.K., Shownkeen, R., Sims, S., Skuce, C.D., Smith, M.L., Soderlund, C., Steward, C.A., Sulston, J.E., Swann, M., Sycamore, N., Taylor, R., Tee, L., Thomas, D.W., Thorpe, A., Tracey, A., Tromans, A.C., Vaudin, M., Wall, M., Wallis, J.M., Whitehead, S.L., Whittaker, P., Willey, D.L., Williams, L., Williams, S.A., Wilming, L., Wray, P.W., Hubbard, T., Durbin, R.M., Bentley, D.R., Beck, S., and Rogers, J. 2001. The DNA sequence and comparative analysis of human chromosome 20. Nature. 414: 865–71.

Deloukas, P., Schuler, G.D., Gyapay, G., Beasley, E.M., Soderlund, C., Rodriguez-Tome, P., Hui, L., Matise, T.C., McKusick, K.B., Beckmann, J.S., Bentolila, S., Bihoreau, M., Birren, B.B., Browne, J., Butler, A., Castle, A.B., Chiannilkulchai, N., Clee, C., Day, P.J., Dehejia, A., Dibling, T., Drouot, N., Duprat, S., Fizames, C., Bentley, D.R., and *et al.* 1998. A physical map of 30,000 human genes. Science. 282: 744–6.

Descartes, A., and Bunce, T. 2000. Programming the Perl DBI. O'Reilly and Associates, Sebastopol, CA.

Deutsch, J.M. 1990. Theoretical Aspects of Electrophoresis. In: Electrophoresis of Large DNA Molecules: Theory and Applications. E. Lai and B.W. Birren eds. Cold Spring Harbor Laboratory Press, Cold Spring Harbor p. 81–99.

Dib, C., Faure, S., Fizames, C., Samson, D., Drouot, N., Vignal, A., Millasseau, P., Marc, S., Hazan, J., Seboun, E., Lathrop, M., Gyapay, G., Morissette, J., and Weissenbach, J. 1996. A comprehensive genetic map of the human genome based on 5,264 microsatellites. Nature. 380: 152–4.

Dubois, P. 2000. MySql. New Riders Publishing, Indianapolis, IN.

Dunham, I., Hunt, A.R., Collins, J.E., Bruskiewich, R., Beare, D.M., Clamp, M., Smink, L.J., Ainscough, R., Almeida, J.P., Babbage, A., Bagguley, C., Bailey, J., Barlow, K., Bates, K.N., Beasley, O., Bird, C.P., Blakey, S., Bridgeman, A.M., Buck, D., Burgess, J., Burrill, W.D., and *et al.* 1999. The DNA sequence of human chromosome 22 [published erratum appears in Nature 2000 Apr 20;404(6780):904]. Nature. 402: 489-95.

Eddy, S.R. 1998. Profile hidden Markov models. Bioinformatics. 14: 755–63.

Ewing, B., and Green, P. 1998. Base-calling of automated sequencer traces using phred. II. Error probabilities. Genome Res. 8: 186–94.

Ewing, B., Hillier, L., Wendl, M.C., and Green, P. 1998. Base-calling of automated sequencer traces using phred. I. Accuracy assessment. Genome Res. 8: 175–85.

Feuerstein, S., and Pribyl, B. 1997. Oracle PL/SQL. Second Edition. O'Reilly and Associates, Sebastopol, CA.

Fichant, G.A., and Burks, C. 1991. Identifying potential tRNA genes in genomic DNA sequences. J Mol Biol. 220: 659–71.

Fiers, W., Contreras, R., Haegemann, G., Rogiers, R., Van de Voorde, A., Van Heuverswyn, H., Van Herreweghe, J., Volckaert, G., and Ysebaert, M. 1978. Complete nucleotide sequence of SV40 DNA. Nature. 273: 113–20.

Fleischmann, R.D., Adams, M.D., White, O., Clayton, R.A., Kirkness, E.F., Kerlavage, A.R., Bult, C.J., Tomb, J.F., Dougherty, B.A., Merrick, J.M., and et al. 1995. Whole-genome random sequencing and assembly of Haemophilus influenzae Rd. Science. 269: 496–512.

Giddings, M.C., Brumley, R.L., Jr., Haker, M., and Smith, L.M. 1993. An adaptive, object oriented strategy for base calling in DNA sequence analysis. Nucleic Acids Res. 21: 4530–40.

Giddings, M.C., Severin, J., Westphall, M., Wu, J., and Smith, L.M. 1998. A software system for data analysis in automated DNA sequencing. Genome Res. 8: 644–65.

Gleeson, T., and Hillier, L. 1991. A trace display and editing program for data from fluorescence based sequencing machines. Nucleic Acids Res. 19: 6481–3.

Golden, J., Abajiian, C., and Tibbetts, C. 1995. Evolutionary Optimization of a Neural Network-based Signal Processor for Photometric Data from an Automated DNA Sequencer. In: Evolutionary Programming IV. Proceedings of the Fourth Annual Conference on Evolutionary Programming. p. 579–601.

Goodman, N., Rozen, S., Stein, L.D., and Smith, A.G. 1998. The LabBase system for data management in large scale biology research laboratories. Bioinformatics. 14: 562–74.

Gordon, D., Abajian, C., and Green, P. 1998. Consed: a graphical tool for sequence finishing. Genome Res. 8: 195–202.

Gordon, D., Desmarais, C., and Green, P. 2001. Automated finishing with autofinish. Genome Res. 11: 614–25.

Green, E.D., and Waterston, R.H. 1991. The human genome project: Prospects and implications for clinical medicine. Jama. 266: 1966–75.

Hattori, M., Fujiyama, A., Taylor, T.D., Watanabe, H., Yada, T., Park, H.S., Toyoda, A., Ishii, K., Totoki, Y., Choi, D.K., Soeda, E., Ohki, M., Takagi, T., Sakaki, Y., Taudien, S., Blechschmidt, K., Polley, A., Menzel, U., Delabar, J., Kumpf, K., Lehmann, R., Patterson, D., Reichwald, K., Rump, A., Schillhabel, M., and Schudy, A. 2000. The DNA sequence of human chromosome 21. The chromosome 21 mapping and sequencing consortium. Nature. 405: 311–9.

Hattori, M., and Taylor, T.D. 2001. Part three in the book of genes. Nature. 414: 854–5.

Hillier, L.D., Lennon, G., Becker, M., Bonaldo, M.F., Chiapelli, B., Chissoe, S., Dietrich, N., DuBuque, T., Favello, A., Gish, W., Hawkins,

M., Hultman, M., Kucaba, T., Lacy, M., Le, M., Le, N., Mardis, E., Moore, B., Morris, M., Parsons, J., Prange, C., Rifkin, L., Rohlfing, T., Schellenberg, K., Marra, M., and *et al.* 1996. Generation and analysis of 280,000 human expressed sequence tags. Genome Res. 6: 807–28.

Huang, X., and Madan, A. 1999. CAP3: A DNA sequence assembly program. Genome Res. 9: 868–77.

Hudson, T.J., Stein, L.D., Gerety, S.S., Ma, J., Castle, A.B., Silva, J., Slonim, D.K., Baptista, R., Kruglyak, L., Xu, S.H., and *et al.* 1995. An STS-based map of the human genome. Science. 270: 1945–54.

Huson, D.H., Reinert, K., Kravitz, S.A., Remington, K.A., Delcher, A.L., Dew, I.M., Flanigan, M., Halpern, A.L., Lai, Z., Mobarry, C.M., Sutton, G.G., and Myers, E.W. 2001. Design of a compartmentalized shotgun assembler for the human genome. Bioinformatics. 17 Suppl 1: S132–9.

International Human Genome Mapping Consortium 2001. A physical map of the human genome. Nature. 409: 934–41.

International Human Genome Sequencing Consortium 2001. Initial sequencing and analysis of the human genome. Nature. 409: 860–921.

Johnston, M., Hillier, L., Riles, L., Albermann, K., Andre, B., Ansorge, W., Benes, V., Bruckner, M., Delius, H., Dubois, E., Dusterhoft, A., Entian, K.D., Floeth, M., Goffeau, A., Hebling, U., Heumann, K., Heuss-Neitzel, D., Hilbert, H., Hilger, F., Kleine, K., Kotter, P., Louis, E.J., Messenguy, F., Mewes, H.W., Hoheisel, J.D., and *et al.* 1997. The nucleotide sequence of Saccharomyces cerevisiae chromosome XII. Nature. 387: 87-90.

Kent, W.J., and Haussler, D. 2001. Assembly of the working draft of the human genome with GigAssembler. Genome Res. 11: 1541-8.

Kernighan, B.W., and Ritchie, D.M. 1988. The C Programming Language. Second Edition. Prentice Hall, Upper Saddle River, NJ.

Koletzke, P., and Dorsey, P. 1999. Oracle Developer Advanced Forms and Reports. McGraw-Hill, Berkeley, CA.

Korf, I., Flicek, P., Duan, D., and Brent, M.R. 2001. Integrating genomic homology into gene structure prediction. Bioinformatics. 17 Suppl 1: S140–8.

Korf, I., and Gish, W. 2000. MPBLAST : improved BLAST performance with multiplexed queries. Bioinformatics. 16: 1052–3.

Krauter, K., Montgomery, K., Yoon, S.J., LeBlanc-Straceski, J., Renault, B., Marondel, I., Herdman, V., Cupelli, L., Banks, A., Lieman, J., and *et al.* 1995. A second-generation YAC contig map of human chromosome 12. Nature. 377: 321–33.

Lander, E.S. 2000. Preface to Volume 1. Ann Rev Genomics Hum Genet. 1: v–vi.

Lander, E.S., and Waterman, M.S. 1988. Genomic mapping by fingerprinting random clones: a mathematical analysis. Genomics. 2: 231–9.

Lawrence, C.B., Honda, S., Parrott, N.W., Flood, T.C., Gu, L., Zhang, L., Jain, M., Larson, S., and Myers, E.W. 1994. The genome reconstruction manager: a software environment for supporting high-throughput DNA sequencing. Genomics. 23: 192–201.

Lawrence, C.B., and Solovyev, V.V. 1994. Assignment of position-specific error probability to primary DNA sequence data. Nucleic Acids Res. 22: 1272–80.

Levene, S.D., and Zimm, B.H. 1989. Understanding the anomalous electrophoresis of bent DNA molecules: a reptation model. Science. 245: 396–9.

Loney, K., and Koch, G. 2000. Oracle8i, The Complete Reference. Osborne McGraw-Hill, Berkeley, CA.

Lowe, T.M., and Eddy, S.R. 1997. tRNAscan-SE: a program for improved detection of transfer RNA genes in genomic sequence. Nucleic Acids Res. 25: 955–64.

Luckey, J.A., and Smith, L.M. 1993. A model for the mobility of single-stranded DNA in capillary gel electrophoresis. Electrophoresis. 14: 492–501.

Lukashin, A.V., and Borodovsky, M. 1998. GeneMark.hmm: new solutions for gene finding. Nucleic Acids Res. 26: 1107–15.

Lutz, M. 2001. Programming Python. Second edition. O'Reilly and Associates, Inc., Sebastopol, CA.

Marra, M., Hillier, L., Kucaba, T., Allen, M., Barstead, R., Beck, C., Blistain, A., Bonaldo, M., Bowers, Y., Bowles, L., Cardenas, M., Chamberlain, A., Chappell, J., Clifton, S., Favello, A., Geisel, S., Gibbons, M., Harvey, N., Hill, F., Jackson, Y., Kohn, S., Lennon, G., Mardis, E., Martin, J., Waterston, R., and et al. 1999. An encyclopedia of mouse genes. Nat Genet. 21: 191–4.

McClelland, M., Florea, L., Sanderson, K., Clifton, S.W., Parkhill, J., Churcher, C., Dougan, G., Wilson, R.K., and Miller, W. 2000. Comparison of the Escherichia coli K-12 genome with sampled genomes of a Klebsiella pneumoniae and three salmonella enterica serovars, Typhimurium, Typhi and Paratyphi. Nucleic Acids Res. 28: 4974–86.

McClelland, M., Sanderson, K.E., Spieth, J., Clifton, S.W., Latreille, P., Courtney, L., Porwollik, S., Ali, J., Dante, M., Du, F., Hou, S., Layman, D., Leonard, S., Nguyen, C., Scott, K., Holmes, A., Grewal, N., Mulvaney, E., Ryan, E., Sun, H., Florea, L., Miller, W., Stoneking, T., Nhan, M., Waterston, R., and Wilson, R.K. 2001. Complete genome sequence of Salmonella enterica serovar Typhimurium LT2. Nature. 413: 852–6.

Myers, E.W. 1999. Whole-genome DNA Sequencing. Computing in Science and Engineering. 1: 33-43.

Myers, E.W., Sutton, G.G., Delcher, A.L., Dew, I.M., Fasulo, D.P., Flanigan, M.J., Kravitz, S.A., Mobarry, C.M., Reinert, K.H., Remington, K.A., Anson, E.L., Bolanos, R.A., Chou, H.H., Jordan, C.M., Halpern, A.L., Lonardi, S., Beasley, E.M., Brandon, R.C., Chen, L., Dunn, P.J., Lai, Z., Liang, Y., Nusskern, D.R., Zhan, M., Zhang, Q., Zheng, X., Rubin, G.M., Adams, M.D., and Venter, J.C. 2000. A whole-genome assembly of Drosophila. Science. 287: 2196–204.

Nakai, K., and Horton, P. 1999. PSORT: a program for detecting sorting signals in proteins and predicting their subcellular localization. Trends Biochem Sci. 24: 34–6.

Olivier, M., Aggarwal, A., Allen, J., Almendras, A.A., Bajorek, E.S., Beasley, E.M., Brady, S.D., Bushard, J.M., Bustos, V.I., Chu, A., Chung, T.R., De Witte, A., Denys, M.E., Dominguez, R., Fang, N.Y., Foster, B.D., Freudenberg, R.W., Hadley, D., Hamilton, L.R., Jeffrey, T.J., Kelly, L., Lazzeroni, L., Levy, M.R., Lewis, S.C., Liu, X., Lopez, F.J., Louie, B., Marquis, J.P., Martinez, R.A., Matsuura, M.K., Misherghi, N.S., Norton, J.A., Olshen, A., Perkins, S.M., Perou, A.J., Piercy, C., Piercy, M., Qin, F., Reif, T., Sheppard, K., Shokoohi, V., Smick, G.A., Sun, W.L., Stewart, E.A., Fernando, J., Tejeda, Tran, N.M., Trejo, T., Vo, N.T., Yan, S.C., Zierten, D.L., Zhao, S., Sachidanandam, R., Trask, B.J., Myers, R.M., and Cox, D.R. 2001. A high-resolution radiation hybrid map of the human genome draft sequence. Science. 291: 1298-302.

Olson, M.V., Dutchik, J.E., Graham, M.Y., Brodeur, G.M., Helms, C., Frank, M., MacCollin, M., Scheinman, R., and Frank, T. 1986. Random-clone strategy for genomic restriction mapping in yeast. Proc Natl Acad Sci U S A. 83: 7826–30.

Pachter, L., Alexandersson, M., and Cawley, S. 2002. Applications of generalized pair hidden markov models to alignment and gene finding problems. J Comput Biol. 9: 389–99.

Parkhill, J., Dougan, G., James, K.D., Thomson, N.R., Pickard, D., Wain, J., Churcher, C., Mungall, K.L., Bentley, S.D., Holden, M.T., Sebaihia, M., Baker, S., Basham, D., Brooks, K., Chillingworth, T., Connerton, P., Cronin, A., Davis, P., Davies, R.M., Dowd, L., White, N., Farrar, J., Feltwell, T., Hamlin, N., Haque, A., Hien, T.T., Holroyd, S., Jagels, K., Krogh, A., Larsen, T.S., Leather, S., Moule, S., O'Gaora, P., Parry, C., Quail, M., Rutherford, K., Simmonds, M., Skelton, J., Stevens, K., Whitehead, S., and Barrell, B.G. 2001. Complete genome sequence of a multiple drug resistant Salmonella enterica serovar Typhi CT18. Nature. 413: 848–52.

Parsons, J.D. 1995. Miropeats: graphical DNA sequence comparisons. Comput Appl Biosci. 11: 615–9.

Pearson, W.R., and Lipman, D.J. 1988. Improved tools for biological sequence comparison. Proc Natl Acad Sci U S A. 85: 2444–8.

Pevzner, P.A., Tang, H., and Waterman, M.S. 2001. An Eulerian path approach to DNA fragment assembly. Proc Natl Acad Sci U S A. 98: 9748–53.

Reddy, V.B., Thimmappaya, B., Dhar, R., Subramanian, K.N., Zain, B.S., Pan, J., Ghosh, P.K., Celma, M.L., and Weissman, S.M. 1978. The genome of simian virus 40. Science. 200: 494–502.

Salamov, A.A., and Solovyev, V.V. 2000. Ab initio gene finding in Drosophila genomic DNA. Genome Res. 10: 516–22.

Salzberg, S.L., Pertea, M., Delcher, A.L., Gardner, M.J., and Tettelin, H. 1999. Interpolated Markov models for eukaryotic gene finding. Genomics. 59: 24–31.

Sanger, F., Coulson, A.R., Barrell, B.G., Smith, A.J., and Roe, B.A. 1980. Cloning in single-stranded bacteriophage as an aid to rapid DNA sequencing. J Mol Biol. 143: 161–78.

Sanger, F., Coulson, A.R., Friedmann, T., Air, G.M., Barrell, B.G., Brown, N.L., Fiddes, J.C., Hutchison, C.A., 3rd, Slocombe, P.M., and Smith, M. 1978. The nucleotide sequence of bacteriophage phiX174. J Mol Biol. 125: 225–46.

Sanger, F., Coulson, A.R., Hong, G.F., Hill, D.F., and Petersen, G.B. 1982. Nucleotide sequence of bacteriophage lambda DNA. J Mol Biol. 162: 729–73.

Sanger, F., Nicklen, S., and Coulson, A.R. 1977. DNA sequencing with chain-terminating inhibitors. Proc Natl Acad Sci U S A. 74: 5463–7.

Searls, D.B. 2000. Bioinformatics tools for whole genomes. Annu Rev Genomics Hum Genet. 1: 251–79.

Shizuya, H., Birren, B., Kim, U.J., Mancino, V., Slepak, T., Tachiiri, Y., and Simon, M. 1992. Cloning and stable maintenance of 300-kilobase-pair fragments of human DNA in Escherichia coli using an F-factor-based vector. Proc Natl Acad Sci U S A. 89: 8794–7.

Smith, T.F., and Waterman, M.S. 1981. Identification of common molecular subsequences. J Mol Biol. 147: 195–7.

Smith, T.M., Abajian, C., and Hood, L. 1997. Hopper: software for automating data tracking and flow in DNA sequencing. Comput Appl Biosci. 13: 175–82.

Sonnhammer, E.L., and Durbin, R. 1994. A workbench for large-scale sequence homology analysis. Comput Appl Biosci. 10: 301–7.

Staden, R. 1977. Sequence data handling by computer. Nucleic Acids Res. 4: 4037–51.

Staden, R. 1996. The Staden sequence analysis package. Mol Biotechnol. 5: 233–41.

Stein, L.D., and Thierry-Mieg, J. 1998. Scriptable access to the Caenorhabditis elegans genome sequence and other ACEDB databases. Genome Res. 8: 1308–15.

Stewart, E.A., McKusick, K.B., Aggarwal, A., Bajorek, E., Brady, S., Chu, A., Fang, N., Hadley, D., Harris, M., Hussain, S., Lee, R., Maratukulam, A., O'Connor, K., Perkins, S., Piercy, M., Qin, F., Reif, T., Sanders, C., She, X., Sun, W.L., Tabar, P., Voyticky, S., Cowles, S., Fan, J.B., Cox, D.R., and *et al.* 1997. An STS-based radiation hybrid map of the human genome. Genome Res. 7: 422–33.

Strauss, E.C., Kobori, J.A., Siu, G., and Hood, L.E. 1986. Specific-primer-directed DNA sequencing. Anal Biochem. 154: 353–60.

Sulston, J., Du, Z., Thomas, K., Wilson, R., Hillier, L., Staden, R., Halloran, N., Green, P., Thierry-Mieg, J., Qiu, L., and *et al.* 1992. The C. elegans genome sequencing project: a beginning. Nature. 356: 37–41.

Tabata, S., Kaneko, T., Nakamura, Y., Kotani, H., Kato, T., Asamizu, E., Miyajima, N., Sasamoto, S., Kimura, T., Hosouchi, T., Kawashima, K., Kohara, M., Matsumoto, M., Matsuno, A., Muraki, A., Nakayama, S., Nakazaki, N., Naruo, K., Okumura, S., Shinpo, S., Takeuchi, C., Wada, T., Watanabe, A., Yamada, M., Yasuda, M., Sato, S., de la Bastide, M., Huang, E., Spiegel, L., Gnoj, L., O'Shaughnessy, A., Preston, R., Habermann, K., Murray, J., Johnson, D., Rohlfing, T., Nelson, J., Stoneking, T., Pepin, K., Spieth, J., Sekhon, M., Armstrong, J., Becker, M., Belter, E., Cordum, H., Cordes, M., Courtney, L., Courtney, W., Dante, M., Du, H., Edwards, J., Fryman, J., Haakensen, B., Lamar, E., Latreille, P., Leonard, S., Meyer, R., Mulvaney, E., Ozersky, P., Riley, A., Strowmatt, C., Wagner-McPherson, C., Wollam, A., Yoakum, M., Bell, M., Dedhia, N., Parnell, L., Shah, R., Rodriguez, M., See, L.H., Vil, D., Baker, J., Kirchoff, K., Toth, K., King, L., Bahret, A., Miller, B., Marra, M., Martienssen, R., McCombie, W.R., Wilson, R.K., Murphy, G., Bancroft, I., Volckaert, G., Wambutt, R., Dusterhoft, A., Stiekema, W., Pohl, T., Entian, K.D., Terryn, N., Hartley, N., Bent, E., Johnson, S., Langham, S.A., McCullagh, B., Robben, J., Grymonprez, B., Zimmermann, W., Ramsperger, U., Wedler, H., Balke, K., Wedler, E., Peters, S., van Staveren, M., Dirkse, W., Mooijman, P., Lankhorst, R.K., Weitzenegger, T., Bothe, G., Rose, M., Hauf, J., Berneiser, S., Hempel, S., Feldpausch, M., Lamberth, S., Villarroel, R., Gielen, J., Ardiles, W., Bents, O., Lemcke, K., Kolesov, G., Mayer, K., Rudd, S., Schoof, H., Schueller, C., Zaccaria, P., Mewes, H.W., Bevan, M., and Fransz, P. 2000. Sequence and analysis of chromosome 5 of the plant Arabidopsis thaliana. Nature. 408: 823–6.

Tatusov, R.L., Natale, D.A., Garkavtsev, I.V., Tatusova, T.A., Shankavaram, U.T., Rao, B.S., Kiryutin, B., Galperin, M.Y., Fedorova,

N.D., and Koonin, E.V. 2001. The COG database: new developments in phylogenetic classification of proteins from complete genomes. Nucleic Acids Res. 29: 22–8.

Venter, J.C., Adams, M.D., Myers, E.W., Li, P.W., Mural, R.J., Sutton, G.G., Smith, H.O., Yandell, M., Evans, C.A., Holt, R.A., Gocayne, J.D., Amanatides, P., Ballew, R.M., Huson, D.H., Wortman, J.R., Zhang, Q., Kodira, C.D., Zheng, X.H., Chen, L., Skupski, M., Subramanian, G., Thomas, P.D., Zhang, J., Gabor Miklos, G.L., Nelson, C., Broder, S., Clark, A.G., Nadeau, J., McKusick, V.A., Zinder, N., Levine, A.J., Roberts, R.J., Simon, M., Slayman, C., Hunkapiller, M., Bolanos, R., Delcher, A., Dew, I., Fasulo, D., Flanigan, M., Florea, L., Halpern, A., Hannenhalli, S., Kravitz, S., Levy, S., Mobarry, C., Reinert, K., Remington, K., Abu-Threideh, J., Beasley, E., Biddick, K., Bonazzi, V., Brandon, R., Cargill, M., Chandramouliswaran, I., Charlab, R., Chaturvedi, K., Deng, Z., Di Francesco, V., Dunn, P., Eilbeck, K., Evangelista, C., Gabrielian, A.E., Gan, W., Ge, W., Gong, F., Gu, Z., Guan, P., Heiman, T.J., Higgins, M.E., Ji, R.R., Ke, Z., Ketchum, K.A., Lai, Z., Lei, Y., Li, Z., Li, J., Liang, Y., Lin, X., Lu, F., Merkulov, G.V., Milshina, N., Moore, H.M., Naik, A.K., Narayan, V.A., Neelam, B., Nusskern, D., Rusch, D.B., Salzberg, S., Shao, W., Shue, B., Sun, J., Wang, Z., Wang, A., Wang, X., Wang, J., Wei, M., Wides, R., Xiao, C., Yan, C., Yao, A., Ye, J., Zhan, M., Zhang, W., Zhang, H., Zhao, Q., Zheng, L., Zhong, F., Zhong, W., Zhu, S., Zhao, S., Gilbert, D., Baumhueter, S., Spier, G., Carter, C., Cravchik, A., Woodage, T., Ali, F., An, H., Awe, A., Baldwin, D., Baden, H., Barnstead, M., Barrow, I., Beeson, K., Busam, D., Carver, A., Center, A., Cheng, M.L., Curry, L., Danaher, S., Davenport, L., Desilets, R., Dietz, S., Dodson, K., Doup, L., Ferriera, S., Garg, N., Gluecksmann, A., Hart, B., Haynes, J., Haynes, C., Heiner, C., Hladun, S., Hostin, D., Houck, J., Howland, T., Ibegwam, C., Johnson, J., Kalush, F., Kline, L., Koduru, S., Love, A., Mann, F., May, D., McCawley, S., McIntosh, T., McMullen, I., Moy, M., Moy, L., Murphy, B., Nelson, K., Pfannkoch, C., Pratts, E., Puri, V., Qureshi, H., Reardon, M., Rodriguez, R., Rogers, Y.H., Romblad, D., Ruhfel, B., Scott, R., Sitter, C., Smallwood, M., Stewart, E., Strong, R., Suh, E., Thomas, R., Tint, N.N., Tse, S., Vech, C., Wang, G., Wetter, J., Williams, S., Williams, M., Windsor, S., Winn-Deen, E., Wolfe, K., Zaveri, J., Zaveri, K., Abril, J.F., Guigo, R., Campbell, M.J., Sjolander, K.V., Karlak, B., Kejariwal, A., Mi, H., Lazareva, B., Hatton, T., Narechania, A., Diemer, K., Muruganujan, A., Guo, N., Sato, S., Bafna, V., Istrail, S., Lippert, R., Schwartz, R., Walenz, B., Yooseph, S., Allen, D., Basu, A., Baxendale, J., Blick, L., Caminha, M., Carnes-Stine, J., Caulk, P., Chiang, Y.H., Coyne, M., Dahlke, C., Mays, A., Dombroski, M.,

Donnelly, M., Ely, D., Esparham, S., Fosler, C., Gire, H., Glanowski, S., Glasser, K., Glodek, A., Gorokhov, M., Graham, K., Gropman, B., Harris, M., Heil, J., Henderson, S., Hoover, J., Jennings, D., Jordan, C., Jordan, J., Kasha, J., Kagan, L., Kraft, C., Levitsky, A., Lewis, M., Liu, X., Lopez, J., Ma, D., Majoros, W., McDaniel, J., Murphy, S., Newman, M., Nguyen, T., Nguyen, N., Nodell, M., Pan, S., Peck, J., Peterson, M., Rowe, W., Sanders, R., Scott, J., Simpson, M., Smith, T., Sprague, A., Stockwell, T., Turner, R., Venter, E., Wang, M., Wen, M., Wu, D., Wu, M., Xia, A., Zandieh, A., and Zhu, X. 2001. The sequence of the human genome. Science. 291: 1304–51.

Venter, J.C., Adams, M.D., Sutton, G.G., Kerlavage, A.R., Smith, H.O., and Hunkapiller, M. 1998. Shotgun sequencing of the human genome. Science. 280: 1540–2.

Wall, L., Christiansen, T., and Orwant, J. 2000. Programming Perl. Third edition. O'Reilly and Associates, Inc., Sebastopol, CA.

Walsh, S., Anderson, M., and Cartinhour, S.W. 1998. ACEDB: a database for genome information. Methods Biochem Anal. 39: 299–318.

Waterman, M.S. 2000. Introduction to Computational Biology. Chapman and Hall — CRC, Boca Raton, FL.

Weber, J.L., and Myers, E.W. 1997. Human whole-genome shotgun sequencing. Genome Res. 7: 401–9.

Wendl, M.C., Dear, S., Hodgson, D., and Hillier, L. 1998. Automated sequence preprocessing in a large-scale sequencing environment. Genome Res. 8: 975–84.

Wendl, M.C., Korf, I., Chinwalla, A.T., and Hillier, L.W. 2001a. Automated processing of raw DNA sequence data. IEEE Eng Med Biol Mag. 20: 41–8.

Wendl, M.C., Marra, M.A., Hillier, L.W., Chinwalla, A.T., Wilson, R.K., and Waterston, R.H. 2001b. Theories and applications for sequencing randomly selected clones. Genome Res. 11: 274–80.

Wheelan, S.J., and Boguski, M.S. 1998. Late-night thoughts on the sequence annotation problem. Genome Res. 8: 168–9.

Wilson, R., Ainscough, R., Anderson, K., Baynes, C., Berks, M., Bonfield, J., Burton, J., Connell, M., Copsey, T., Cooper, J., and et al. 1994. 2.2 Mb of contiguous nucleotide sequence from chromosome III of C. elegans [see comments]. Nature. 368: 32–8.

Wootton, J.C., and Federhen, S. 1993. Statistics of local complexity in amino acid sequences and sequence databases. Computers and Chemistry. 17: 149–163.

Worden, D. 2000. Sybase System 11 Develpoment Handbook. Morgan Kaufmann Publishers, San Francisco, CA.

From: *Genome Mapping and Sequencing*
© 2003 Horizon Scientific Press, Wymondham, UK

13

Annotating Mammalian Genome Sequence

John E. Collins and David M. Beare

Abstract

As more mammalian genomes have become sequenced attention has turned to sequence annotation. An annotated sequence provides a wealth of information about the organism not directly obvious from the sequence alone. It also acts as a standard, allowing investigators around the world to work on the same basic gene structures and to compare subsequent findings. In this chapter we show how to assemble finished sequence clones in a contiguous reference sequence and then subject this to an array of sequence analysis tools. By examining these analyses we show how to annotate exon/intron structures on the genomic sequence to define genes. In regions where there is insufficient evidence to draw a complete gene structure or where further evidence is required, we suggest methods to identify the necessary sequence from mRNA sources. Finally, we show how these data can be compiled into simple flat files of coordinates on a reference sequence and transferred between investigators.

1. Introduction

The sequence of a genome is the key to understanding the biology of the organism. The sequence itself provides a scaffold or reference to which functional features can be attached or annotated. The annotation can take many forms, such as genes, repeats, variations or disease association. Complete and standardised sequence annotation is the foundation for subsequent functional experiments, and accurate annotation is fundamental for rapid progress. This chapter describes a curated annotation system where the available evidence is used to align gene structures to the reference sequence. Here each annotation is checked against all the existing data, an assessment made, further directed laboratory work performed where necessary and a comprehensive transcript map produced. This differs from the automated annotation systems, such as ENSEMBL (Hubbard *et al.*, 2002) and the UCSC (Kent *et al.*, 2002), which are currently a first pass genome annotation.

Annotating a cDNA on a reference sequence places it in a genomic context and generates new and useful data. A gene structure in the genomic landscape leads to information such as promoter regions, defined exons, intron sequence, splice junctions and 3' control elements, all unknown in a cDNA sequence. It also assists with the design of experiments to detect alternative transcripts, to establish a tissue expression profile and to express the full length protein. Furthermore, predicting the coding portion of each gene and comparison to protein or protein domain databases can suggest possible functions. Studying the genomic sequence in and around a gene for sequence variation reveals useful polymorphic markers as well as biologically significant alterations in exons and control regions. Using such markers to screen populations can demonstrate links between phenotype and genotype, both within normal variation and disease causing alterations. Pin-pointing a gene within the genome provides valuable positional data required for linkage analysis, polymorphic marker association studies and correlation with chromosome translocations and duplications.

Before beginning the annotation process it is useful to clarify what is to be annotated. Considering the gene content, mammalian genes can be divided into 3 main groups, protein coding, non-protein coding (Mattick, 2001) and pseudogenes. In a protein coding gene you expect to find a cDNA sequence which matches the genomic sequence, normally splicing at consensus splice sites. Within this cDNA is an open reading frame (ORF) beginning with an ATG start codon and finishing with a stop codon. Prior to the 5' end of the start of transcription site are promoter sequences and control elements. At the 3' end there is a polyA sequence in the cDNA but not in the genomic DNA and a polyA addition signal hexamer. The majority of non-coding genes comprise ribosomal RNAs, transfer RNAs and small nuclear RNAs, and can be annotated using similarity searches of databases of experimentally derived RNA sequence. A further set of non-coding RNA has been described as antisense RNA with a possible regulatory role (Kumar and Carmichael, 1998). There is increasing evidence that other non-coding RNAs may also be common (Mattick, 2001). Pseudogenes are non-functional copies of a coding gene found elsewhere in the genome. The majority of pseudogenes are believed to have occurred by retrotransposition of a coding mRNA. As a general rule they do not contain introns, often have a polyA tail at the 3' end in genomic sequence and have a disrupted ORF caused by a premature stop codon or an insertion or deletion leading to a frame shift. Pseudogenes can also be produced by genome duplication. These generally retain remnants of the exon/intron structure but again have a disrupted ORF.

Genes can be identified within genomic sequence using sequence similarity searches, computational gene structure prediction or through cross-species conservation. First, similarity searches of expressed sequence databases, such as ESTs and cDNAs, provide all the information required to define a complete gene structure. This also carries the assurance of genuine gene expression, but is confounded by cDNA library artefact and limited spatial or temporal expression patterns (Guigo et al., 2000; Collins et al., 2003). Matching cDNA sequence and its translation products also provides data for annotation of pseudogenes. Second, exons and gene structures can be predicted computationally from the genomic sequence by searching for statistical characteristics of genes (Solovyev et al., 1995; Burge and Karlin, 1997). Whilst these programs have proven relatively successful in predicting exons in mammalian genomes, they do not accurately define entire gene structures and have an unacceptably high rate of false positive and false negative exon prediction. Third, as the sequence of related genomes becomes available, matching conserved regions between species can be used to indicate the presence of a transcribed sequence (Dubchak et al., 2000; Mayor et al., 2000; Korf et al., 2001). However, this approach can result in additional noise in the system, as seen in the level of similarity between the human and mouse genome not associated with genes (Deloukas et al., 2001; Frazer et al., 2001; Collins et al., 2003; Kondrashov and Shabalina, 2002). In practise a combination of these three methods is used to establish the presence of a transcribed sequence and then the EST/cDNA databases are combined with additional directed cDNA sequencing to provide the necessary information to define the complete exon structure (Collins et al., 2003).

This chapter aims to provide a guide to annotating part or a whole mammalian genome. It will describe how to assemble the finished sequenced clones into a contiguous reference sequence. Initially this is used to search the numerous public databases to find matching cDNA or protein sequences. The addition of data from exon prediction programs and cross species homology matches highlights further potential transcripts. These data are displayed in a graphical viewer, the features examined and the complete genes annotated. Incomplete genes or regions that display the characteristics of genes but with no existing cDNA sequence are confirmed with directed laboratory work. The completed annotation and reference sequence can then be transferred wherever necessary in a flat file of coordinates.

2. 'Golden Path' Sequence Assembly

2.1 Strategy

Using the 'clone by clone' method for mapping and sequencing the human genome (International Human Genome Sequencing Consortium, 2001) a minimally overlapping tiling path of cloned genomic sequence fragments (clones) are sequenced for each chromosome. When the sequencing phase for a chromosome is completed an assembly or 'golden path' can be constructed and a consensus, or reference sequence, derived from it. This reference sequence can then be annotated without the constraints of artificial clone boundaries and each gene or feature can be mapped within the context of the whole chromosome.

The ACeDB database (http://www.acedb.org ; Durbin and Thierry-Mieg, 1991) has been developed for genome mapping and sequencing projects and it serves as an ideal display environment for an assembled whole chromosome reference sequence, associated subsequences and annotation. In an ACeDB database data is organised within objects belonging to defined classes. A model defines each class of object, and how the data associated with each class of object is stored. Within each class, tags are used to organise the data. For example, in the standard model for the Sequence class there are several homology tags including EST_homol and DNA_homol. For genomic sequence objects in the Sequence class the EST_homol tag is used to define the alignment of matching EST sequences. Subsequence and Source_Exon tags can be used to define gene structures and the Feature tag is used to define CpG islands, predicted exons and tandem repeats. The ability of ACeDB to store data within objects and graphically display sequence related data makes it an ideal choice. The capability to export feature tables in GFF format is also very useful (see section 5.1).

Although there may be other comparable database solutions, for the methodologies discussed here ACeDB will be the database of choice. For help with ACeDB see http://www.acedb.org/Software/whelp.

2.2 Sequence Alignments

To assemble a golden path reference sequence for any given chromosome, overlapping clones are aligned to form sequence contigs (sub links). Establishing these alignments is largely straightforward, as before submitting a sequence to the public databases (EMBL, GenBank, DDJB) many sequencing centres establish the overlaps with neighbouring clones and clip the finished sequence to leave 100bp, 200bp or 2kb of sequence overlap. However, this is not always the case and insertions/deletions, repeats or both can complicate the alignment of sequences, particularly for unclipped sequences with many kilobases of overlap.

During sequence assembly a problem may arise where insertions or deletions within the overlap result in an alternative way to align the sequences. As a general rule, it is best to select the alignment which yields the longest consensus sequence, avoiding the exclusion by deletion of functional genomic sequence. Once the alignments have been established, clones are trimmed so that each overlapping region in the reference sequence is contributed by one or other of the clones, but not both. The position on the reference sequence where the source sequence switches from one clone to another is referred to as a 'switch point'.

In the assembly methods using ACeDB described here, the SuperLink sequence represents a whole chromosome reference sequence which is assembled from a series of sub links. Sub links are contigs of sequenced clones where alignments have been made using a pre-defined assembly (see Protocol 1) or by automated methods using cross_match (Green, unpublished; http://www.phrap.com) where no assembly is currently available (Protocol 2). In the SuperLink the sub links are separated by gaps where no sequence is available. Gap lengths are estimated by the best available mapping data. Figure 1 shows the structure of the golden path reference sequence (SuperLink) and Table 1 defines the hierarchy and nomenclature in ACeDB.

It is important to note that once an individual clone becomes a subsequence of a sub link it may inherit sequence from a neighbouring clone in the overlapping region and therefore it may not exactly match the original sequence. For this reason, the reference sequence (or its component

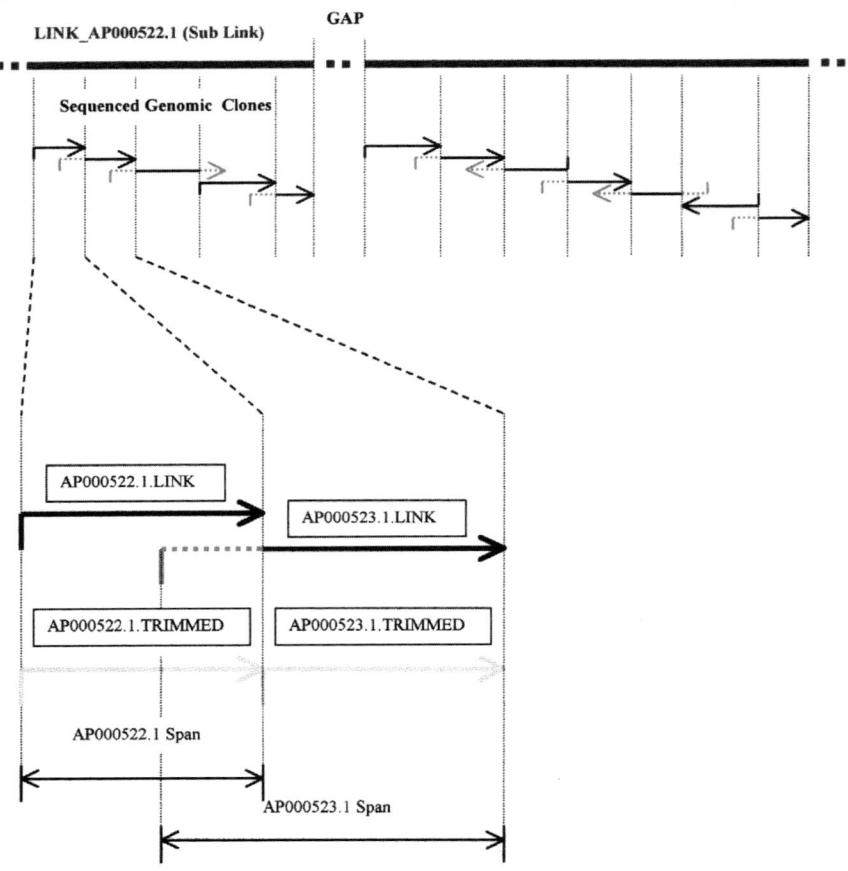

Figure 1. Structure of the golden path reference sequence. See text for description.

subsequences) should be used for analyses and annotation rather than clone sequences retrieved directly from the public databases.

Until the sequencing of all human chromosomes is completed there will be two alternative routes to the assembly of a whole chromosome reference sequence in ACeDB. The finished or almost finished chromosomes have publicly available AGP (Assembly Golden Path) format files which contain all the information required to construct whole chromosome reference sequences. At http://genome.cse.ucsc.edu/goldenPath/datorg.html a full description of the format is given, but briefly it contains the following information:-

Table 1. The hierarchy of sequences in a SuperLink assembly in ACeDB. See also Figure 1.

Level	Name	Type	Subsequence of
1	SuperLink	Whole Chromosome (reference sequence)	-
2	LINK_*	Sub Link (sequence contig)	Level 1
3	*.LINK	Clone (cloned genomic fragment)	Level 2
4	*.TRIMMED	Trimmed Clone (overlaps trimmed)	Level 3

* = accession number and version, *e.g.* AC005500.2

1. Sequence order:	based on the tiling path (normally pTel to qTel)
2. Sequence definitions:	unique database accession and version number
3. Sequence orientation:	'+' pTel to qTel, or '-'qTel to pTel
4. Sequence trimming:	coordinate range to be included in the reference
5. Gap definitions:	size and type

However, if the chromosome of interest is not finished and does not have an AGP file, the TPF (Tile Path Format) file should be used instead. These files are also publicly available and a description of the format is given at http://cvsweb.sanger.ac.uk/cgi-bin/cvsweb.cgi/genome-map/formats/TPF. TPFs give the order of sequences and positions of gaps (and sizes where known), but no assembly information. Consequently the alignment of each of the clones in the tile path must be established and this complicates and lengthens the process. It should also be noted that a TPF represents a snapshot of a genome mapping and sequencing project and it will lack the stability and reliability of a finished sequence map.

3. Computational Sequence Analyses

3.1 Automation

Once clones have been assembled into a reference sequence, a number of computational analyses can be performed to identify known and putative genes. Where possible these analyses should be run against sub link sequences rather than individual clones in order to eliminate potential clone boundary effects. However, if computational resources are limited it may be necessary to run against the clone sequences. As mentioned earlier in section 2 the clone sequences used for analysis must be derived from the reference sequence and not directly from the public databases.

For large scale projects an 'analysis pipeline' can be established, using an automated system for running database searches and gene prediction programs. It should be designed to maximise the use of available resources and also track the progress of the analyses. The handling of any failed processes is particularly important as a full set of analyses is a crucial first stage in the annotation process. It is also essential to log details such as the source and version of software and databases, and the date and time when analyses were started and completed. Ideally, for any size of project a batch job processing system is required. With such a system, resource intensive jobs can be batch processed and scheduled to run during 'off-peak' times to maximise efficiency. For small projects with relatively few analyses to be run it is possible to batch process jobs using remote web servers. For example, the bioperl package (see http://www.bioperl.org) contains a RemoteBlast module which allows multiple blast searches to be run remotely on the NCBI blast server via HTTP. There are web versions of many bioinformatics programs and The Genome Analysis Pipeline website at http://compbio.ornl.gov/tools/pipeline has many options and allows multiple gene finding analyses to be run on a query sequence. Another option is Genotator, described as a workbench for sequence annotation (Harris, 1997) http://www.fruitfly.org/~nomi/genotator and has a 'back end' for running multiple sequence analyses (see also section 5.2).

3.2 Analysis Software and Databases

A variety of software systems and databases are available for analysing genomic sequence. Described here are the analyses which proved most useful for identifying and annotating genes on chromosome 22 (Dunham et al., 1999; Collins et al., 2003). As mentioned in section 2.2, before running any analysis the reference sequence and its associated genomic subsequences should be exported from ACeDB in FASTA format. This can be done by creating a keyset of these sequence objects and clicking the 'Export' button from the keyset window and then selecting the 'DNA in FASTA format' option.

The first analysis to run against the exported sequences is RepeatMasker http://repeatmasker.genome.washington.edu/cgi-bin/RepeatMasker. This is run with the -ace option to generate ACE output files as well as repeat masked sequence which will be used for the analyses detailed in tables 2 to 5. This avoids multiple matches to repeats in all the subsequent analyses.

The masked sequence is then run through an analysis pipeline as shown in Tables 2 to 5 to gather evidence for gene discovery from a variety of sources. The more sources of evidence which support an annotation the more confidence there is in that annotation. *Ab initio* gene prediction (Table 2) by programs such as GenScan indicates the presence of potential

Table 2. Ab *initio* Gene Prediction.

Rationale :	Use exon and gene prediction algorithms.
Methods :	Genscan
	http://www.molbiol.ox.ac.uk/documentation/gene_predict/genscan.htm
	ftp://ftp.sanger.ac.uk/pub/T62/Software/gs2ace
	Fgenesh
	http://www.softberry.com/berry.phtml?topic=gfind
	ftp://ftp.sanger.ac.uk/pub/T62/Software/fh2ace

genes but require additional supporting evidence. Similarity searches against EST and cDNA databases (Table 3) are the most effective method of identifying genes and also for annotating exon structures. Evidence of expression can also come from cross-species comparison. Sequence regions which are highly conserved between species are strong indicators of genes (Table 4). While the previous methods are effective gene discovery tools, finding protein similarities by searching protein databases such as SWISS-PROT, and the protein families database PFAM can also prove useful in identifying potential function for novel genes (Table 3.). Scanning genomic sequence for CpG islands and promoter regions can be used to identify the 5' ends of genes, and can be useful not only for gene discovery but also in confirming if an annotation is partial or complete (Table 5).

For DNA similarity searches the percent identity cut-off should be 90 - 95%, and for protein similarity searches this should be lowered to 40%.

3.3 Parsing Data

After the completion of computational analyses the data is parsed into ACeDB. Most features will be attached to the reference sequence using the

Table 3. Similarity searches.

Rationale :	Find similarities with expressed sequence and peptides	
Methods :	blastn (DNA to DNA)	http://blast.wustl.edu/
	blastx (DNA to Peptide)	http://blast.wustl.edu/
Databases :	EST	http://www.ncbi.nlm.nih.gov/dbEST
	UniGene clusters	ftp://ftp.ncbi.nih.gov/repository/UniGene
	SWISS-PROT (Protein)	http://www.ebi.ac.uk/swissprot
	Pfam (Protein Domains)	http://www.sanger.ac.uk/Software/Pfam
	Vertebrate mRNA	Subset of the EMBL database
		http://www.ebi.ac.uk/embl
		(defined by mRNA in DE line and
		Vertebrata in the OC line)

Table 4. Cross-species Sequence Conservation.

Rationale :	Find cross-species sequence conservation which may correspond to expressed sequence.
Methods :	Cross Species MegaBLAST
	http://www.ncbi.nlm.nih.gov/blast/tracemb.html
	tblastx
	http://blast.wustl.edu
	Blat
	http://www.soe.ucsc.edu/~kent/exe
	Exonerate
	http://www.ensembl.org/Docs/wiki/html/EnsemblDocs/Exonerate.html
	Exofish (Exon Finding by Sequences Homology) finds homology to Tetraodon *nigrovridis*
	http://www.genoscope.cns.fr/proxy/cgi-bin/exofish.cgi
Databases :	Mus *musculus* Genome Sequence
	ftp://ftp.ncbi.nih.gov/genbank/genomes/M_musculus
	Drosophila *melanogaster* Genome Sequence
	ftp://ftp.ncbi.nih.gov/genbank/genomes/D_melanogaster
	Caaenorhabditis *elegans* Genome Sequence
	ftp://ftp.ncbi.nih.gov/genbank/genomes/C_elegans
	Tetraodon *nigrovridis* Genome Sequence
	http://www.genoscope.cns.fr/externe/tetraodon/Ressource.html
	Human ecores (Evolutionary Conserved Regions) available at http://www.genoscope.cns.fr/externe/tetraodon/Data/Ecores

Table 5. Gene 5' End Prediction.

Rationale :	Find features associated with the 5' ends of genes.
Methods :	PromoterInspector
	http://www.genomatix.de/software_services/software/PromoterInspector/PromoterInspector.html
	Eponine transcription start prediction
	http://www.sanger.ac.uk/Users/td2/eponine
	CpG Islands
	http://www.hgmp.mrc.ac.uk/Software/EMBOSS/Apps/cpgreport.html
	ftp://ftp.sanger.ac.uk/pub/T62/Software/cpg2ace

Feature tag or one of the "Sequence"_homol tags for sequence similarities. The format of ACE files conforming to the standard ACeDB sequence model for each of these types is shown in the examples below where the object in ?class Sequence is in bold, and the tag in the class is in italics.

Features
Sequence : source sequence
Feature   method start   end   score   "text comments"

Similarities
Sequence : source sequence
"Sequence"_homol seq method score source_start source_end seq_start seq_end

For example below is an ACE file defining features and sequence similarities for genomic sequence AL050312.LINK.

Sequence : "AL050312.LINK"

EST_homol		A622692	dbEST	2055	470	880	1	411
vertebrate_mRNA_homol	AB016768	vertRNA	5	11308	11318	1361	1371
Pep_homol		P02463	Blastx	56	28615	28 532	474	501
Motif_homol	AluSg	RepeatMasker_SINE	8.8	39	335	1	296
Feature	Predicted_CpG_island	5754	8017	0.760000	"68.6 \%GC,	#CpGs= 204"		

Gene structures are subsequences of the reference sequence with multiple Source_Exon tags used to define the spliced exons. The following example is an ACE format file defining the gene AL050312.C22.1.mRNA. First the span of the gene is defined in the genomic sequence (base 1 of the gene at base 10060 of the genomic and base 11599 of the gene at base 21658 of the genomic), and then in a new sequence object the regions which are exons within this span are described as the Source_Exon.

Sequence :	AL050312.LINK		
Subsequence	AL050312.C22.1.mRNA	10060	21658
Sequence :	AL050312.C22.1.mRNA		
Source	AL050312.LINK		
Source_Exon	1	124	
Source_Exon	7245	7361	
Source_Exon	8409	8547	
Source_Exon	11101	11599	

RepeatMasker has ACE and GFF output options which allow the data to be parsed directly into ACeDB using the 'Read .ace files' option from the main window. For blast output, the post-processing program MSPcrunch (Sonnhammer and Durbin, 1994) can be used with the -4 option to output the data in ACE format (The '-4' refers to compatibility with ACeDB version 4). For more information on MSPcrunch see http://www.cgr.ki.se/cgr/groups/sonnhammer/MSPcrunch.html. However, not all programs have built in support for ACeDB and the output must be converted to ACE format. The Perl programming language (http://www.perl.com) is a very powerful language for regular expression matching and text processing and can be used to reformat output for parsing into ACeDB. Example Perl scripts can be found in ftp://ftp.sanger.ac.uk/pub/T62/Software (Table 6).

Table 6. Perl scripts to convert program output into ACE format. Avialble from ftp://ftp.sanger.ac.uk/pub/T62/Software.

fh2ace	Converts output from the fgenesh gene prediction program to gene structures in ACE format
gs2ace	Cgi script which runs the gene prediction program 'genscan' on the MIT web server and processes the output to create gene structutres in ACE format
cpg2ace	Runs the EMBOSS program 'newcpgreport' and converts the output to features in ACE format.

cDNA sequences can be mapped to the reference sequence using 'est2genome'; (http://www.hgmp.mrc.ac.uk/Software/EMBOSS/Apps/est2 genome.html; Mott, 1997). This program performs a gapped alignment with splice junction recognition, which greatly improves performance compared to other sequence alignment programs such as blast. The output from est2genome is processed by the Perl script 'estg2ace' (ftp://ftp.sanger. ac.uk/pub/T62/Software/estg2ace) which creates files for gene structures in ACE format. For GFF format conversion (and cautionary note) see Section 5.2.

4. From Analysis to Laboratory Confirmed Genes

4.1 Assessing Analysis Data

4.1.1 Definition of a Gene

When annotating genomes a gene definition is useful to standardise data and prevent misunderstanding. As described in the introduction, mammalian genes include a number of features. Below a set of rules is used to establish whether a gene displays these features and is therefore considered complete, or requires further laboratory work. Each annotated gene structure can be placed into one of three categories, coding gene, non-coding gene or pseudogene.

- The basis of a coding gene is a 5' untranslated region, an open reading frame (ORF) and a 3' untranslated region. This is normally, but not necessarily, split into exons. The ORF of a coding gene must start with the translation initiation codon ATG and finish with one of the three stop codons TAA, TAG or TGA. The start codon is often, but not always, proceeded by an in frame stop codon. At the 3' end of the cDNA a polyA tail is added which is not present in genomic DNA. This coincides with a polyA addition signal hexamer within

50 bases of the 3' end. The most frequent hexamer is AATAAA, followed by ATTAAA and a series of additional variations from AATAAA (Beaudoing *et al.*, 2000). Other features of mammalian genes to consider include possibly a CpG island at the beginning of the 5' untranslated region (Bird, 1986), a predicted promoter region (Scherf *et al.*, 2000; Down and Hubbard, 2002) and a possible Kozak sequence in strong consensus (AnnATGN or GnnATGG), or adequate consensus (GnnATGY or YnnATGG) around the translation start codon (Kozak, 1999).

- Non-coding RNA genes include small RNAs such as tRNA, small nuclear RNA (snRNA) and small nucleolar RNA (snoRNA). Other non-coding sequences represent control RNAs such as antisense sequences (Kumar and Carmichael, 1998). These either overlap with an exon of a coding gene or lie adjacent to the 5' end of a coding gene and are always on the opposite strand.
- A pseudogene matches a known gene or protein but has evidence of disrupted function, either from a disrupted open reading frame, or because the structure is disrupted relative to an expressed gene or cDNA and there is no evidence of expression of RNA from the annotated structure, or because it has been previously identified as a pseudogene in the literature.

4.1.2 Viewing Genomic Analysis and Defining Gene Structure

In order to annotate genomic sequence the first step is to conduct electronic analysis of the sequence, to identify sequence features. In general the strategy for mammalian genomes is two pronged. In the first part similarity to other known DNA and protein sequences is identified by searching against all available and appropriate databases. In general this will be conducted using the Blast suite of programs (Altschul *et al.*, 1990) against the DNA and protein databases available from Genbank/EMBL/DDBJ. In the second part algorithms that aim to predict features such as exons, genes, CpG islands, etc. will be run on the sequence and the results displayed. In mammals a complication is the presence of a high number of interspersed or tandem repeat sequences, and to avoid the problems associated with these sequences, they will be masked in the genomic sequence using a program such as RepeatMasker. The next step is to assess the evidence for a particular feature within the sequence, and to produce models of gene structures supported by evidence. Alignment of the similarity searches and feature prediction program outputs against genomic DNA in a display tool such as ACEDB, allows the evidence for a gene to be assessed. Our aim is to define complete coding and non-coding genes structures using 100% match cDNA from experimentally derived RNA sequence, and to define pseudogenes from similarity to functional genes found elsewhere in the genome. cDNA or EST sequence from the organism which matches the genomic DNA perfectly, allowing for the possibility of polymorphisms, can be pieced together to form a gene containing the required features of a coding or non-coding gene. Without complete cDNA or EST coverage a gene can not be considered to be supported by expression data and therefore is not finished. However other features can help predict a preliminary structure. Gene paralogues or homologues detected by Blastx searches can indicate possible exons. Similarly gene prediction programs such as GenScan or fgenesh provide an approximate gene structure. A predicted CpG island or a Unigene cluster can indicate the beginning or end of a gene in the genomic landscape, respectively. In some cases a gene may be supported by sufficient evidence over its full length from the electronic analysis to enable a complete annotation. However in many cases this will not be the case and additional data will be required to complete the structure. For instance it may be necessary to obtain additional sequence from cDNA to cover some of the presumed exons, or to join up two structures which might in reality be part of the same gene. As the new sequence is obtained it can be added back into the analysis to aid the annotation. The process of completing a gene annotation can involve several cycles of electronic analysis followed by experimental work as is illustrated in the example below.

Figure 2 shows a hypothetical graphical view of a typical example of sequence analysis aligned against genomic sequence and the addition of further data progressing toward a completed coding gene in 3 rounds of annotation. Initially a cluster of ESTs from dbEST and partial cDNA match indicate the possible 3' end of a gene. This can be confirmed by the presence of a polyA tail at the 3' end of an EST or cDNA sequence which is not found in the genomic DNA, with a polyA signal hexamer within 50 bases of the 3' end as additional proof. Looking at the sequence immediately 5' for evidence of the start of this gene reveals a predicted CpG island. A single EST situated 5' of this gene indicates two exons, but with the intervening exons missing. Two structures have been annotated in the first round (see Figure 2 filled boxes), one for the 3' end and the other for the 5' exons, which have yet to be shown to be part of the same gene. The blastx matches show the coding exons of a related gene stretching the across the region suggesting that the two structures are from the same gene. The gene prediction program also confirms this hypothesis, but again only indicates coding exons. To establish whether these two annotations

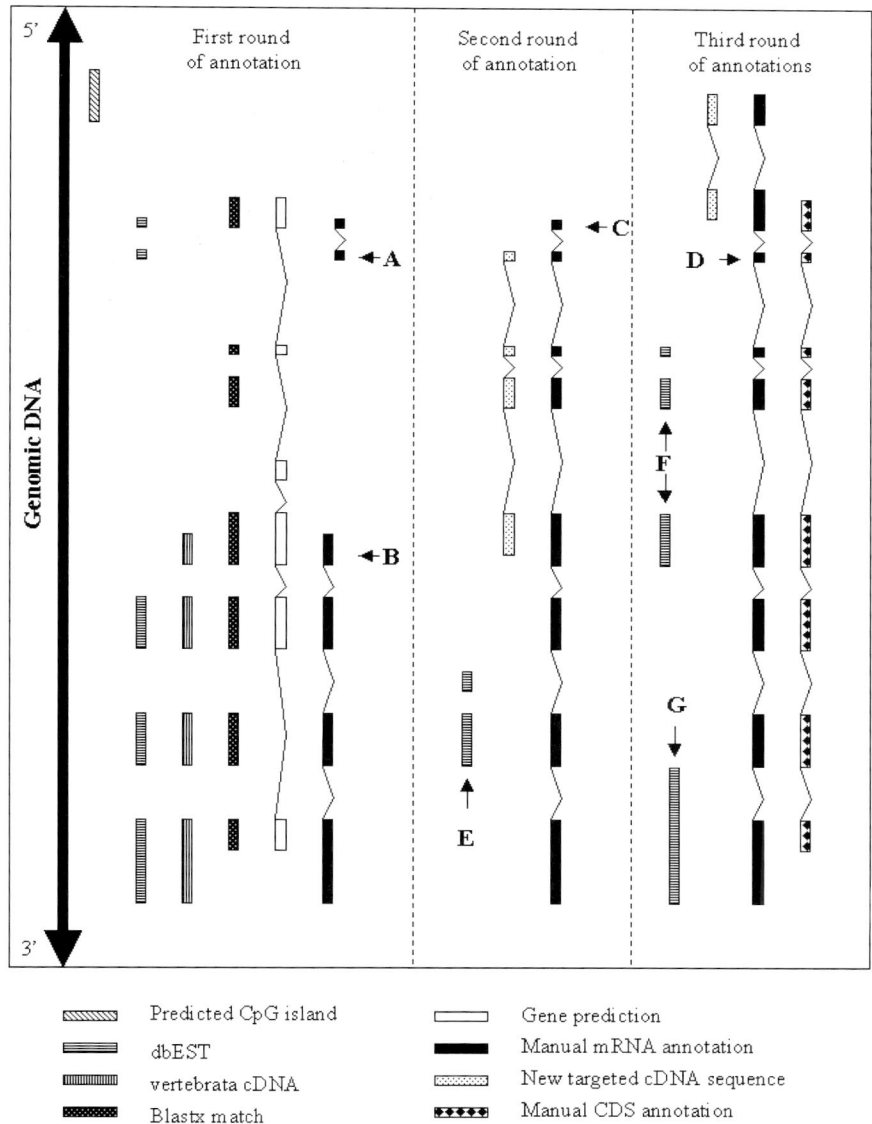

First round of annotation

Second round of annotation

Third round of annotations

5′

3′

Genomic DNA

	Predicted CpG island		Gene prediction
	dbEST		Manual mRNA annotation
	vertebrata cDNA		New targeted cDNA sequence
	Blastx match		Manual CDS annotation

Figure 2. Using genome analysis to annotate a complete gene. See text for description.

are indeed a single gene and to identify the intervening exon/intron structure, primers are designed to amplify a fragment between the annotations using an appropriate cDNA source as template (at A and B). The fragment is then sequenced, matched back to the genomic DNA and the annotation updated. In the second round of annotation this new targeted cDNA sequence shows that the two annotations are in fact a single gene and a new structure has been replaced by a single annotation showing the newly defined intervening exon/intron structure. However, the gene may still not be complete. The ORF extends outside the annotation and both the blastx and gene prediction program suggest the start is further 5'. A predicted CpG island is also indicated immediately 5', possibly signalling the 5' end of the gene. To extend the gene 5' the cDNA vectorette bubble PCR method (Riley *et al.*, 1990) can be used to amplify and sequence a fragment from a cDNA library (see Figure 2 5' of C). In the third round of annotation the new cDNA sequence amplified from a vectorette cDNA library shows the extension of the gene beyond the blastx and gene prediction program start codon and into the CpG island. The gene now has an annotated mRNA with 5 and 3' untranslated regions and an ORF.

The example in Figure 2 shows a number of interesting features relating to interpreting genomic DNA analysis. Comparing the blastx match of a related gene to the final annotation it is noticeable that exon 3 is missing (labelled D). In such cases the combination of a short exon and a diverged amino acid sequence from the related genes prove insufficient for a match detected by blastx. Alternatively this exon genuinely may not be present in the related gene, but it is worth checking that the lost exon does not alter the ORF of subsequent exons. This illustrates the problems encountered if an annotation is not defined by perfect match cDNA or EST sequence. Similarly, comparing the gene prediction program structure to the final annotation it is evident that although the prediction is good, it is not completely the same. The program failed to predict one exon and predicted an additional exon. However, it would be improper to call this incorrect as these exons may represent alternative splice forms of the gene. During the second and third rounds of analysis the region was used to search EST and cDNA databases to reveal any new deposited sequences. As is not uncommon, new ESTs were discovered (see E, F and G). The EST labelled E shows an alternative exon splicing onto the 5' end of exon 8 and the EST labelled F confirms part of the targeted cDNA sequence from the second round. The EST labelled G shows a sequence which has not spliced relative to the genomic sequence. This is just one example of the many EST artefact sequences observed in the human EST databases (Wolfsberg and Landsman, 1997) and can be a serious problem for any automatic annotation system.

4.2 Isolating Full Length cDNA

Once the initial annotation is complete the preliminary structures can be confirmed by cross exon PCR (Protocol 3) and the partial genes extended by vectorette PCR cDNA library screening (Protocols 4 and 5). Here a cDNA library is split into pools, amplified and the insert excised with a restriction enzyme (RE) (Figure 3). The vectorette bubble linker is then ligated onto either end of the insert. An initial round of screening with a pair of gene specific primers (GSP) establishes which pools contain an insert/bubble construct of the gene of interest. Amplification of these pools with a single gene specific primer and the bubble primer 224 yields the part of the cDNA insert, while the other gene specific primer and 224 give the remained of the insert. These PCR fragments are sequenced to provide the sequence of the original cDNA clone.

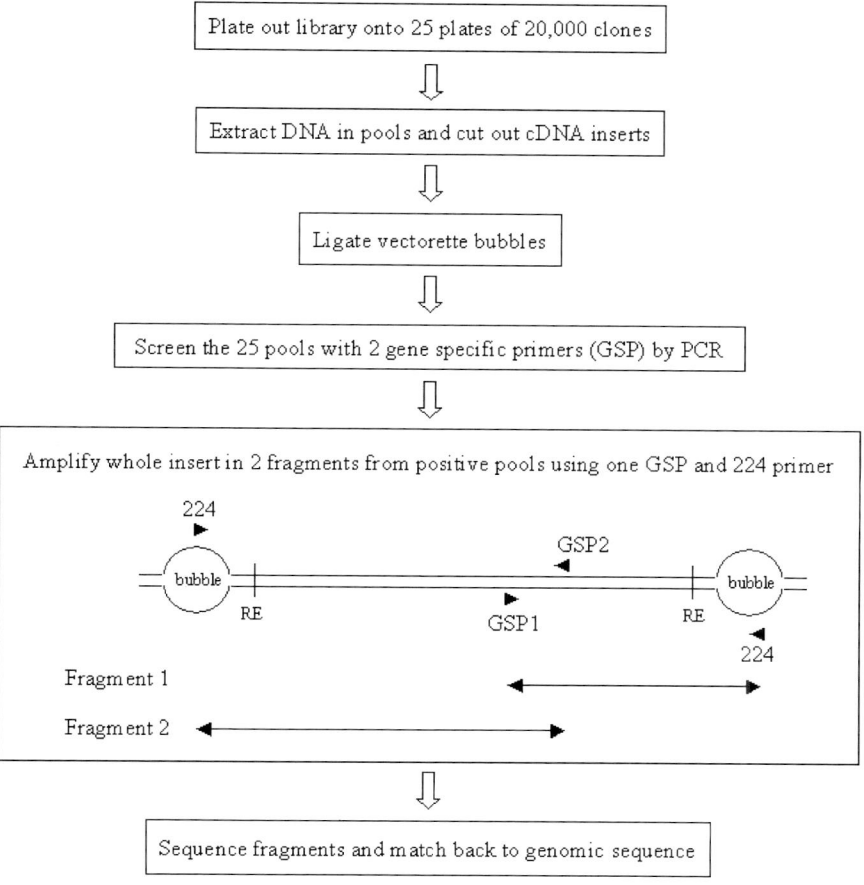

Figure 3. Isolation of a cDNA library clone insert by vectorette bubble PCR. See text for description.

Initially primers are designed to complete the annotation of the gene. When used with cDNA as templates the primers must be within exon sequence and a primer design program such as primer3 (Rozen, 2000) can help identify the oligonucleotide sequences. It is useful to design a sense and antisense primer within the same exon to allow genomic DNA to act as a positive control for the PCR. When attempting to amplify between exons, two pairs of primers are designed, one pair from each surrounding exon (see figure 2 A and B). Once it has been established that both pairs of primers amplify genomic DNA and cDNA from a particular tissue, then the appropriate primers, one from each pair, may be used together to amplify between exons using the same cDNA source (Protocol 3). No amplification would therefore indicate that the two exons are from different genes or the intervening exon or exons are too large to be amplified. When designing primers to screen vectorette cDNA library pools, amplification of genomic DNA as a positive control is also critical for proving the PCR is working correctly. If no primers are found, then the next exon 5' or 3' may be used. It is preferable to make the amplification product over 100 bases for the vectorette cDNA library screening to assist in interpretation after gel electrophoresis.

After completion of directed cDNA sequencing and update of the annotation it is necessary to assess whether the gene is finished. By considering the features that make up a gene a decision is made on a gene by gene basis and the annotation is either declared finished or further laboratory work undertaken (Protocol 6).

4.3 Problems with Annotating Genes

Imposing a set of rules on a biological system is not a simple task. Our understanding of the complexities of the mammalian cell is limited and new feature of the genome are continually discovered. Additionally the sequence databases are not perfect and are incomplete. The following list includes some of the problems which have been observed during the annotation of the gene content of human chromosome 22.

1. Sequencing errors in cDNA or genomic DNA sequence can prevent accurate gene annotation. This is a particular problem if a base is added or missing in the ORF, causing a frame shift. PCR primers designed from the genomic DNA sequence around the problem region can be used to amplify fragments from unrelated individuals to establish the true genomic sequence. Similarly EST sequences from different cDNA libraries may indicate the correct sequence.

2. Insertion and deletion polymorphisms observed between the cDNA and genomic sequence can cause a similar problem to sequencing errors. Again amplifying fragments from unrelated individuals can help to confirm a polymorphism or reject it as a sequencing error.

3. Unfinished regions in the genomic sequence can often lead to gene annotation continuing into genomic sequence gaps. These can be tagged as start or end incomplete and await further genomic sequence. Such cDNA sequences can also be used to help isolate clones to fill sequence gaps.

4. Chromosome 22q11 contains several genome duplications and which has led to gene amplifications (Dunham *et al.*, 1999). These are often observed as partial gene duplications, sometimes with a disrupted ORF compared to the expressed copy and therefore classed as a pseudogene, but sometimes retaining all or part of the ORF and therefore not meeting the criteria for a pseudogene. These partial gene duplications could be expressed genes, however the surrounding genomic landscapes suggest that this is unlikely (Collins, 2003). This has also been observed in other parts of the chromosome, such as a fragment of the immunoglobulin lambda locus dispersed on the long arm of chromosome 22 distant from the IgL locus itself.

5. A polyA sequence in the genomic DNA immediately 3' of a gene annotation is an indication of an incorrect 3' end. During the reverse transcription of cDNA from mRNA using a polyT oligonucleotide, priming from an internal polyA sequence can lead to a truncated cDNA molecule. cDNA or EST sequences 3' of the annotation which have a polyA signal hexamer associated with polyA in the cDNA but not genomic DNA are evidence of the true 3' end.

6. Cloning artefacts are known to occur in cDNA libraries and these are regularly observed in ESTs (Wolfsberg and Landsman, 1997; Collins *et al.*, 2003). This can lead to conflicting data when annotating a gene. A common observation is incomplete mRNA splicing which can be excluded during manual annotation. Chimeric cDNA clones have also been found matching partially to chromosome 22 and partially to elsewhere in the genome. Having the complete genome sequence available can help identify such clones.

7. As more genomes are annotated the quantity of gene sequences in various databases increases. This is generally helpful, assuming all the previous annotation is correct. Incorrect gene structures have found their way through the public databases and eventually return to the original sequence as a perfect match. Careful filtering of data is required to prevent such circular bogus confirmation.

8. Several chromosome 22 genes appear to be a collection of exons splicing together in repeat sequences. These do not contain ORFs and can not be classed as coding genes. It remains to be seen if these transcripts have any function.

5. Data Exchange

5.1 Exporting Annotation Data

One of the strengths of ACeDB is its graphical display of sequence information. However, in order to analyse the annotation data stored in the database it is desirable to export the data in a simple coordinate based format. This allows easy data flow between collaborating laboratories, data release into the public domain, and also the comparison of data sets (see 5.2) which can aid the annotation process.

Feature tables can be exported from ACeDB by selecting the 'Export features' option from the graphical view of a sequence object when running xace, or using the gif submenu's seqfeatures command when running giface. For help with ACeDB's xace and giface interfaces see http://www. acedb.org/Software/whelp. Data is exported in GFF format which has a simple one line per feature record format. The format is structured but does allow some flexibility. The official definition of the GFF acronym is now 'General Feature Format' (although Gene Finding Format and Gene Feature Format are still commonly used) which reflects the extended usage of the format to include nucleotide and peptide sequences.

The format is described in detail at http://www.sanger.ac.uk/Software/ formats/GFF/GFF_Spec.shtml, but briefly, each line contains a feature record with tab delimited fields as follows:-

1. Reference sequence name
2. The source or method which generated the feature
3. The feature (intron, exon, CDS, etc)
4. Start coordinate relative to the reference sequence
5. End coordinate relative to the reference sequence
6. Numeric score (if there is no score a "." is used)
7. Strand (+/–), or "." if not stranded
8. Frame (0, 1 or 2) or "." for features that have no translation frame
9. Attributes - or other information as a semi-colon separated list (*e.g.* the id of the feature such as 'Sequence AC005500.C22.1')

Comment lines beginning with the '#' character are often used to define a header with date, author and other useful information about the file. Another related format commonly in use is GTF (Gene Transfer Format). For the full specification see http://genes.cs.wustl.edu/GTF2.html. This format has the same structure as GFF but usage of some fields is forced and extended.

5.2 Comparison of Datasets

The ability to compare GFF formatted data sets has a number of applications. New matches from searching updated databases can be directly compared to the current annotation set and potential new genes or extensions to existing structures can be flagged for attention. The proximity of gene 5' ends to predicted CpG islands or promoters can also be established. Prediction methods can be evaluated by comparison of the set for a given method against the experimentally confirmed set.

At http://www.sanger.ac.uk/Software/formats/GFF there are numerous useful links to GFF utility programs and Perl modules which can be used for manipulating and analysing GFF data files. In addition there is extensive support for GFF in the bioperl library (see http://doc.bioperl.org/releases/bioperl-1.0.2/Bio/DB/GFF.html) and also in the AcePerl library (http://stein.cshl.org/AcePerl). The latter contains many useful functions which can be used for fast and easy access to ACeDB data. For a simple Perl script example of GFF conversion to ACE format download the file 'example_gff2ace' at ftp://ftp.sanger.ac.uk/pub/T62/Software.

For viewing data in the context of the genome the UCSC Genome Browser supports the upload of GFF annotation, for either private viewing or for general access. For details see http://genome.ucsc.edu/goldenPath/help/customTrack.html. Alternatively there are programs such as Genotator (Harris, 1997) and Apollo (http://www.ensembl.org/apollo) which are annotation browsers and editing tools. Genotator supports ACE format and Apollo has direct links to Ensembl data and has a GFF adaptor for data import. However, it should be noted that the flexibility of the GFF format can be a disadvantage. Although data can easily be exported, ultimately the success in transferring data between databases, viewers or annotation tools will be determined by the compatibility of the source and target data structures.

5.3 Future Developments

In order to accurately find and annotate all the genes across an entire genome, experimental investigation is needed not only for the verification of computationally predicted genes but also to expand the knowledge base for known genes. This is a considerable undertaking on a genome wide scale and with multiple laboratories intensively investigating specific genes or genomic regions there is unlimited scope for duplicated efforts and inconsistency. In addition, although the collating and presentation of genome sequence data has been successful (Ensembl Genome Browser (Hubbard *et al.*, 2002), NCBI MapViewer http://www.ncbi.nlm.nih.gov/mapview/map_search.cgi, UCSC Genome Browser (Kent *et al.*, 2002) there is a real danger that experimental annotation data will be detached from the genome sequence in a multitude of locations and presentation formats.

The DAS (Distributed Annotation System, http://biodas.org) is an attempt to address this issue with a system enabling experimental annotation data from multiple sources to be collated and displayed at a central site. Distant laboratories would be able to contribute experimental annotation data relative to a common reference sequence in a common format. There are a number of difficulties with this approach, not least of which is the agreement of a common format. Other complications arise from the instability of the genomic sequence in the databases which affect the curation of the reference sequence. Multiple sequence version numbers as a result of repeated editing and re-submission are common and the genome sequence is unlikely to be static for some years to come. However, if the problems can be overcome and if the system is widely accepted, DAS could be invaluable for assimilating annotation data as a common and unified resource.

6. Protocols

6.1 Protocol 1. Link Building from AGP

Requires: An internet connection. A UNIX operating system with cross_match (Green, unpublished) and Perl (version 5) installed with the LWP.pm and Getopts.pm Perl modules.

1. Firstly, create a directory called LinkTools in your home directory. Then download the LinkTools.tar.gz package (or LinkTools_only.tar.gz if you already have ACeDB installed) from ftp://ftp.sanger.ac.uk/pub/T62/Software

   Save the downloaded file to the LinkTools directory. Then type 'set path = ($home/LinkTools $path)'. You should also add this line to your .cshrc startup file (home directory). cd to LinkTools and unpack the downloaded file using the following command -

> gunzip -c LinkTools.tar.gz | tar -xvf -
> or, gunzip -c LinkTools_only.tar.gz | tar -xvf –

Note: The scripts and programs in the LinkTools directory will need to be executable, so you may need to run the command 'chmod u+x *' to do this.

2. Run perl script 'setup'. This creates a directory tree for link building, and an ACeDB database (linkace). The root directory created will be $home/links, where $home is the users home directory as defined in the HOME environment variable. If you wish to change this you will need to edit the script, and use the -root option in all subsequent scripts. However, it is better to use the default and then rename the links directory when the build is complete.

3. Go to the following URL to download latest AGP:-

   http://cvsweb.sanger.ac.uk/cgi-bin/cvsweb.cgi/genome-map/data/agp/?cvsroot=Enseml

   click the desired Chr<chr>.agp
   Download the latest agp file to the 'links' directory
   This file can be edited and cut down to a smaller region of interest if desired.

4. Run perl script 'AGP2Link'. This script reads the downloaded Chr<chr>.agp file and retrieves all the required Sequences from the NCBI website. These are saved into the $root/genomic directory as 'sequence.version.seq' files. It will take a long time to complete for a whole chromosome and will also require a lot of disk space. Allow at least ~100 kilobytes of space for each accession retrieved.

   Once all the sequences have been retrieved, AGP2Link builds a SuperLink sequence object and trims all the sequences according to the coordinates specified in the AGP file. The complete reference sequence (SuperLink) is composed of each of the trimmed sequences concatenated in order. The SuperLink and all the trimmed sequences are saved in ACE format in $root/linkace/rawdata.

5. Run perl script 'ImpAce'. This will import the ACE files generated by AGP2Link files (in $root/linkace/rawdata) to the Linkace ACeDB database. If you have downloaded the LinkTools_only.tar.gz package you will need to use the -path option to set the path to the ACeDB executables. See the log file (in $root/linkace) for any parse errors.

6.2 Protocol 2. Link Building from TPF

Requires: An internet connection. A UNIX operating system with cross_match (Green, unpublished) and perl (version 5) installed with the LWP.pm and Getopts.pm perl modules.

1. Firstly, create a directory called LinkTools in your home directory. Then download the LinkTools.tar.gz package (or LinkTools_only. tar.gz if you already have ACeDB installed) from ftp://ftp.sanger. ac.uk/pub/T62/Software

   Save the downloaded file to the LinkTools directory. Then type 'set path = ($home/LinkTools $path)'. You should also add this line to your .cshrc startup file (home directory). cd to LinkTools and unpack the downloaded file using the following command -

   gunzip -c LinkTools.tar.gz | tar -xvf -
   or, gunzip -c LinkTools_only.tar.gz | tar -xvf –

Note: The scripts and programs in the LinkTools directory will need to be executable, so you may need to run the command 'chmod u+x *' to do this.

2. Run perl script 'setup'. This creates a directory tree for link building, and an ACeDB database (linkace). The root directory created will be $home/links, where $home is the users home directory as defined in the HOME environment variable. If you wish to change this you will need to edit the script, and use the -root option in all subsequent scripts. However, it is better to use the default and then rename the links directory when the build is complete.

3. Download latest TPF at:-

   Go to http://cvsweb.sanger.ac.uk/cgi-bin/cvsweb.cgi/genome-map/data select latest data directory click Chr<chr>. Download the latest tpf.txt file to the 'links' directory. This file can be edited and cut down to a smaller region of interest if desired. Alternatively, if a TPF is not available you can create one yourself, providing you know the EMBL accession numbers of the sequences and their order.

4. Run perl script 'LinkPrep'. This script reads the downloaded tpf.txt file and retrieves all the required sequences. These are saved into the $root/genomic directory as 'sequence.version.seq' files. It will take a long time to complete for a whole chromosome and will also require a lot of disk space. Allow at least ~100 kilobytes of memory for each sequence.

A new file ($root/tpf.table) is also created which records the sequence version number and length of the sequence retrieved. This file is required by the LinkBuild script.

Often, though not always, estimated gap sizes are given in the TPF. If this is the case a file of gap sizes ($root/gap_sizes) is saved. This file can be read by LinkMan later (automatically), when building the SuperLink from sub links.

Sequences which could not be retrieved will be logged in the error file $root/LinkPrep.errors

All sequences in the TPF must be retrieved, so try searching the nucleotide databases from http://www.ncbi.nlm.nih.gov:80/entrez/query.fcgi

5. Run perl script 'LinkBuild'. LinkBuild reads the file '$root/tpf.table' made by LinkPrep. For each sequence in the table it calls the 'match' perl script which uses cross_match to align sequence1 to sequence2. If an overlap cannot be determined, an additional gap is inserted and the failure is reported in the file '$root/LinkBuild. failed'. Any warnings issued (for example the presence of indels) are logged in the file '$root/LinkBuild.warnings'.

On completion, read the LinkBuild.failed and LinkBuild.warnings files. Follow up any alignments listed in these files by examining the relevant part of LinkBuild.log (use logreader –id <sequence> to do this). It is important to note that due to the complexity of some overlaps this method does not achieve 100% accuracy.

If a correction is to be made, edit the LinkBuild.lbf file accordingly. To close a gap remove the blank line and insert the overlap size in the second comma separated field. To change an overlap size, edit the second field and to change the orientation of a sequence edit the fourth field (+ or –). The .lbf file format is as follows:-

sequence.version,overlap_size,sequence_length,orientation

For example, here is part of the LinkBuild.lbf for human chromosome 6:-

AL024506.1,100,71653,+
AL137220.8,,126482,+

AL138884.10,1956,10063,+
AL031785.1,100,152553,–
AL139387.12,100,83352,+

6. Run perl script 'LinkAssemble'. This generates ACeDB format link objects using the table of alignments in the file 'LinkBuild.lbf' (output from LinkBuild). Each link object is saved as '$root/LINK_<firstsequence.version>.ace'.

7. Check the alignments in each link by running the perl script 'LinkCheck'. This will save a report for each link in the $root/reports directory and also ACE format files defining the trim tags (required later by LinkTrim).

Look through the reports, follow up those alignments listed and correct any misalignments (edit the LinkBuild.lbf file; see 5.). When fully checked, if any edits have been made, it will be necessary to run LinkAssemble and LinkCheck again.

8. Run the perl script 'LinkMan'. This generates the 'SuperLink' (link of links). If a file of gap sizes is found (as output by LinkPrep) it will be used to map the sub links onto the SuperLink, or if not, 50kb will be used as the default gap size.

LinkMan will also convert all sequence names to the form '<sequence.version>.LINK', to distinguish between EMBL/Gen Bank versions of the sequences and those which will constitute part of the link; in some cases containing regions of neighbouring clones in overlapping segments.

9. Run the perl script 'LinkTrim'. LinkTrim reads the $root/links.ace file and the trim tags files created by LinkCheck and creates a trim table (saved as $root/<linkname>.TrimTable) defining how each sequence should be trimmed. It then checks for conflicts *e.g.* for two + orientation clones, a sequence1 trim right and sequence2 trim left are conflicting, or conversely no sequence1 trim right and no sequence2 trim left are also conflicting. Conflicts are reported in the file $root/LinkTrim.errors. If no conflicts are found LinkTrim creates <sequence.version>.TRIMMED sequence objects for all the sequences, saving each in ace format in $root/linkace/rawdata. The complete reference sequence is composed of each of the trimmed sequences concatenated in order.

If errors are found, examine the <linkname>.TrimTable file(s) in $root/links. Fortunately errors are usually rare but they can happen where there is redundancy, or in short overlaps which are complicated by indels. In such cases edit the trim tags file using the perl script 'trimfix'. This has options to trim or include left and right overlaps for a specified sequence id. For example, there is a conflict in the TrimTable shown below which can be resolved by trimming the second sequence at the left end:-

sequence,length,link_start,link_end,orientation,include_left, include_right
AL009176.1.LINK,87903,1,87903,+,no,yes
AL135910.22.LINK,42938,87804,130741,+,yes,yes

To correct this use the command:-

trimfix -id AL135910.22.LINK -trim_left

Or, in this example the conflict can be resolved by including the left end:-

sequence,length,link_start,link_end,orientation,include_left, include_right
AL009176.1.LINK,87903,1,87903,+,yes,no
AL135910.22.LINK,42938,87804,130741,+,no,yes

To correct this use the command:-

trimfix -id AL135910.22.LINK -include_left

The trimfix options -trim_right and -include_right can be used in a similar way.

Note: For clones in reverse (-) orientation, left and right refer to the 'true' left and right of the clone sequence, not left and right relative to the link.

10. Run perl script 'ImpAce'.
This will import all the ACE format files created during the link building process (all located in $root/linkace/rawdata) to the Linkace ACeDB database. If you have downloaded the LinkTools_only.tar.gz package you will need to use the -path option to set the path to the ACeDB executables. See the log file (in $root/linkace) for any parse errors.

If ACE format files need to be imported to Linkace (*e.g.* trim corrections) after the first set of files have been imported, it is sensible to rename $root/linkace/rawdata to a different name (for example, 'imported<date>') and then create a new empty rawdata directory. This archives the imported ACE files and when ImpAce is run again it will only import the new ACE files.

6.3 Protocol 3. Confirming Preliminary Exons and Amplification Between Exons

Equipment and Reagents

- Separate gene specific primers (200 ng/µl)
- 10x PCR buffer: 670 mM Tris-HCl (pH 8.3), 166 mM (NH₄)₂SO₄, 67 mM MgCl₂
- 0.5 mg/ml BSA
- 1/20 2-mecaptoethanol (714 mM)
- Cresol Red/sucrose: 0.008% Cresol Red, 28% sucrose
- 5mM dNTP
- Advantage 2 Taq Polymerase mix Clontech 8430-1 (50x)
- Ampli-Taq (5 units/µl)
- PerfectMatch PCR Enhancer Stratagene 600129 (1 unit/µl)
- TaqExtender PCR additive Stratagene 600148 (5 units/µl)
- cDNA template
- Genomic DNA (12.5 µl)
- 10 mg/ml ethidium bromide
- agarose
- DNA fragment size marker
- QIAquick Gel Extraction Kit QIAgen Cat. no. 28706
- Shrimp alkaline phosphatase (SAP) (1 unit/µl) Amersham E70092Y
- ExonucleaseI (ExoI) (10 unit/µl) Amersham E70073Z
- 10xSAP buffer (supplied with the SAP)
- Access to DNA sequencing reagents and equipment

Method

A. Primer testing and cDNA source validation
1. Design two pairs of gene specific primers within exons either side of region to be confirmed.
2. To chose cDNA source, check existing data for an indication of where the gene is expressed, for example EST data or expression of a homologue.
3. To confirm that both pairs of primers are represented in the cDNA source, amplify each pair using cDNA template and a genomic DNA

positive control.

4. For each 15 μl PCR reaction use 1.5 μl of 10x PCR buffer, 1.5 μl of 5mM dNTP, 0.495 μl of 0.5 mg/ml BSA, 0.21 μl of 1/20 2-mecaptoethanol, 5.425 μl of Cresol Red/sucrose, 0.375 μl of 200 ng/μl of each gene specific primer and 0.12 μl of Ampli-Taq.

5. Add 5 μl of 0.1 ng/μl cDNA template or 5 μl of 12.5 ng/μl genomic DNA to give a total of 15 μl.

6. Place in a PCR machine and heat to 94°C for 5 minutes.

7. Start cycling program with 94°C for 5 seconds, 60°C for 30 seconds, 72°C for 30 seconds for 35 cycles and finishing with 72°C for 5 minutes. Annealing temperatures may need to be altered.

8. Run products with a size marker on a 2.5% agarose gel containing 0.4 μl/ml ethidium bromide (no loading buffer is required when using cresol red/sucrose).

B. Amplification between exons

1. To amplify between exons, pick the sense primer from the 5' primer pair and the antisense primer from the 3' primer pair to make an intra-exon primer pair.

2. For each 15 μl PCR reaction use 1.5 μl of PCR buffer, 1.5 μl of 5mM dNTP, 0.495 μl of 0.5 mg/ml BSA, 0.21 μl of 1/20 2-mecaptoethanol, 5.185 μl of Cresol Red/sucrose, 0.375 μl of 200 ng/μl of each gene specific primer, 0.12 μl of Advantage 2 polymerase mix, 0.12 μl of TaqExtender, 0.12 μl of PerfectMatch and 5 μl of 0.1 ng/μl cDNA template.

3. Start cycling program with 94°C for 2 minutes for one cycle, followed by 94°C for 5 seconds, 55-65°C for 30 seconds, 72°C for 5 minutes for 35 cycles and finishing with 72°C for 5 minutes. Annealing temperatures may need to be altered depending on the primers and PCR machine used. Long extension times may be required for larger fragments.

Note: We have found that Advantage 2 polymerase mix works best with short cycle times at 94°C.

4. Run products with a size marker on a 1.2% agarose gel containing 0.4 μl/ml ethidium bromide (no loading buffer is required).

5. Excise fragments for further analysis from the gel and store in 100 μl of sterile water at 4°C.

6. To amplify the fragment for sequencing or further analysis use 5 μl of the water around the gel slice as a template in a further PCR amplication using the reagents as above with cycling of 94°C for 5

seconds, 60°C for 30 seconds, 72°C for 3 minutes for 20 cycles. 4 × 15 μl reactions will provide plenty of product. PerfectMatch need not be used unless the region contains local short repeated sequences.

7. Re-amplifications which give single fragments can be cleaned with 1 μl SAP, 0.1 μl ExoI, 1 μl 10xSAP buffer and 8 μl of PCR product at 37°C for 30 minutes followed by 80°C for 15 minutes. However, re-amplifications which gives multiple fragments need to be separated on an agarose gel, excised, cleaned with a QIAquick Gel Extraction Kit and eluted in 30 μl buffer as supplied.

8. Sequence fragments with the gene specific primers.

9. Match the sequence to genomic DNA to confirm or extend annotation.

6.4 Protocol 4. cDNA Vectorette Library Construction

Equipment and Reagents

- Titrated cDNA library in a plasmid vector where the insert can be excised by digesting with a single restriction enzyme.

Note: It is best to use an enzyme which is not common to prevent cutting the insert in addition to excising it from the vector.

- LB medium (sterile)
- Appropriate antibiotic for cDNA library
- QIAfilter Plasmid Midi Kit. Qiagene Cat. No. 12245
- Sterile water
- Appropriate restriction enzyme and buffer
- 10 mg/ml BSA
- 100 mM spermidine
- Phenol/chloroform 1:1 v/v
- 3M Na-acetate
- Ethanol
- $T_{0.1}E$ buffer: Tris 10 mM Tris-HCl, 0.1 mM EDTA pH 8.0
- Ligation buffer: 50 mM Tris-HCl pH 7.6, 10 mM $MgCl_2$, 1mM DDT
- Annealed vectorette bubbles (1 pmole/μl) (Table 7)
- rATP (100mM)
- T4 DNA ligase (2.5 units/μl)
- Gene specific primers from known genes in the cDNA library for titrating and testing.

Table 7. Primers for vectorette PCR.

224	CGAATCGTAACCGTTCGTACGAGAATCGCT
BPBI	CAAGGAGAGGACGCTGTCTGTCGAAGGTAAGGAACGGAC GAGAGAAGGGAGAG
BPBII	CTCTCCCTTCTCGAATCGTAACCGTTCGTACGAGAATCGCT GTCCTCTCCTTG

BPBI and II produce a blunt end bubble, sticky ends may be added to the 5' end of BPBI and the 3' end of BPBII during synthesis as required.

Method

1. Set up 25 tubes of 20 ml LB with appropriate antibiotic.
2. Inoculate each with 20,000 colonies from a cDNA library glycerol stock.
3. Grow shaking overnight at temperature and shaking speed suggested for the cDNA library.
4. Pellet the cells at 4000 rpm for 10 minutes at room temperature.
5. Pour off the media and invert to drain the excess.

Note: It is important to remove all the media as this can effect the subsequent enzyme reactions.

6. Miniprep using QIAfilter Plasmid Midi Kit and resuspend in 30 µl.
7. Digest approximately 1 mg of DNA (5 µl) in 30 µl with the chosen restriction enzyme: use 3 µl of 10x restriction buffer, 0.3 µl of 10 mg/ml BSA, 0.3 µl of 100 mM spermidime, 2 µl of enzyme (10 U/µl) and make up with sterile water. Incubate at appropriate temperature for 2 hours.

Note: Adding 1 µl of enzyme at the start followed by 1 µl after 1 hour gives more efficient cutting.

8. Add 70 µl of sterile water and extract with phenol/chloroform.
9. Add 10 µl of 3M Na-acetate and 200 µl of cold ethanol, shake well and freeze for 1 hour at –20°C.
10. Spin in a microfuge for 15 minutes, remove the supernatant, wash the pellet with 70% ethanol and air dry.
11. Resuspend in 100 µl of ligation buffer.
12. Add 10 µl of 1 pmol/µl annealed vectorette bubbles which are compatible with the restriction enzyme used (see Table 7). To

anneal bubbles add 1 μmole of each primer to 25 μl of 1M NaCl and make up to 1 ml with sterile water. Heat to 65°C for 1 minute and air cool.

13. Add 1.1 μl of rATP and 2.5 units of T4 DNA ligase. And incubate overnight at 16°C.

14. Dilute the vectorette library to 500 μl with $T_{0.1}E$ and store at −20°C.

15. Titrate the amount of DNA required for cDNA library screening and vectorette PCR using primer from several known genes. This is approximately 1:1000 for screening and 1:100 for vectorette PCR.

6.5 Protocol 5. Screening and Isolating Fragments from Vectorette cDNA Libraries

Equipment and Reagents

- cDNA vectorette libraries
- Gene specific primers (200 ng/μl)
- PCR buffer: 670 mM Tris-HCl (pH 8.3), 166 mM $(NH_4)_2SO_4$, 67 mM $MgCl_2$
- 0.5 mg/ml BSA
- 1/20 2-mecaptoethanol (714 mM)
- Cresol Red/sucrose: 0.008% Cresol Red, 28% sucrose
- 224 vectorette bubble primer (200 ng/μl) (Table 7)
- 5mM dNTP
- Ampli-Taq (5 units/μl)
- PerfectMatch PCR Enhancer Stratagene 600129
- TaqExtender PCR additive Stratagene 600148
- 10 mg/ml ethidium bromide
- agarose
- DNA fragment size marker
- QIAquick Gel Extraction Kit Cat. no. 28706
- Shrimp alkaline phosphatase (SAP) (1 unit/μl) Amersham E70092Y
- ExonucleaseI (ExoI) (10 unit/μl) Amersham E70073Z
- 10xSAP buffer (supplied with the SAP)
- Access to DNA sequencing reagents and equipment

Method

1. Identify the pool or pools which amplify the fragment of the gene of interest by PCR on diluted cDNA vectorette libraries. Use the

PCR method in Protocol 3 method A. Either screen all 25 libraries or 5 pools of 5 libraries followed by a second round of screening. If the primers give clean results it is possible to use the pool of 5 libraries (100,000 clones) for the vectorette PCR.

2. For each vectorette PCR reaction use 1.5 μl of PCR buffer, 1.5 μl of 5mM dNTP, 0.495 μl of 0.5 mg/ml BSA, 0.21 μl of 1/20 2-mecaptoethanol, 4.545 μl of Cresol Red/sucrose, 0.375 μl of 200 ng/μl gene specific primer and 0.375 μl of 200 ng/μl 224 bubble primer.

3. Add 5 μl of diluted cDNA vectorette library to give a total of 14 μl.

4. Place in a PCR machine and heat to 94°C and add 1 μl of enzyme mix comprising 0.12 μl of Ampli-Taq, 0.12 μl of TaqExtender, 0.12 μl of PerfectMatch and 0.64 μl of Cresol Red/sucrose. Attempt to add the enzyme mix to all samples within 5 minutes.

5. Start cycling program with 94°C for 5 seconds, 68°C for 30 seconds, 72°C for 3 minutes for 17 cycles, followed by 94°C for 5 seconds, 65°C for 30 seconds, 72°C for 3 minutes for 18 cycles and finishing with 72°C for 5 minutes. Annealing temperatures may need to be altered depending on the PCR machine used. Long extension times may be required for larger fragments.

6. Run products with a size marker on a 1% agarose gel containing 0.4 μl/ml ethidium bromide (no loading buffer is required when using cresol red/sucrose).

7. Excise fragments for further analysis from the gel and store in 100 μl of sterile water at 4°C.

8. To amplify the fragment for sequencing see Protocol 3 method B.

9. Sequence fragments with the gene specific primer and the 224 primer.

10. Match the sequence to genomic DNA to extend annotation.

6.6 Protocol 6. Features of Finished Genes

Equipment and Reagents

- Graphical view system for genomic DNA analysis, such as ACeDB
- Genomic sequence analysis (see Section 3)

Method

1. Features to look for in a coding gene
   i) Does the cDNA match exactly across the entire length of the gene, allowing for the possibility of polymorphisms?

ii) Does the cDNA contain an open reading frame (ORF) of over 300 bases?

iii) Does the cDNA finish with a polyA tail not seen in the genomic DNA?

iv) Is there a polyA hexamer within 50 bases of the 3' end of the cDNA?

v) Does the ATG start codon conform to the Kozak consensus sequence?

vi) Does the cDNA splice compared to genomic DNA (not essential) and do these splices conform to the consensus?

vii) Is the 5' end of the gene in a predicted CpG island (not essential)?

viii) Is there an in frame stop codon before the start ATG codon at the 5' end of the gene (not essential but useful to confirm the entire ORF is annotated)?

ix) Is there a promoter prediction at the 5' end of the gene?

2. Features to look for in a non-coding gene

i) Does the annotation match a sequence in a non-coding RNA database, such as tRNA, snRNA, snoRNA?

ii) Does the annotation contain a selection of the above coding gene features but no ORF > 300 bases?

iii) Is the gene unspliced or with short introns and lie directly 5' of an annotated gene but on the opposite strand, suggesting possible antisense transcript?

3. Features to look for in a pseudogene

i) Does the annotation match the amino acid sequence of a coding gene but with a disrupted ORF?

ii) If the region has no protein match but several non-exact ESTs matches, do the ESTs match best to another part of the genome suggesting an unannotated coding gene?

iii) If unspliced, is there a polyA tail in the genomic at the 3' end suggesting retrotransposon?

iv) If spliced is the coding gene close in the genomic sequence suggesting a local genome duplication?

Conclusion

The annotation of a reference sequence will continue to evolve as new data becomes available. In this chapter we have concentrated mostly on coding genes, but there is increasing evidence that there are a variety of other transcripts that are involved in control systems, as well as features which at first sight appear to be genes but are really remnants of the evolution process (Mounsey *et al.*, 2002). The goal of annotating a reference sequence is to define all the parts of the genome that contribute to the function of that organism. It is these data that provides the basic information to unravel the complicated processes involved in cell biology.

Acknowledgements

We would like to thank Charlotte Cole and Ian Dunham for critically reading this manuscript.

References

Altschul, S.F., Gish, W., Miller, W., Myers, E.W., and Lipman, D.J. 1990. Basic local alignment search tool. J Mol Biol. 215: 403–10.

Beaudoing, E., Freier, S., Wyatt, J.R., Claverie, J.M., and Gautheret, D. 2000. Patterns of variant polyadenylation signal usage in human genes. Genome Res. 10: 1001–10.

Bird, A.P. 1986. CpG-rich islands and the function of DNA methylation. Nature. 321: 209–13.

Burge, C., and Karlin, S. 1997. Prediction of complete gene structures in human genomic DNA. J Mol Biol. 268: 78–94.

Collins, J.E., Goward M.E., Cole C.G., Smink L.J., Huckle E.J., Bye J.M., Beare D.M., and Dunham, I. 2003. Re-evaluating human gene annotation: a second generation analysis of chromosome 22. Genome Res. 13: 27–36.

Deloukas, P., Matthews, L.H., Ashurst, J., Burton, J., Gilbert, J.G., Jones, M., *et al* 2001. The DNA sequence and comparative analysis of human chromosome 20. Nature. 414: 865–71.

Down, T.A., and Hubbard, T.J. 2002. Computational detection and location of transcription start sites in mammalian genomic DNA. Genome Res. 12: 458–61.

Dubchak, I., Brudno, M., Loots, G.G., Pachter, L., Mayor, C., Rubin, E.M., and Frazer, K.A. 2000. Active conservation of noncoding sequences revealed by three-way species comparisons. Genome Res. 10: 1304–6.

Dunham, I., Hunt, A.R., Collins, J.E., Bruskiewich, R., Beare, D.M., Clamp, M., *et al.* 1999. The DNA sequence of human chromosome 22. Nature. 402: 489–95.

Durbin, R. and J. Thierry-Mieg. 1991. A C. elegans database.http://www.sanger.ac.uk/Sofeware/Acedb/

Frazer, K.A., Sheehan, J.B., Stokowski, R.P., Chen, X., Hosseini, R., Cheng, J.F., Fodor, S.P., Cox, D.R., and Patil, N. 2001. Evolutionarily conserved sequences on human chromosome 21. Genome Res. 11: 1651–9.

431

Guigo, R., Agarwal, P., Abril, J.F., Burset, M., and Fickett, J.W. 2000. An assessment of gene prediction accuracy in large DNA sequences. Genome Res. 10: 1631–42.

Harris, N.L. 1997. Genotator: a workbench for sequence annotation. Genome Res. 7: 754–62.

Hubbard, T., Barker, D., Birney, E., Cameron, G., Chen, Y., Clark, L. *et al.* 2002. The Ensembl genome database project. Nucleic Acids Res. 30: 38–41.

International Human Genome Sequencing Consortium. 2001. Initial sequencing and analysis of the human genome. Nature. 409: 860–921.

Kent, W.J., Sugnet, C.W., Furey, T.S., Roskin, K.M., Pringle, T.H., Zahler, A.M., and Haussler, D. 2002. The human genome browser at UCSC. Genome Res. 12: 996–1006.

Kondrashov, A.S., and Shabalina, S.A. 2002. Classification of common conserved sequences in mammalian intergenic regions. Hum Mol Genet. 11: 669–74.

Korf, I., Flicek, P., Duan, D., and Brent, M.R. 2001. Integrating genomic homology into gene structure prediction. Bioinformatics. 17 Suppl 1: S140–8.

Kozak, M. 1999. Initiation of translation in prokaryotes and eukaryotes. Gene. 234: 187–208.

Kumar, M., and Carmichael, G.G. 1998. Antisense RNA: function and fate of duplex RNA in cells of higher eukaryotes. Microbiol Mol Biol Rev. 62: 1415–34.

Mattick, J.S. 2001. Non-coding RNAs: the architects of eukaryotic complexity. EMBO Rep. 2: 986–91.

Mayor, C., Brudno, M., Schwartz, J.R., Poliakov, A., Rubin, E.M., Frazer, K.A., Pachter, L.S., and Dubchak, I. 2000. VISTA : visualizing global DNA sequence alignments of arbitrary length. Bioinformatics. 16: 1046–7.

Mott, R. 1997. EST_GENOME: A program to align spliced DNA sequences to unspliced genomic DNA. Comput. Appl. Blosci. 13: 477–478.

Mounsey, A., Bauer, P., and Hope, I.A. 2002. Evidence suggesting that a fifth of annotated Caenorhabditis elegans genes may be pseudogenes. Genome Res. 12: 770–5.

Riley, J., Butler, R., Ogilvie, D., Finniear, R., Jenner, D., Powell, S., Anand, R., Smith, J.C., and Markham, A.F. 1990. A novel, rapid method for the isolation of terminal sequences from yeast artificial chromosome (YAC) clones. Nucleic Acids Res. 18: 2887–90.

Rozen, S. and Skaletsky H. J. 2000. Primer3 on the WWW for general users and for biologist programmers. In: Krawetz S, Misener S (eds)

Bioinformatics Methods and Protocols: Methods in Molecular Biology. Humana Press, Totowa, NJ, pp 365–386

Scherf, M., Klingenhoff, A., and Werner, T. 2000. Highly specific localization of promoter regions in large genomic sequences by PromoterInspector: a novel context analysis approach. J Mol Biol. 297: 599–606.

Solovyev, V.V., Salamov, A.A., and Lawrence, C.B. 1995. Identification of human gene structure using linear discriminant functions and dynamic programming. Proc Int Conf Intell Syst Mol Biol. 3: 367–75.

Sonnhammer, E.L., and Durbin, R. 1994. A workbench for large-scale sequence homology analysis. Comput Appl Biosci. 10: 301–7.

Wolfsberg, T.G., and Landsman, D. 1997. A comparison of expressed sequence tags (ESTs) to human genomic sequences. Nucleic Acids Res. 25: 1626–32.

From: *Genome Mapping and Sequencing*
© 2003 Horizon Scientific Press, Wymondham, UK

14

Sequence Databases

Gregory D. Schuler

Abstract

Biological sequence databases represent indispensable tools for scientific discovery. A comprehensive database of all publicly available DNA sequences has been maintained through a longstanding collaboration among informatics groups in the United States, Europe, and Japan. Over the past decade, extraordinary gains have been made in the extensive sequencing of genomes and transcriptomes. Consequently, the databases have developed new procedures for bulk submission of new classes of data in both intermediate and finished forms. Additional databases have been established to track clone sequencing progress and to capture the original sequencing traces. More recent developments have focused on tools for browsing and analyzing whole genomes.

1. Introduction

In the universe of sequence databases, the event sometimes described as the "Big Bang" was the development of rapid DNA sequencing in 1977 (Maxam and Gilbert, 1977; Sanger *et al.*, 1977a). In this same year, the first genome was sequenced—the 5386 bp genome of coliphage ΦX174 (Sanger *et al.*, 1977b). As sequencing technology has continued to improve, the number of sequence entries in the public databases has swelled to the point that by the end of 2001, nearly 15 million sequence entries constituting 15.8 Gb of DNA have been deposited in the sequence databases. Concomitantly, there has been a steady increase in the size and complexity of genomes whose sequence has been determined (see Figure 1). In 1995, the first genome of a free-living organism was sequenced—the 1.8 Mb genome of the bacterium *H. influenzae* (Fleischmann *et al.*, 1995). The 12 Mb genome of *S. cerevisiae* was the first eukaryotic genome to be sequenced (Goffeau *et al.*, 1996). The genomes of multicellular organisms represented an order of magnitude increase in size, with *C. elegans* at 97 Mb (The *C. elegans* Sequencing Consortium 1998) and *D. melanogaster* at 137 Mb (Adams *et al.*, 2000). Finally, the first vertebrate genome sequenced, albeit in "working draft" form (finishing is now underway), is the 3.2 Gb sequence of *H. sapiens* (The International Human Genome Sequencing Consortium 2001).

In the early history of the sequence databases, releases were made on floppy diskettes or even printed in book form! Today, the database is large enough that some institutions are choosing not to devote the hard disk space to a storing a complete copy and are instead depending upon Internet resources for searching and retrieving sequence entries.In this chapter, we discuss the basic structure of sequence databases, mechanisms for contributing data, and how to make effective use of the online resources.

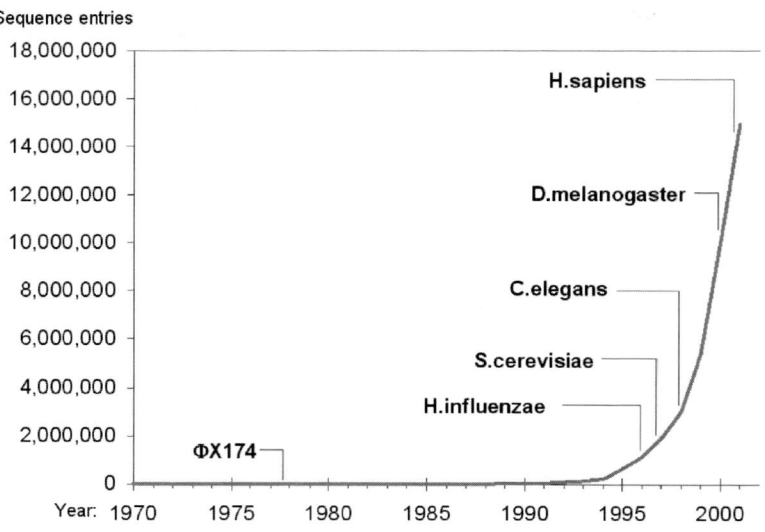

Figure 1. Milestones in DNA Sequencing. Both the number of sequence entries in GenBank and the sizes of genomes sequenced have been steadily increasing since the development of rapid DNA sequencing in the late 1970s.

2. Nucleotide Sequence Databases

2.1 International Sequence Database Consortium

The worldwide comprehensive collection of public sequence data is maintained by a longstanding collaboration among database groups in the UK, Japan, and the USA. The European Bioinformatics Institute (EBI) manages the sequence database of the European Molecular Biology Laboratory (EMBL), headquartered in Heidelberg, Germany (Stoesser *et al.*, 2002). The EBI is an EMBL outstation located on the grounds of the Wellcome Trust Genome Campus in Hinxton, UK, next to the Wellcome Trust Sanger Institute and the UK Human Genome Mapping Project Resource Centre. The DNA Data Bank of Japan (DDBJ) is located at the National Institute of Genetics in Mishima, Japan (Tateno *et al.*, 2002). The American sequence database, GenBank, is managed by the National Center for Biotechnology Information (NCBI) (Benson *et al.*, 2002). The NCBI is a division of the National Library of Medicine, located on the campus of the US National Institutes of Health (NIH) in Bethesda, Maryland. Data are independently collected by the three groups and redistributed to the others on a daily basis. Consequently, data producers are free to submit to any of the databases and users of any of the sites are assured of a comprehensive collection. Through regular meetings and communication, the collaborators have achieved uniformity of entry accessioning and versioning, taxonomic nomenclature, and feature annotation. Table 1 provides Internet addresses for data submission and retrieval at the three database sites.

Sequence entries consist of an original DNA sequence, biological source information, literature citations, annotated features, and the protein sequences derived by translation of annotated coding sequences. An example of a typical mRNA sequence entry as it would appear in GenBank or DDBJ is shown in Figure 2, and as it would appear in EMBL in Figure 3. Despite the use of different presentation formats, the underlying data is the same and entries can be freely converted back and forth between the two formats without loss of information. All three databases have authority to assign accession numbers to sequences received. Because entries may later be updated to extend the sequence or to correct errors, identifying a precise sequence of bases requires both an accession number and version number. As normally written, the two would be separated by a dot, for example AY055475.1.

The database is partitioned into divisions, many of which are based on crude taxonomic groups. For example in Figure 2, the sequence from Chinese scorpion *B. martensii* is in the INV (invertebrate) division (indicated

Table 1. Useful Internet Addresses

DDBJ

Home	http://www.ddbj.nig.ac.jp
Submission overview	http://www.ddbj.nig.ac.jp/ddbjing/data-in/dataflow_doc-e.html
Sakura	http://sakura.ddbj.nig.ac.jp
Entry retrieval	http://srs.ddbj.nig.ac.jp/index-e.html
FTP downloads	ftp://ftp.ddbj.nig.ac.jp/dna

EMBL

Home	http://www.ebi.ac.uk/embl
Submission overview	http://www.ebi.ac.uk/embl/Submission
Webin	http://www.ebi.ac.uk/embl/Submission/webin.html
Entry retrieval	http://srs6.ebi.ac.uk
FTP downloads	ftp://ftp.ebi.ac.uk/pub/databases/embl/release

GenBank

Home	http://www.ncbi.nih.gov
Submission overview	http://www.ncbi.nih.gov/Genbank
BankIt	http://www.ncbi.nih.gov/BankIt
Sequin	http://www.ncbi.nih.gov/Sequin
Taxonomy	http://www.ncbi.nih.gov/Taxonomy
Entry retrieval	http://www.ncbi.nih.gov/Entrez
FTP downloads	ftp://ftp.ncbi.nih.gov/genbank

in the top line of the entry). Similarly, *H. sapiens* sequences would be in the PRI (primate) division and those of *M. musculus* would be in the ROD (rodent) division. Other special-purpose divisions are more reflective of the information source or some special property of the data. For example, sequences derived from patent applications are in the PAT division and expressed sequence tag (EST) sequences are in the EST division.

2.2 The GenBank Taxonomy

The portion of the tree of life represented in the sequence databases has been steadily increasing. At the time of this writing, over 120,000 individual species have at least one sequence in GenBank. When looking at any sequence entry, it is clearly of interest to know from which of these many species the sequence was derived and how that organism is related to other forms of life. Furthermore, it is desirable to use a set of controlled organism names that are the same across sequence databases. To serve this purpose, a taxonomic classification system has been developed by the NCBI in collaboration with EBI and DDBJ and with the valuable assistance of external advisors and curators. As such, the GenBank Taxonomy is an amalgam of data from multiple sources and should not be used as a phylogenetic authority.

The GenBank Taxonomy is a molecular-based classification system that contains the names of all organisms that are represented in the sequence database. Each species, as well as each superspecies level (*e.g.* class, phylum, etc.) and each subspecies level (*e.g.* strain) has been given a distinct taxon identifier (in Figure 2, this identifier appears in a /db_xref qualifier of the source feature). Within the database, each taxon (with the exception of the root) has exactly one predecessor given such that the dataset can formally be represented as a tree.

```
LOCUS       AY055475               242 bp    mRNA    linear    INV 16-AUG-2002
DEFINITION  Buthus martensii venom peptide beta-CT (beta-Ct) mRNA, complete
            cds.
ACCESSION   AY055475
VERSION     AY055475.1  GI:22267422
KEYWORDS    .
SOURCE      Chinese scorpion.
  ORGANISM  Buthus martensii
            Eukaryota; Metazoa; Arthropoda; Chelicerata; Arachnida; Scorpiones;
            Buthoidea; Buthidae; Mesobuthus.
REFERENCE   1  (bases 1 to 242)
  AUTHORS   Zeng,X.C. and Wang,S.X.
  TITLE     Evidence that BmTXK beta-BmKCT cDNA from Chinese scorpion Buthus
            martensii Karsch is an artifact generated in the reverse
            transcription process
  JOURNAL   FEBS Lett. 520 (1-3), 183-184 (2002)
  MEDLINE   22040283
   PUBMED   12044895
REFERENCE   2  (bases 1 to 242)
  AUTHORS   Zeng,X.-C., Li,W.-X. and Zhu,S.-Y.
  TITLE     Direct Submission
  JOURNAL   Submitted (10-SEP-2001) Biotechnology Department, Wuhan University,
            Virology Institute, Luojia Street, Wuhan, Hubei 430072, P.R.China
FEATURES             Location/Qualifiers
     source          1..242
                     /organism="Buthus martensii"
                     /db_xref="taxon:34649"
                     /tissue_type="venom gland"
                     /note="authority: Buthus martensii Karsch"
     gene            1..242
                     /gene="beta-Ct"
                     /note="from recombination between BmTXKbeta and BmKCT"
     CDS             35..121
                     /gene="beta-Ct"
                     /note="BmKbeta-CT; no disulfide bridge"
                     /codon_start=1
                     /product="venom peptide beta-CT"
                     /protein_id="AAL24435.1"
                     /db_xref="GI:22267423"
                     /translation="MMKQQFFLFLAVIVMISSVIEAGRGKEM"
BASE COUNT       81 a     33 c     53 g     75 t
ORIGIN
        1 tacaatttta catagccccg aagattatcg gtaaatgatg aaacaacagt tcttcttgtt
       61 cttagcggtg attgtgatga tttcttctgt cattgaggcg ggaagaggca aggaaatgta
      121 gggaatgttg cggaggtatt ggaaaatgct ttggcccaca atgtctgtgt aaccgtatat
      181 gaataattaa aaatgtacac ctgaacagat catttaatga ataataaata ttaataagca
      241 tt
//
```

Figure 2. A Sample Sequence Entry. Sequence AY055475.1 in GenBank format showing the mRNA sequence of *B. martensii* venom peptide beta-CT.

A web-based interface allows users to browse through the tree and, optionally show the numbers of sequence entries that are present for each level. For example, suppose one was interested in knowing how many sea

```
ID    AY055475    standard; RNA; INV; 242 BP.
XX
AC    AY055475;
XX
SV    AY055475.1
XX
DT    16-AUG-2002 (Rel. 72, Created)
DT    16-AUG-2002 (Rel. 72, Last updated, Version 1)
XX
DE    Buthus martensii venom peptide beta-CT (beta-Ct) mRNA, complete cds.
XX
KW    .
XX
OS    Buthus martensii (Chinese scorpion)
OC    Eukaryota; Metazoa; Arthropoda; Chelicerata; Arachnida; Scorpiones;
OC    Buthoidea; Buthidae; Mesobuthus.
XX
RN    [1]
RP    1-242
RX    MEDLINE; 22040283.
RX    PUBMED; 12044895.
RA    Zeng X.C., Wang S.X.;
RT    "Evidence that BmTXK beta-BmKCT cDNA from Chinese scorpion Buthus martensii
RT    Karsch is an artifact generated in the reverse transcription process";
RL    FEBS Lett. 520(1-3):183-184(2002).
XX
RN    [2]
RP    1-242
RA    Zeng X.-C., Li W.-X., Zhu S.-Y.;
RT    ;
RL    Submitted (10-SEP-2001) to the EMBL/GenBank/DDBJ databases.
RL    Biotechnology Department, Wuhan University, Virology Institute, Luojia
RL    Street, Wuhan, Hubei 430072, P.R.China
XX
FH    Key             Location/Qualifiers
FH
FT    source          1..242
FT                    /db_xref="taxon:34649"
FT                    /note="authority: Buthus martensii Karsch"
FT                    /organism="Buthus martensii"
FT                    /tissue_type="venom gland"
FT    CDS             35..121
FT                    /codon_start=1
FT                    /note="BmKbeta-CT; no disulfide bridge"
FT                    /gene="beta-Ct"
FT                    /product="venom peptide beta-CT"
FT                    /protein_id="AAL24435.1"
FT                    /translation="MMKQQFFLFLAVIVMISSVIEAGRGKEM"
XX
SQ    Sequence 242 BP; 81 A; 33 C; 53 G; 75 T; 0 other;
      tacaattttta catagccccg aagattatcg gtaaatgatg aaacaacagt tcttcttgtt        60
      cttagcggtg attgtgatga tttcttctgt cattgaggcg ggaagaggca aggaaatgta       120
      gggaatgttg cggaggtatt ggaaaatgct ttggcccaca atgtctgtgt aaccgtatat       180
      gaataattaa aaatgtacac ctgaacagat catttaatga ataataaata ttaataagca       240
      tt                                                                      242
//
```

Figure 3. A Sample Sequence Entry. Sequence AY055475.1 in EMBL format showing the mRNA sequence of *B.martensii* venom peptide beta-CT.

urchin sequences have been generated. Figure 4 shows some excerpts from a session in the Taxonomy Browser attempting to explore this question. *Echinoidea* is a broad term for sea urchins that includes many species, which at the time of the search, collectively had 89,170 sequence entries in GenBank. As it happens, nearly all of the work has been done on one particular species of sea urchin. To see which one, the user would successively follow the branches of the tree that have the largest numbers of sequences, first to *Echinacea,* then to *Strongylocentrotidae,* and finally to the purple urchin *Strongylocentrotus purpuratus* (in this example, stepping three levels at a time because this is the default number of levels shown in the browser).

Each taxonomic node has a number of attributes. One that is particularly important for computational analysis is the genetic code that should be used for protein translation. Other attributes include various aliases that may come from retired taxonomic names, common names, and common misspellings. For example, the puffer fish *Takifugu rubripes* was previously known as *Fugu rubripes*, a name which is still widely used among researchers. The use of aliases allow sequences to be retrieved using either the old or the new names.

2.3 Sequence Annotation

The dictionary definition of annotation is the addition of critical commentary or explanatory notes to a body of text, for example that of a literary work. Sequence annotation involves the attachment of explanatory notes—and more structured information—to specific coordinates on a sequence. Each sequence feature consists of a feature key, a location, and any number of qualifiers. An example from Figure 2 is a source feature, attached to location 1.242 (the entire length of the sequence), and carrying (among others) a qualifier /tissue_type that has "venom gland" as its value. The source feature with an /organism qualifier is mandatory for all sequence entries.

Arguably, one of the most important feature types is CDS, the coding sequence. The bases identified by the feature location can be translated to produce the sequence of the protein product. The fact that some species use non-standard genetic codes illustrates the importance of knowing the taxonomic origin of the sequence. Unless a CDS feature is flagged as being partial, it should begin with a start codon, end with a stop codon, and have no internal stops. Useful qualifiers on the CDS feature are /translation, which gives the protein sequence, and /protein_id, which allows the entire annotated protein entry to be fetched. In some rare cases, the amino acid

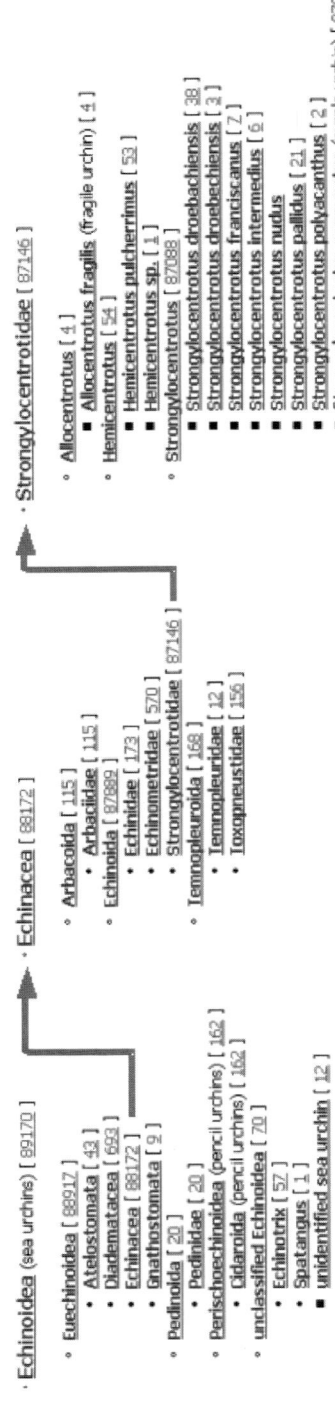

Figure 4. Browsing the Taxonomy. Portions of the display pages from a TaxBrowser session exploring the sea urchin lineage. By default, three taxonomic levels are shown per page. By enabling a display option, the numbers of nucleotide sequence from each taxon are shown in brackets. Arrows indicate the names that were clicked during the session.

sequence given in /translation may differ from what is obtained by translation. Examples would include proteins containing selenocystein residues or those translated with the aid of suppressor tRNAs. It is possible to add a "translation exception" to CDS features in order to document such phenomena.

The sequence databases maintain a master list of valid feature keys, their meanings, and the qualifiers that are allowed for each. Table 2 shows a selected list of the more commonly used feature keys. For a complete listing plus additional details, the reader is referred to online feature table documentation (www.ncbi.nih.gov/projects/collab/FT).

2.4 Data Submission

At the time when the sequence databases were first established, significant effort was expended by database staff in scanning journals and keying in sequences from figures. Clearly, this was a rather time-consuming and error-prone process and did not scale with the exponential increase in the production of sequence data. Moreover, it did not take long before journals stopped devoting space to sequences. Instead, virtually all journals now require that authors deposit sequences directly into the database and that the accession numbers be listed in the article.

With the responsibility for submission shifting to the authors, it was crucial that the databases provide tools for entry preparation. One option is the use of a standalone program developed at the NCBI called Sequin, which interactively guides the user through the process and produces a file that can be sent to the database (Kans and Ouellette, 2001). In addition, each of the database web sites provides a form-based facility for submitting data (Webin for EMBL, Sakura for DDBJ, and BankIt for NCBI). Submitted sequences are subjected to a number of automated tests such as checking for contamination with cloning vector and verifying that the CDS can be translated. In addition, entries are manually reviewed by trained indexers who may contact submitters in order to resolve any questions that may arise.

The interactive approaches described above work well for occasional submitters but would be completely impractical for a genome center producing thousands of sequence entries per day. To address this need, special bulk submission databases have been established, such as dbEST, dbSTS, and dbGSS. Centers are urged to contact the database staff to discuss the best solution. For example, in one case programmers at a sequencing center wrote a program using function libraries provided by the

Table 2. Commonly Used Sequence Features

Feature key	Description
CDS	Coding sequence. Sequence of bases that are translated to make the primary protein product (location includes stop codon).
exon	Range of contiguous bases that are retained in the spliced mature transcript (may include untranslated regions).
gene	Region of biological interest identified as a gene and for which a name has been assigned.
intron	Bases that are transcribed but removed from the mature transcript by splicing.
mat_peptide	Subsequence of the CDS that corresponds to the mature peptide following from proteolytic cleavage (location does not include stop codon).
misc_binding	Site which covalently or non-covalently binds another moiety that cannot be described by any other feature key.
misc_feature	Region of interest which cannot be described by another feature key.
mRNA	DNA sequence corresponding to a mature mRNA (includes both untranslated and coding sequences, but not introns).
polyA_signal	Signal required for polyadenylation of an mRNA (most commonly AATAAA).
polyA_site	Site on an mRNA to which the poly(A) tail is added.
primer_bind	Binding sites for oligonucleotide primers. Often used to annotate PCR or overgo primers.
promoter	Region on a DNA molecule involved in RNA polymerase binding to initiate transcription.
protein_bind	Non-covalent protein binding site on nucleic acid.
repeat_region	Region of genome containing repeating units.
repeat_unit	Single repeat element.
rRNA	DNA sequence corresponding to a ribosomal RNA.
scRNA	DNA sequence corresponding to a small cytoplasmic RNA.
sig_peptide	Signal peptide coding sequence involved in attaching nascent polypeptide to the membrane leader sequence.
snRNA	DNA sequence corresponding to a small nuclear RNA.
snoRNA	DNA sequence corresponding to a small nucleolar RNA.
source	Identifies the biological source of sequenced material (this is a mandatory feature). In addition to taxonomic species, attributes such as strain, chromosome, and map position may be indicated.
STS	Sequence tagged site. A short, single-copy DNA sequence detected by PCR and used as a mapping landmark.
tRNA	DNA sequence corresponding to a transfer RNA.

Continued

unsure	Submitter is unsure of exact sequence in this region.
variation	Site(s) at which some other strain or haplotype contains stable differences compared to the presented sequence.

For a complete listing of allowed feature keys, together with required and optional qualifiers, see http://www.ncbi.nih.gov/projects/collab/FT.

database in order to dump sequences directly from their relational database into the required format.

3. Genomic Sequences

3.1 Sequencing Strategies

All genome sequencing projects being undertaken today involve random shotgun sequencing at some level. In the whole genome shotgun (WGS) approach, the genome is assembled (or "built") from sequencing reads generated from across the entire genome. This strategy has been used in the sequencing of most small genomes, such as bacteria. The clone-by-clone approach, also known as hierarchical shotgun, was pioneered for the sequencing of medium-sized genomes such as *C.elegans*. Under this scenario, a clone map of the genome is generated, shotgun and finishing is undertaken for each individual clone, and finally the genome is assembled from the overlapping clone sequences. For larger genomes such as *H.sapiens*, the choice of strategy was hotly debated (Green, 1997; Weber and Myers, 1997) and both approaches were attempted. For several genome projects currently underway, such as *M.musculus* and *R.norvegicus*, different combinations of the two approaches have been devised.

3.2 HTG Sequence Entries

Genome sequencing centers work closely with GenBank, EMBL, and DDBJ to ensure smooth submission of high throughput genomic (HTG) sequences. Early in the course of the Human Genome Project, the participants agreed that all sequence assemblies of at least 2 kb should be released to the public immediately because of their critical importance to the advancement of science (this is often called the Bermuda Agreement after the site of the meeting at which it was crafted). Consequently, database groups developed procedures for representing unfinished clone entries which may consist of multiple sequence contigs separated by gaps. Each of these entries carries the keyword HTG plus any number of additional keywords have been used to detail the status of unfinished clones (see Table 3). In order to make effective use of the data, it is important for users to understand the meaning of these keywords because they have a bearing on data quality (Jang *et al.*, 1999). For example, HTGS_DRAFT indicates a level of quality that was adopted for the draft stage of the Human Genome Project in which the sequence of a clone was assembled from roughly half the number of shotgun reads as would normally be used for a finished sequence (4-5x redundancy, also known as "half shotgun"). At this level of coverage, the reads may fall together into about 10-40 sequence contigs, typically without ordering information. In contrast, entries carrying the HTGS_FULLTOP keyword have about twice as many reads and typically assemble into 1-4 contigs. To maintain a clear distinction, unfinished sequences are segregated into the HTG database division, whereas finished entries are included in the traditional division for the species (PRI for *H. sapiens,* ROD for *M. musculus*, and so forth). As a clone progresses though different stages of completion and is eventually regarded as finished, the keywords and division will change from time to time, but the accession number remains unchanged.

An accessory database that has been used in clone-by-clone projects is the Clone Registry, a system developed at the NCBI to store information about BACs and PACs being used in genome sequencing projects. Sequencing centers have the ability to register their intent to sequence a clone, the fact that it is in the pipeline, or the accession number of a sequence that has been deposited. In addition, connections to other types of clone-oriented information, such as BAC end-sequences or BAC fingerprint contigs, have been included. Results from large-scale FISH analysis of BACs (Cheung *et al.*, 2001) have been added to provide connections to cytogenetic resources. Associations between clones and mapped markers, mostly derived by PCR or overgo screening of BAC libraries, provide additional mapping data.

3.3 WGS Assemblies

An increasing number of larger genomes are being attempted using a WGS approach. Just as with clone-based sequences, the research community benefits from having data made available in unfinished form and updated periodically as it becomes more complete. However, many of the conventions that have been used for clone simply don't work for WGS assemblies. For example, there is no entity (like a clone) that would be stable from assembly to assembly and would be the natural choice to carry the accession number.

Table 3. HTGS Keywords

Keyword	Description
HTG	Keyword present on all high-throughput genomic sequencing projects.
HTGS_PHASE0	Phase-0 sequence. Unassembled reads (simply concatenated end-to-end) used to test the quality of a shotgun library or determine overlaps with other clones.
HTGS_PHASE1	Phase-1 sequence. Unfinished sequence with unordered sequence contigs.
HTGS_PHASE2	Phase-2 sequence. Unfinished sequence in which the contigs have been ordered and oriented.
HTGS_DRAFT	A project assembled from roughly half the number of reads (4-5x redundancy) that would be used in a finished project.
HTGS_FULLTOP	A project with the full depth of reads normally produced for the shotgun phase (8-10x).
HTGS_ENRICHED	A project assembled from some clone reads plus some WGS reads recruited by sequence matching.
HTGS_ACTIVEFIN	Active project in the finishing process.
HTGS_CANCELLED	Project no longer being sequenced. Common reasons for cancellation are that the clone is redundant with others already finished or is thought to be rearranged.

Instead, accession numbers must be given to a large number of sequence contigs. However, it is difficult (and possibly meaningless) to track their history from assembly to assembly. Another difference is that, whereas clones are sequenced independently of one another, the contigs of a WGS assembly are generated over the whole genome at once and it would be useful to group together and retrieve contigs derived from the same assembly. To address these needs, the databases have recently established procedures for accessioning WGS assemblies, which have so far been applied to assemblies of *O. sativa*, *A. gambiae*, and *M. musculus*.

WGS accession numbers consist of 12 letters and digits that can be divided into several fields. The first four letters use a code assigned to each sequencing project. This is followed by two digits defining the revision of the assembly and six digits to identify the individual contigs. For example, the Beijing Genomics Institute project to sequence the genome of *O.sativa* indica cultivar (Yu *et al.*, 2002) was given the abbreviation AAAA (simply because it was the first to be processed in this system). It is still in its initial version, which consists of 103044 sequence contigs having accession numbers AAAA01000001 through AAAA01103044. If this assembly were to be updated, the digits identifying the build would change from 01 to 02 and the contigs would be renumbered, starting again from 1.

3.4 The Trace Archive

One disadvantage of the WGS approach is that an assembly may not even be attempted until a significant portion of the reads have been generated. The Trace Archive was established to provide public access to the raw sequencing traces, together with base calls, quality scores, and additional ancillary information. Although clearly not as desirable as a genomic assembly, traces have a variety of uses, such as identifying exon/intron boundaries by comparison with a cDNA sequence or discovering putative SNPs. Having a permanent archive of the original data generated by a sequencing project additionally allows for reassembly at some future point in order to take advantage of improvements in assembly software. The Trace Archive was established to support the mouse genome sequencing project and quickly expanded to include traces from several additional species. At the time of this writing, more than 200 million traces have been submitted. Moreover, although originally intended for WGS reads, the Trace Archive can also accommodate traces from other types of large-scale sequencing projects, such as BAC-ends and ESTs.

The Trace Archive is administered jointly by the NCBI (www.ncbi.nih. gov/Traces) and Ensembl (trace.ensembl.org). Detailed data submission procedures may be found at either of the two web sites. In general, traces may be submitted either by FTP or on high-density tapes and several file formats are supported (SCF is the most commonly used). Data collected at either site are mirrored to the other. Submissions must also include an ancillary data file with a number of essential pieces of information not provided by the trace file itself, such as the trace type (WGS, BAC-end, EST, etc.) and the organism that was the source of the material. Ancillary fields that are particularly valuable for WGS reads are the plasmid library name, the mean and the standard deviation of insert size for that library, the template name, and template end (forward or reverse). These data are used by WGS assemblers to form connections among sequence contigs and impose distance constraints. Upon receipt, the database assigns a unique trace identifier, the TI, that can be used as a retrieval key. Both the NCBI and Ensembl web sites provide multiple ways to access the data. Sequence similarity searching can be performed to find traces matching a query sequence provided by the user. As shown in Figure 5, a web-based interface several graphical and textual views of traces and related information. Finally, trace data may be downloaded in bulk for assembly or analysis on the user's local computer.

4. Transcribed Sequences

4.1 Database of Expressed Sequence Tags (dbEST)

For most species, large-scale sequencing of cDNAs preceded genomic sequencing by several years. The most common approach involves the generation of "expressed sequence tags" (ESTs), a term that was coined in a landmark study from Venter and his colleagues (Adams *et al.*, 1991). The basic strategy involves selecting cDNA clones at random and performing single-pass sequencing reads from one or both ends of their inserts. There is no initial attempt to identify or characterize the clones and it is fully expected that many clones will be redundant with others already sampled. The resulting sequences are short (around 500 bp) and relatively inaccurate (about 2% error). Nevertheless, ESTs were found to be an invaluable resource for the discovery of new genes, particularly those involved in human disease processes (Sikela and Auffray, 1993; Boguski *et al.*, 1994).

Following the initial demonstration of the utility and cost-effectiveness of the EST approach, many similar projects were initiated, resulting in an ever-increasing numbers of human ESTs. In 1992, a database called dbEST was established (Boguski *et al.*, 1993) to serve as a collection point for ESTs, which are then distributed to the scientific community as the EST division of GenBank. The fact that this division accounts for approximately two-thirds of all sequences in the database is a testament to the popularity of the EST approach. One appealing feature of many EST projects is that the physical clones are archived and made available to the community. Thus, it is possible to use sequence database searching as a means of identifying clones that may be used for experimental purposes.

4.2 Full-Insert cDNA Sequencing

As useful as ESTs are, they are no substitute for having the complete cDNA sequence. In the last few years, a number of projects aiming to produce full-length cDNA sequences have been launched (Strausberg *et al.*, 1999; Kawai *et al.*, 2001; Wiemann *et al.*, 2001; Seki *et al.*, 2002; Stapleton *et al.*, 2002). One strategy is to construct cDNA libraries using methods designed to enrich for cDNAs with 5'termini. In other cases, libraries are size-selected to favor longer cDNAs. Yet another approach is to use completely standard library construction methods and to rely on EST analysis to identify clones likely to be full-length. However, despite all of these measures, not all of the clones chosen for sequencing will be truly full-length. For this reason, we will refer to these sequences as full-insert cDNA (FI-cDNA) sequences. Many of them have been deposited into the newly created HTC (high-throughput cDNA) division.

4.3 UniGene

The sheer number of cDNA sequences in the databases is extraordinary, indeed for most organisms much larger than the number of genes. A major

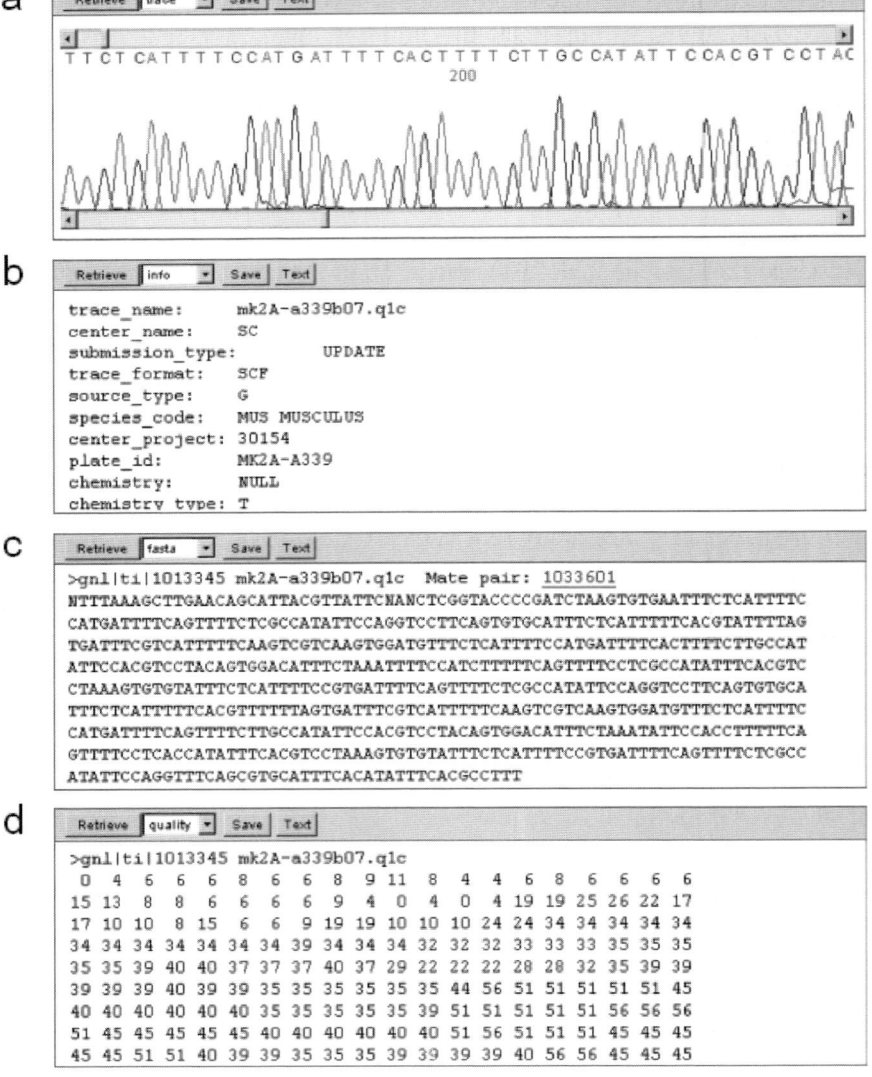

Figure 5. Trace Archive. Views of the trace identified by TI 1013345 showing (a) the chromatograph, (b) ancillary information, (c) base calls, and (d) quality scores.

goal is to reduce the obvious redundancy inherent in this dataset. Computationally, this is a clustering problem in which the sequences are vertices that may be coalesced into clusters representing genes by establishing connections among them.

UniGene is a system developed at the NCBI for clustering transcribed sequences through largely automated means (Schuler, 1997). Sequences included in the computation include ESTs, FI-cDNAs, and other well characterized mRNA sequences found in GenBank. Experience has shown that it is particularly important to eliminate low-quality and apparently artifactual sequences prior to clustering because even a small level of noise can have a large corrupting effect on result. Thus, procedures are in place to eliminate sequences of foreign origin (most commonly *E.coli*) and identify regions that are derived from the cloning vector or artificial primers or linker. At present, UniGene focuses on protein-coding genes of the nuclear genome, so those identified as rRNA or mitochondrial sequence are eliminated. The most common artifacts among these sequences are chimerism, incomplete splicing, and internal priming. Multiple approaches have been devised to identify these cases and lists of sequences for exclusion from UniGene are maintained.

Given a set of sequences, a variety of different sources of information may be used as evidence that any pair of them are or are not derived from the same gene. For well studied sequences, it is useful to make use of locus identifiers because they are based on expert judgment. Sequences with the same clone identifier often represent the two ends of the same cDNA, but clone tracking errors can be a significant issue. Strong sequence similarity provides the most direct evidence of two sequences being derived from the same gene. However, one dilemma is that the level of sequencing error that must be tolerated rivals the level of divergence among gene family members. Values of specific parameters governing acceptable sequence alignments are chosen by examining ratios of true to false connections in curated test sets.

5. Genome Annotation

5.1 Reference Sequences

A number of strategies for automated genome annotation have been devised and improvement of these methods remains an active area of research. Ultimately, experimental validation of all annotation is needed, but this remains an expensive and time-consuming proposition. In order to compare the output of different annotation methods and capture experimental results, a common coordinate system is needed in the form of a stable reference sequence.

RefSeq is an NCBI project aimed at collecting a comprehensive set of reference DNA, RNA, and protein sequences for all organisms (Pruitt and Maglott, 2001). The collection currently contains genomic sequences for over 1000 viruses, 83 bacteria, 16 archaea, and 8 eukaryotes (as of August, 2002). RefSeq staff interact with data producers on the preparation and maintenance of sequence records and, in some cases, the annotation. In the case of the human genome, the NCBI has collaborated with the sequencing centers on the maintenance of several finished chromosomes and the assembly of the remaining draft chromosomes. These reference sequences are used as a substrate for annotation by the University of California at Santa Cruz (UCSC) and Ensembl, in addition to annotation performed at the NCBI. Each of these three sites also maintains a browser for the genomic information.

5.2 Human Genome Browsers

Scientists at the UCSC have been heavily involved in collaborative projects to analyze the human and mouse genomes. The UCSC Genome Browser was initially developed for data exchange among the project participants and has come to be very widely used by the research community as a whole (Kent *et al.*, 2002). About half of the tracks that may be viewed are computed at the UCSC, including alignments of mRNA sequences, the output of gene prediction programs, and the results of cross-species genome comparisons. Additional tracks have been contributed by external scientists. In Figure 6, a small number of these tracks are shown for the CFTR gene. At present, the genomes of *H. sapiens* and *M. musculus* may be viewed and more species are planned for the future.

The genome database produced by Ensembl, a joint project of the EBI and the Sanger Institute, provides consistent annotation and online browsing for several large genomes (Hubbard *et al.*, 2002). By the middle of 2002, the list of genomes included *H. sapiens*, *M. musculus*, *T. rubripes,* and *A.gambiae*. The Ensembl gene prediction pipeline incorporates a wide range of methods, including *ab initio* gene predictions, homology, and gene prediction HMMs. A view of the CFTR gene in the Ensembl browser is shown in Figure 7. As with other browsers, there are many tracks that may be independently turned on or off. Some features have hypertext links to genome resources maintained elsewhere.

The NCBI MapViewer presents a large number of genomes for which there are reference genomic sequences (see above) as well as other genomes (several plants, for example) that are represented only by STS maps (Wheeler *et al.*, 2002). An example showing the CFTR locus is shown in Figure 8. The NCBI gene prediction pipeline relies heavily on alignment of cDNAs, including fragmentary ESTs whose alignments must be chained to form longer gene models. Transcript-based models are augmented with GenomeScan model in regions for which there was no mRNA-based gene. Additional types of features such as SNPs, STS markers, and FISH-mapped BAC clones are also shown.

6. Concluding Remarks

A public archive of all sequence data, whether generated for a small question-oriented study or a large infrastructure-building project, is critical resource for accelerating the pace of scientific investigation. One reason that sequence data is particularly valuable, compared to other types of information, is that they are amenable to many forms of computational analysis. The analysis of sequences has provided the answers to many questions which no one would have thought to ask at the time the sequences were originally generated. Although it is important to capture all public sequence data, the sheer size of the database is making it increasingly unwieldy and there is every reason to expect the exponential rate of growth to continue well into the future. Similarly, genome browsers are experiencing data overload problems as the addition of more

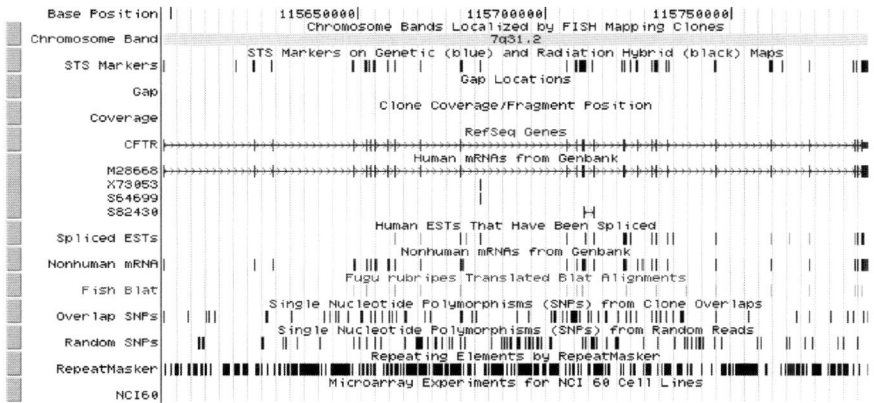

Figure 6. The UCSC Genome Browser. The CFTR gene as it appears in the UCSC genome browser

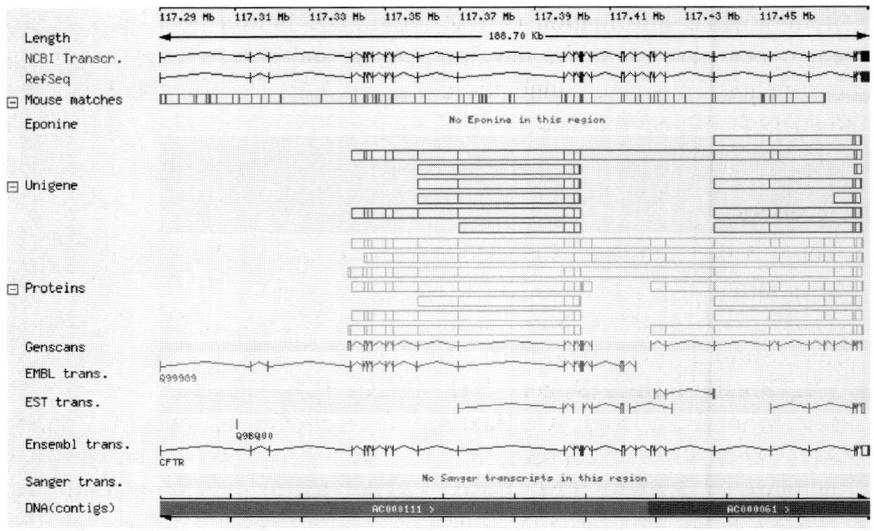

Figure 7. The Ensembl Genome Browser. The CFTR gene as it appears in the Ensembl genome browser.

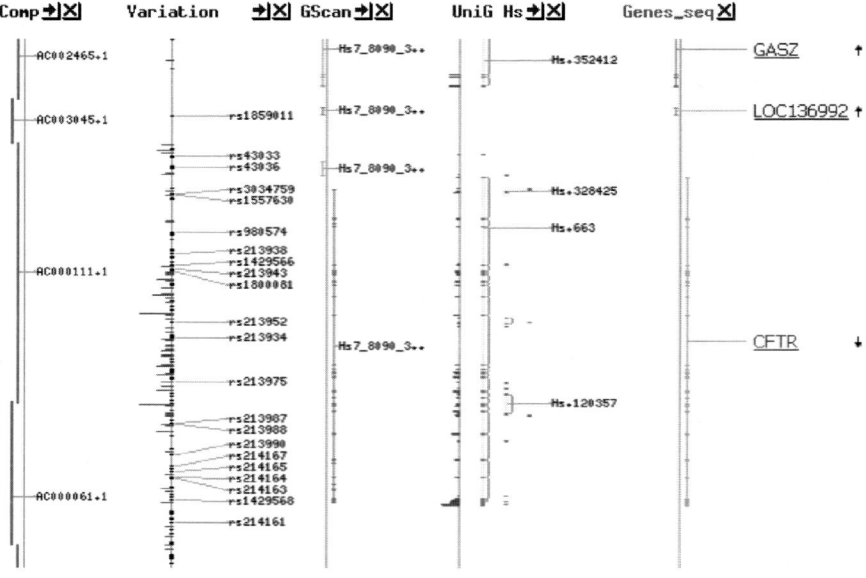

Figure 8. The NCBI MapViewer. The CFTR gene as it appear in the NCBI MapViewer.

annotation tracks is having the counterintuitive effect of making it more difficult to find information of interest. In the future, additional efforts need to be focused on the development of stable reference sequences, experimentally validated annotation, and software for effective searching and analysis of genome data.

References

Adams, M.D., Celniker, S.E., Holt, R.A., Evans, C.A., Gocayne, J.D., Amanatides, P.G., Scherer, S.E., Li, P.W., Hoskins, R.A., Galle, R.F., George, R.A., Lewis, S.E., Richards, S., Ashburner, M., Henderson, S.N., Sutton, G.G., Wortman, J.R., Yandell, M.D., Zhang, Q., Chen, L.X., Brandon, R.C., Rogers, Y.H., Blazej, R.G., Champe, M., Pfeiffer, B.D., Wan, K.H., Doyle, C., Baxter, E.G., Helt, G., Nelson, C.R., Gabor, G.L., Abril, J.F., Agbayani, A., An, H.J., Andrews-Pfannkoch, C., Baldwin, D., Ballew, R.M., Basu, A., Baxendale, J., Bayraktaroglu, L., Beasley, E.M., Beeson, K.Y., Benos, P.V., Berman, B.P., Bhandari, D., Bolshakov, S., Borkova, D., Botchan, M.R., Bouck, J., Brokstein, P., Brottier, P., Burtis, K.C., Busam, D.A., Butler, H., Cadieu, E., Center, A., Chandra, I., Cherry, J.M., Cawley, S., Dahlke, C., Davenport, L.B., Davies, P., de Pablos, B., Delcher, A., Deng, Z., Mays, A.D., Dew, I., Dietz, S.M., Dodson, K., Doup, L.E., Downes, M., Dugan-Rocha, S., Dunkov, B.C., Dunn, P., Durbin, K.J., Evangelista, C.C., Ferraz, C., Ferriera, S., Fleischmann, W., Fosler, C., Gabrielian, A.E., Garg, N.S., Gelbart, W.M., Glasser, K., Glodek, A., Gong, F., Gorrell, J.H., Gu, Z., Guan, P., Harris, M., Harris, N.L., Harvey, D., Heiman, T.J., Hernandez, J.R., Houck, J., Hostin, D., Houston, K.A., Howland, T.J., Wei, M.H., Ibegwam, C., Jalali, M., Kalush, F., Karpen, G.H., Ke, Z., Kennison, J.A., Ketchum, K.A., Kimmel, B.E., Kodira, C.D., Kraft, C., Kravitz, S., Kulp, D., Lai, Z., Lasko, P., Lei, Y., Levitsky, A.A., Li, J., Li, Z., Liang, Y., Lin, X., Liu, X., Mattei, B., McIntosh, T.C., McLeod, M.P., McPherson, D., Merkulov, G., Milshina, N.V., Mobarry, C., Morris, J., Moshrefi, A., Mount, S.M., Moy, M., Murphy, B., Murphy, L., Muzny, D.M., Nelson, D.L., Nelson, D.R., Nelson, K.A., Nixon, K., Nusskern, D.R., Pacleb, J.M., Palazzolo, M., Pittman, G.S., Pan, S., Pollard, J., Puri, V., Reese, M.G., Reinert, K., Remington, K., Saunders, R.D., Scheeler, F., Shen, H., Shue, B.C., Siden-Kiamos, I., Simpson, M., Skupski, M.P., Smith, T., Spier, E., Spradling, A.C., Stapleton, M., Strong, R., Sun, E., Svirskas, R., Tector, C., Turner, R., Venter, E., Wang, A.H., Wang, X., Wang, Z.Y., Wassarman, D.A., Weinstock, G.M., Weissenbach, J., Williams, S.M., WoodageT, Worley, K.C., Wu, D., Yang, S., Yao, Q.A., Ye, J., Yeh, R.F., Zaveri, J.S., Zhan, M., Zhang, G.,

Zhao, Q., Zheng, L., Zheng, X.H., Zhong, F.N., Zhong, W., Zhou, X., Zhu, S., Zhu, X., Smith, H.O., Gibbs, R.A., Myers, E.W., Rubin, G.M., and Venter, J.C. 2000. The genome sequence of Drosophila melanogaster. Science. 287: 2185–95.

Adams, M.D., Kelley, J.M., Gocayne, J.D., Dubnick, M., Polymeropoulos, M.H., Xiao, H., Merril, C.R., Wu, A., Olde, B., Moreno, R.F., *et al.* 1991. Complementary DNA sequencing: expressed sequence tags and human genome project. Science. 252: 1651–6.

Benson, D.A., Karsch-Mizrachi, I., Lipman, D.J., Ostell, J., Rapp, B.A., and Wheeler, D.L. 2002. GenBank. Nucleic Acids Res. 30: 17–20.

Boguski, M.S., Lowe, T.M., and Tolstoshev, C.M. 1993. dbEST—database for "expressed sequence tags". Nature Genet. 4: 332–333.

Boguski, M.S., Tolstoshev, C.M., and Bassett, D.E., Jr. 1994. Gene discovery in dbEST. Science. 265: 1993–4.

Cheung, V.G., Nowak, N., Jang, W., Kirsch, I.R., Zhao, S., Chen, X.N., Furey, T.S., Kim, U.J., Kuo, W.L., Olivier, M., Conroy, J., Kasprzyk, A., Massa, H., Yonescu, R., Sait, S., Thoreen, C., Snijders, A., Lemyre, E., Bailey, J.A., Bruzel, A., Burrill, W.D., Clegg, S.M., Collins, S., Dhami, P., Friedman, C., Han, C.S., Herrick, S., Lee, J., Ligon, A.H., Lowry, S., Morley, M., Narasimhan, S., Osoegawa, K., Peng, Z., Plajzer-Frick, I., Quade, B.J., Scott, D., Sirotkin, K., Thorpe, A.A., Gray, J.W., Hudson, J., Pinkel, D., Ried, T., Rowen, L., Shen-Ong, G.L., Strausberg, R.L., Birney, E., Callen, D.F., Cheng, J.F., Cox, D.R., Doggett, N.A., Carter, N.P., Eichler, E.E., Haussler, D., Korenberg, J.R., Morton, C.C., Albertson, D., Schuler, G., de Jong, P.J., and Trask, B.J. 2001. Integration of cytogenetic landmarks into the draft sequence of the human genome. Nature. 409: 953–8.

Fleischmann, R.D., Adams, M.D., White, O., Clayton, R.A., Kirkness, E.F., Kerlavage, A.R., Bult, C.J., Tomb, J.F., Dougherty, B.A., Merrick, J.M., *et al.* 1995. Whole-genome random sequencing and assembly of Haemophilus influenzae Rd. Science. 269: 496–512.

Goffeau, A., Barrell, B.G., Bussey, H., Davis, R.W., Dujon, B., Feldmann, H., Galibert, F., Hoheisel, J.D., Jacq, C., Johnston, M., Louis, E.J., Mewes, H.W., Murakami, Y., Philippsen, P., Tettelin, H., and Oliver, S.G. 1996. Life with 6000 genes. Science. 274: 546, 563–7.

Green, P. 1997. Against a whole-genome shotgun. Genome Res. 7: 410–7.

Hubbard, T., Barker, D., Birney, E., Cameron, G., Chen, Y., Clark, L., Cox, T., Cuff, J., Curwen, V., Down, T., Durbin, R., Eyras, E., Gilbert, J., Hammond, M., Huminiecki, L., Kasprzyk, A., Lehvaslaiho, H., Lijnzaad, P., Melsopp, C., Mongin, E., Pettett, R., Pocock, M., Potter, S.,

Rust, A., Schmidt, E., Searle, S., Slater, G., Smith, J., Spooner, W., Stabenau, A., Stalker, J., Stupka, E., Ureta-Vidal, A., Vastrik, I., and Clamp, M. 2002. The Ensembl genome database project. Nucleic Acids Res. 30: 38–41.

Jang, W., Chen, H.C., Sicotte, H., and Schuler, G.D. 1999. Making effective use of human genomic sequence data. Trends Genet. 15: 284–6.

Kans, J.A., and Ouellette, B.F. 2001. Submitting DNA sequences to the databases. Methods Biochem Anal. 43: 65–81.

Kawai, J., Shinagawa, A., Shibata, K., Yoshino, M., Itoh, M., Ishii, Y., Arakawa, T., Hara, A., Fukunishi, Y., Konno, H., Adachi, J., Fukuda, S., Aizawa, K., Izawa, M., Nishi, K., Kiyosawa, H., Kondo, S., Yamanaka, I., Saito, T., Okazaki, Y., Gojobori, T., Bono, H., Kasukawa, T., Saito, R., Kadota, K., Matsuda, H., Ashburner, M., Batalov, S., Casavant, T., Fleischmann, W., Gaasterland, T., Gissi, C., King, B., Kochiwa, H., Kuehl, P., Lewis, S., Matsuo, Y., Nikaido, I., Pesole, G., Quackenbush, J., Schriml, L.M., Staubli, F., Suzuki, R., Tomita, M., Wagner, L., Washio, T., Sakai, K., Okido, T., Furuno, M., Aono, H., Baldarelli, R., Barsh, G., Blake, J., Boffelli, D., Bojunga, N., Carninci, P., de Bonaldo, M.F., Brownstein, M.J., Bult, C., Fletcher, C., Fujita, M., Gariboldi, M., Gustincich, S., Hill, D., Hofmann, M., Hume, D.A., Kamiya, M., Lee, N.H., Lyons, P., Marchionni, L., Mashima, J., Mazzarelli, J., Mombaerts, P., Nordone, P., Ring, B., Ringwald, M., Rodriguez, I., Sakamoto, N., Sasaki, H., Sato, K., Schonbach, C., Seya, T., Shibata, Y., Storch, K.F., Suzuki, H., Toyo-oka, K., Wang, K.H., Weitz, C., Whittaker, C., Wilming, L., Wynshaw-Boris, A., Yoshida, K., Hasegawa, Y., Kawaji, H., Kohtsuki, S., and Hayashizaki, Y. 2001. Functional annotation of a full-length mouse cDNA collection. Nature. 409: 685–90.

Kent, W.J., Sugnet, C.W., Furey, T.S., Roskin, K.M., Pringle, T.H., Zahler, A.M., and Haussler, D. 2002. The human genome browser at UCSC. Genome Res. 12: 996–1006.

Maxam, A.M., and Gilbert, W. 1977. A new method for sequencing DNA. Proc Natl Acad Sci U S A. 74: 560–4.

Pruitt, K.D., and Maglott, D.R. 2001. RefSeq and LocusLink: NCBI gene-centered resources. Nucleic Acids Res. 29: 137–40.

Sanger, F., Air, G.M., Barrell, B.G., Brown, N.L., Coulson, A.R., Fiddes, C.A., Hutchison, C.A., Slocombe, P.M., and Smith, M. 1977b. Nucliotide sequence of bacteriophage phi X174 DNA. Nature. 265: 687–95.

Sanger, F., Nicklen, S., and Coulson, A.R. 1977a. DNA sequencing with chain-terminating inhibitors. Proc Natl Acad Sci U S A. 74: 5463–7.

Schuler, G.D. 1997. Pieces of the puzzle: expressed sequence tags and the catalog of human genes. J Mol Med. 75: 694–8.

Seki, M., Narusaka, M., Kamiya, A., Ishida, J., Satou, M., Sakurai, T., Nakajima, M., Enju, A., Akiyama, K., Oono, Y., Muramatsu, M., Hayashizaki, Y., Kawai, J., Carninci, P., Itoh, M., Ishii, Y., Arakawa, T., Shibata, K., Shinagawa, A., and Shinozaki, K. 2002. Functional annotation of a full-length Arabidopsis cDNA collection. Science. 296: 141–5.

Sikela, J.M., and Auffray, C. 1993. Finding new genes faster than ever. Nat Genet. 3: 189–91.

Stapleton, M., Liao, G., Brokstein, P., Hong, L., Carninci, P., Shiraki, T., Hayashizaki,Y., Champe, M., Pacleb, J., Wan, K., Yu, C., Carlson, J., George, R., Celniker, S., and Rubin, G.M. 2002. The Drosophila Gene Collection: Identification of Putative Full-Length cDNAs for 70% of D. melanogaster Genes. Genome Res. 12: 1294–300.

Stoesser, G., Baker, W., van den Broek, A., Camon, E., Garcia-Pastor, M., Kanz, C., Kulikova, T., Leinonen, R., Lin, Q., Lombard, V., Lopez, R., Redaschi, N., Stoehr, P., Tuli, M.A., Tzouvara, K., and Vaughan, R. 2002. The EMBL Nucleotide Sequence Database. Nucleic Acids Res. 30: 21–6.

Strausberg, R.L., Feingold, E.A., Klausner, R.D., and Collins, F.S. 1999. The mammalian gene collection. Science. 286: 455–7.

Tateno, Y., Imanishi, T., Miyazaki, S., Fukami-Kobayashi, K., Saitou, N., Sugawara, H., and Gojobori, T. 2002. DNA Data Bank of Japan (DDBJ) for genome scale research in life science. Nucleic Acids Res. 30: 27–30.

The C. elegans Sequencing Consortium 1998. Genome sequence of the nematode C. elegans: a platform for investigating biology. Science. 282: 2012–8.

The International Human Genome Sequencing Consortium 2001. Initial sequencing and analysis of the human genome. Nature. 409: 860–921.

Weber, J.L., and Myers, E.W. 1997. Human whole-genome shotgun sequencing. Genome Res. 7: 401–9.

Wheeler, D.L., Church, D.M., Lash, A.E., Leipe, D.D., Madden, T.L., Pontius, J.U., Schuler, G.D., Schriml, L.M., Tatusova, T.A., Wagner, L., and Rapp, B.A. 2002. Database resources of the National Center for Biotechnology Information: 2002 update. Nucleic Acids Res. 30: 13–6.

Wiemann, S., Weil, B., Wellenreuther, R., Gassenhuber, J., Glassl, S., Ansorge, W.,Bocher, M., Blocker, H., Bauersachs, S., Blum, H., Lauber, J., Dusterhoft, A., Beyer, A., Kohrer, K., Strack, N., Mewes, H.W., Ottenwalder, B., Obermaier, B., Tampe, J., Heubner, D., Wambutt, R.,

Korn, B., Klein, M., and Poustka, A. 2001. Toward a catalog of human genes and proteins: sequencing and analysis of 500 novel complete protein coding human cDNAs. Genome Res. 11: 422–35.

Yu, J., Hu, S., Wang, J., Wong, G.K., Li, S., Liu, B., Deng, Y., Dai, L., Zhou, Y., Zhang, X., Cao, M., Liu, J., Sun, J., Tang, J., Chen, Y., Huang, X., Lin, W., Ye, C., Tong, W., Cong, L., Geng, J., Han, Y., Li, L., Li, W., Hu, G., Li, J., Liu, Z., Qi, Q., Li, T., Wang, X., Lu, H., Wu, T., Zhu, M., Ni, P., Han, H., Dong, W., Ren, X., Feng, X., Cui, P., Li, X., Wang, H., Xu, X., Zhai, W., Xu, Z., Zhang, J., He, S., Xu, J., Zhang, K., Zheng, X., Dong, J., Zeng, W., Tao, L., Ye, J., Tan, J., Chen, X., He, J., Liu, D., Tian, W., Tian, C., Xia, H., Bao, Q., Li, G., Gao, H., Cao, T., Zhao, W., Li, P., Chen, W., Zhang, Y., Hu, J., Liu, S., Yang, J., Zhang, G., Xiong, Y., Li, Z., Mao, L., Zhou, C., Zhu, Z., Chen, R., Hao, B., Zheng, W., Chen, S., Guo, W., Tao, M., Zhu, L., Yuan, L., and Yang, H. 2002. A draft sequence of the rice genome (Oryza sativa L. ssp. indica). Science. 296: 79–92.

Index

ΦX174, 294, 435-436

3' end, 398-399, 408-410, 415, 430-431

5' end 286, 322, 398, 405, 407, 409, 412, 430

Ab initio gene prediction, 380, 404, 452

Accession, 161, 403, 419-420, 437, 443, 446-447

Accession number. *See* Accession

ACeDB 35, 203, 221, 362, 379-380, 400-402, 404-407, 416-420, 422, 424, 430

    giface 416

    xace 416

AGP 402-403, 418-419

Alpha satellite 131, 136, 157

Alu repeat 11, 98, 322, 344

Aneuploidy 136

Annotation 349, 361, 377-380, 397-398, 400, 402, 404-405, 410-412, 414-418, 426, 429-431, 437, 441, 451-452

    automatic 379, 398, 412

    data 416, 418

    evidence 404

    feature 437

    gene 404, 410, 412, 414-415, 430

    genome 349, 398, 451

    manual 415

    sequence 441

    tags 319, 326, 368-369, 377

Annotation tags. *See* Annotation, tags

Anopheles gambiae genome. *See* Genome, *Anopheles gambiae*.

Apollo 417

Arabidopsis thaliana genome. *See* Genome, *Arabidopsis thaliana*

ARACHNE 11

Array CGH 131, 134

Array painting 142-143

ARS 101

ASN.1 378

Assembly Golden Path. *See* AGP

Autofinish 346, 351, 375-376, 383

Automation 26, 32, 33, 144, 173, 201-202, 303-306, 346, 351, 358-359, 362, 374, 379, 382, 403

BAC 8, 14, 52, 55-58, 61, 71-74, 76-77, 79, 82-83, 85-88, 95, 135, 138, 142-144, 168, 172, 180, 195, 203, 208, 217, 242-243, 246, 248-251, 264, 267-268, 287, 292, 294, 316, 319, 321, 325-326, 329-330, 333, 343, 345, 347, 349, 359, 365, 369, 373, 381-382, 446, 448, 453

Bacterial Artificial Chromosome. *See* BAC

Base calling 366-368, 370, 377, 450

Bermuda Rules 376, 445

BigDye™ terminators 302, 322-323, 327-330, 346

BigDye™ primer 329-330

Bioperl 404, 417

Blast 12, 30, 170, 195, 203, 233, 249-251, 287, 309, 368-369, 379, 404-405, 407-409

    blastn 12, 405

    blastx 379, 405, 407, 410-412

Blast some sequence. *See* BSS

Breakpoints 129, 141-144

BSS 203, 233

Buried markers 42-44

Buried clones 178, 210, 213, 228, 233

Bury 178

Caenorhabditis elegans genome. See Genome, Caenorhabditis elegans

Canonical clones 233

Canonical marker 42

Cap3 373

cDNA 379, 398-400, 405, 408-415, 424-430, 448-451, 453